Biomethanization of the organic fraction of municipal solid wastes

Edited by
J. Mata-Alvarez
Department of Chemical Engineering, University of Barcelona,
Barcelona, Spain

Published by IWA Publishing, Alliance House, 12 Caxton Street, London SW1H 0QS, UK
Telephone: +44 (0)20 7654 5500; Fax: +44 (0)20 7654 5555; Email: publications@iwap.co.uk
Web: **www.iwapublishing.com**

First published 2003

Printed by TJ International (Ltd), Padstow, Cornwall, UK

British Library Cataloguing in Publication Data
A CIP catalogue record for this book is available from the British Library

Library of Congress Cataloging-in-Publication Data
A catalog record for this book is available from the Library of Congress

ISBN: 1 900222 14 0

Contents

Preface

J. Mata-Alvarez

This book is intended as an introduction to the fundamentals and the more practical aspects of the anaerobic digestion of organic solid wastes, particularly those derived from households; that is, the organic fraction of municipal solid wastes (OFMSW). In this sense it is useful as a textbook for specialized courses and also for practitioners, as it also gives practical methods to follow-up digesters.

In the first part, the book covers the most relevant aspects of anaerobic digestion (AD) of organic wastes. In first and second chapters the fundamentals and kinetic aspects of AD are reviewed, paying special emphasis to the aspects related to solid wastes. Chapter 1 very briefly reviews the microbial and environmental aspects affecting AD process. This introduction is necessary to give a comprehensive view of the process and to understand the practical principles given in the second part, as well as the origin of possible problems arising from the management of the process. Chapter 2 emphasizes the role of kinetics in designing the reactor, paying special attention to existing models, particularly dynamic ones. The intention of this introduction is to facilitate technology transfer from laboratory or pilot-plant experience to full-scale process, so that improvements can be implemented in current digesters.

Chapter 3 describes some methods for analysis and optimization of reactor performance to be carried out in laboratory, such as methanogenic activity tests or experimental evaluation of the biodegradation kinetics of solid organic wastes. Thus hydrolysis, a key step in the biodegradation of complex wastes, as pointed out in Chapter 2, is considered thoroughly here, giving methods to estimate the kinetic constant involved.

Chapter 4 offers a perspective of the evolution of research and application of AD to the treatment of OFMSW. The history of the process is reviewed, pointing out the significant findings in laboratory and pilot-plant research that gave rise to the first industrial application of this technology. Comments are made on the significance of yield reports in the literature, as they are very much dependent on chosen units. Commercial references are also provided although this topic is tackled more thoroughly in Chapter 5. This chapter deals with the different reaction patterns applied to industrial reactors; in particular, the aspects related to the more simple, small-scale batch reactors compared with the usual high-capacity continuous reactors. Also, a review is carried out of the aspects related to the one- and two-phase digesters. Industrial reactors are classified in accordance with the system they use, pointing out advantages and drawbacks of each one. The first part of the book ends with Chapter 6, in which a detailed description of the characteristics related to the different sources of OFMSW is given. This topic is quite significant as obviously AD yields are directly related to these characteristics. In Chapter 6 practical methods to follow-up the digestion process are given, and guidelines are offered for assessing the correct performance of the digester.

The second part of the book is devoted to issues of particular interest. Thus in Chapter 7 a review of what is and the advantages that represent to digest in the same reactor different types of wastes is provided. Co-digestion, the name of this procedure, enables the co-treatment of organic wastes of different origin in a more economically feasible way. Some examples of co-digestion are given, always having OFMSW as a base-substrate. Some complementary issues about co-digestion related to the topic of the modelling aspects explained in Chapter 2 are examined here. Finally, some full-scale co-digestion plants are briefly discussed. Chapter 8 introduces a promising area for improving the anaerobic digestion process of solid wastes. Various types of pre-treatment (mechanical, biological, physico-chemical) for increasing the biodegradability and thus the yields of the process are reviewed in detail.

Chapter 9 deals with another practical aspect, related to integrated waste management: the use of fermentation products of anaerobic digesters for biological nutrient removal processes in wastewater treatment plants. This chapter gives an example of integrated waste management, a field in which both economic and technical advances can be obtained. Balances are given to justify the approach and the Chapter ends with a full-scale case study.

Chapter 10 focuses on the important topic of economics and ecological advantages of the process. The use of compost, integration with composting technology, and advantages over it and other technologies are detailed in the framework of an environmental impact assessment of biowaste treatment.

Lastly, Chapter 11 gives a deep review on the anaerobic digestion of MSW in landfills, with special emphasis on landfill process enhancement and strategies for its application.

J Mata-Alvarez
July 2002

Contributors

B.K. Ahring
Biocentrum-DTU, Technical University of Denmark, Building 227, DK-2800, Lyngby, Denmark.

A.B. Al-Yousfi
UNEP/ROWA, P.O. Box 10880, Manama, Bahrain

I. Angelidaki
IWA Anaerobic Digestion Modelling Task Group, Advanced Wastewater Management Centre, The University of Queensland, St. Lucia, QLD 4072, Australia

L. De Baere
Organic Waste Systems (O.W.S), Dok Noord 4, 9000 Gent, Belgium

Professor P. Battistoni
Engineering Faculty, Institute of Hydraulics, University of Ancona, Via Brecce Bianche, 60123 Ancona, Italy

D. Bolzonella
Department of Science and Technology, University of Verona, Strada Le Grazie 15, 37134 Verona, Italy

Professor F. Cecchi
Facolta Di Scienze MMFFNN, Dip. Scientifico e Technologico, Strada le Grazie 15, 37134 Verona, Italy

J.P. Delgenès
Laboratoire de Biotechnologie de l'Environnement, Institut National de la Recherche Agronomique, Avenue des Etangs, 11100 Narbonne, France.

Dr W. Edelmann
arbi GmbH (Arbeitsgemeinschaft Bioenergie), Lättichstr 8, CH-6340 Baar, Switzerland

J.L. García-Heras
Environmental Engineering Section, CEIT – Technological Research Centre, Apartado 1555, San Sebastian 20018, Spain

H. Hartmann
The Anaerobic Microbiology/Biotechnology Research Group, Department of Biotechnology, The Technical University of Denmark, 2800 Lyngby, Denmark

L. Innocenti
Department of Science and Technology, University of Verona, Strada Le Grazie 15, 37134 Verona, Italy

Professor J. Mata-Alvarez
Department of Chemical Engineering, University of Barcelona, Marti i Franques 1 p6, 08028 Barcelona, Spain

Professor R. Moletta
Laboratoire de Biotechnologie de l'Environnement, Institut National de la Recherche Agronomique, Avenue des Etangs, 11100 Narbonne, France.

P. Pavan
Department of Environmental Sciences, University of Venice, Dorsoduro, 2137, 30123 Venice, Italy

V. Penaud
Laboratoire de Biotechnologie de l'Environnement, Institut National de la Recherche Agronomique, Avenue des Etangs, 11100 Narbonne, France.

Professor F.G. Pohland
Department of Civil & Environmental Engineering, University of Pittsburgh, 1140 Benedum Hall, Pittsburgh, PA 15261, USA

Dr D. R. Reinhart
Department of Civil & Environmental Engineering, University of Central Florida, FL 32816-2993, USA

W.T.M. Sanders
Sub-department of Environmental Technology, Wageningen Agricultural University, PO Box 8129, 6700 EV, Wageningen, The Netherlands

P. Traverso
Department of Environmental Sciences, University of Venice, Calle Larga S.Marta 2137, 30123 Venice, Italy

P. Vandevivere
Organic Waste Systems (O.W.S), Dok Noord 4, 9000 Gent, Belgium

Dr J.B. Van Lier
EP & RC Foundation, PO Box 500, Wageningen 6700 AM, Netherlands

A.H.M. Veeken
Department of Agricultural, Environmental and Systems Technology, Wageningen Agricultural University, P.O. Box 8129 6700 EV Wageningen The Netherlands

Professor W. Verstraete
University Gent, Laboratory for Microbial Ecology, Coupure Links 653, B-9000 Gent, Belgium

Dr G. Zeeman
Sub-department of Environmental Technology, Wageningen Agricultural University, PO Box 8129, 6700 EV, Wageningen, The Netherlands

1

Fundamentals of the anaerobic digestion process

J. Mata-Alvarez

1.1. INTRODUCTION

Anaerobic digestion is a process which, in the absence of oxygen, decomposes organic matter. The main product is biogas – a mixture of approximately 65% methane and 35% carbon dioxide – along with a reduced amount of a bacterial biomass. The development of anaerobic digestion technology took place at the beginning of the 19th century, although after the Second World War biological aerobic treatments and tertiary treatments were the main features of the incipient waste-treatment processes. However, owing to the energy crises of the 1970s, anaerobic digestion technology underwent significant growth.

Traditionally, anaerobic digestion (AD) has been used to treat liquid wastes with or without suspended solids, such as manures, domestic or industrial wastewaters, sludges from biological or physico-chemical treatments, etc.

Edited by J. Mata-Alvarez. ISBN: 1 900222 14 0

References ranging from scientific studies to full-scale experiences are very extensive in this field. However, there are large quantities of solid wastes, such as agricultural and municipal (MSW), that had attracted comparatively little attention on the part of specialists in anaerobic digestion until recently. Because of the large organic matter content, these wastes offer great potential for biogas production.

One of the first studies on the topic of anaerobic digestion of solid wastes was conducted by G.C. Golueke and P.H. McGauhen (1970, 1971), at the University of California. These researchers examined the technical feasibility of the anaerobic digestion of MSW mixed with other wastes such as manure and sewage sludge. After this study, many others were published (see, for instance, Kispert et al., 1975), devoted to the study of MSW as a whole; that is, before any sorting and mainly addressed to the problem of landfill biomethanization (see Chapter 11). In Europe the beginning of the study of anaerobic digestion of solid waste took place later, in the 1980s, and was dedicated directly to the organic fraction of municipal solid waste (OFMSW).

Anaerobic digestion of solid wastes includes many aspects. OFMSW is a specific waste, characterized by its complexity. In this book, many of these aspects are examined: from the more theoretical to the more practical. In this chapter, and next one, the fundamental issues are reviewed. Some description of the metabolic pathways is presented, with special attention to the thermodynamics, so as to see the importance of some intermediates of the process. Environmental factors such as the presence of nutrients and the presence of inhibitors or toxicants, together with the effect of temperature, are also considered. This is important to understand the phenomena going on inside the reactor and to help in interpreting the values of the measured variables. Finally some description of the control strategy for anaerobic digesters is also given.

1.1.1. Bioreaction paths in anaerobic digestion

OFMSW is a complex substrate and obviously requires a more complex metabolic pathway to be degraded as it involves a more intricate series of metabolic reactions before final conversion to methane. Some of the important intermediates of methane formation have been identified and will be described here to help in the understanding of the overall process. This can be of importance in the effective control of the start-up and operation of anaerobic digesters.

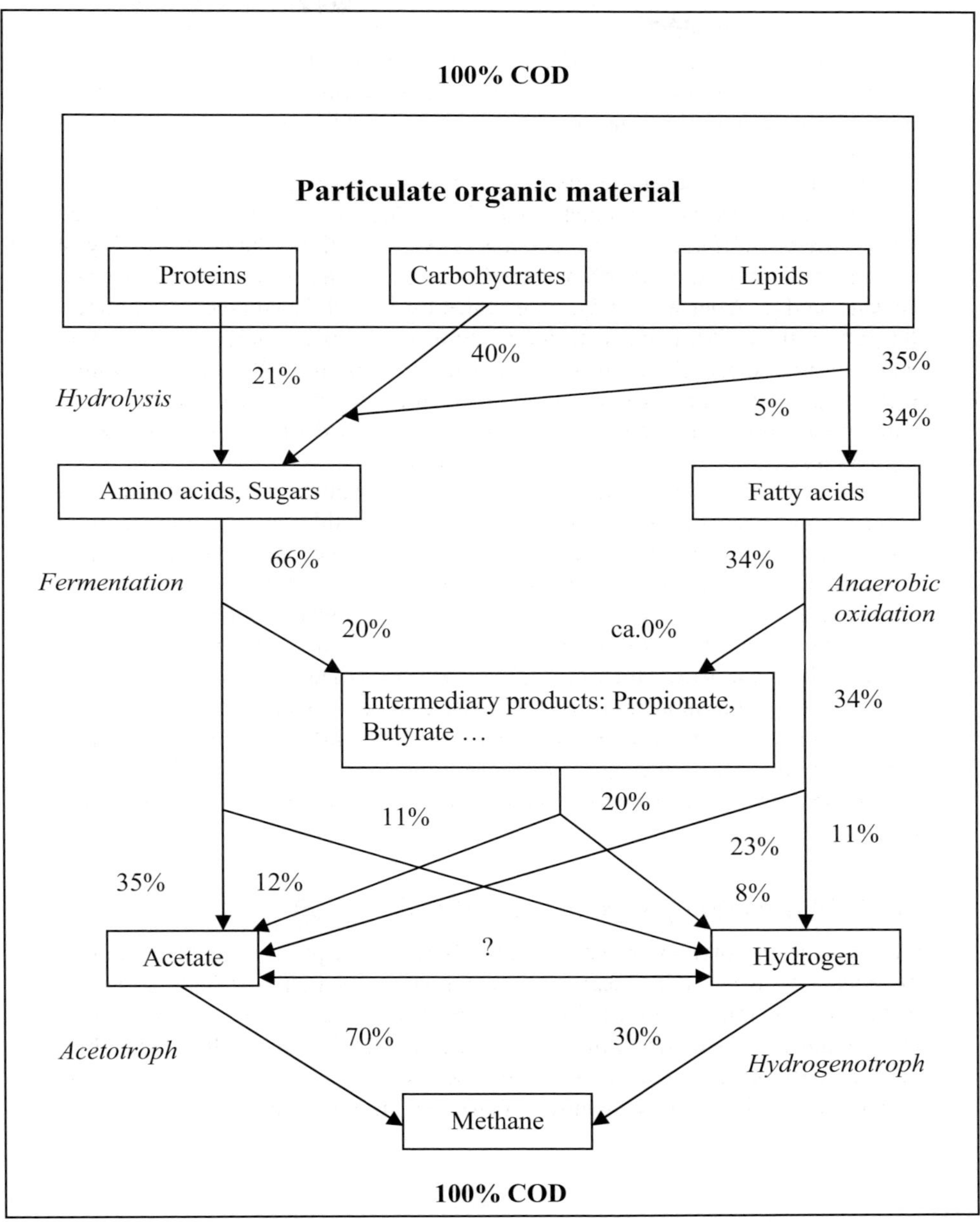

Figure 1.1. Scheme of the biodegradation steps of complex matter (Siegrist et al., 1993).

It is clear that microorganism population distribution is highly dependent on substrate and product concentration as well as on environmental conditions (pH, temperature, hydrogen concentration, etc.). However, knowledge of the microbial ecology of the microorganisms involved will arise from the use of modern molecular techniques which complement traditional cultivation-dependent and microscopic identification techniques with a non-destructive *in situ* analysis (Merkel et al., 1999).

Anaerobic degradation of the organic fraction of municipal solid waste requires the concerted action of a highly varied microbial population, consisting of several groups of strict and facultative bacteria strains. Figure 1.1 shows a scheme of the anaerobic digestion path, from long-chain organic compounds (proteins, carbohydrates and lipids) to the final products, methane and carbon dioxide (Siegrist et al, 1993; Gujer and Zehnder, 1983). The figure also shows COD flow percentages, which come from a balance made for sewage sludge digestion (also a complex waste). To clarify this, another diagram is presented in Chapter 2 (section 2.3). Also in Chapter 2, Figure 2.4 summarises the reaction patterns in a matrix. These patterns make up the identified steps of the anaerobic process and have been specifically described by Pohland (1992) as follows:

1) Hydrolysis of polymers to organic monomeric or dimeric substances (sugars, organic acids, amino acids, etc.) by hydrolytic microorganisms. These bacteria decompose long-chain organic polymers such as proteins, polysaccharides and lipids to monomeric compounds. Hydrolytic bacteria have the ability to produce hydrolytic enzymes, which degrade both soluble and non-soluble high-molecular weight organic compounds. These bacteria can be classified in accordance with the type of exo-enzyme produced and can be inhibited by the accumulation of sugars and amino acids.
2) Fermentation of monomeric organic substances to hydrogen or formate, carbon dioxide, pyruvate, volatile fatty acids, and other organic products, such as ethanol, ketones or lactic acid. The bacteria involved in these transformations are called generically fermentative bacteria. Once solubilised, these monomeric substances are available to be transported to the inside of fermentative bacteria where they are converted to the cited substances.
3) Oxidation of reduced compounds to hydrogen, carbon dioxide and acetate by obligate hydrogen-producing acetogens (OHPAs).
4) Acetogenic respiration of bicarbonates by homoacetogenic bacteria (HA). HA catabolism is mixotropic and these bacteria catabolise mixtures of carbon dioxide and hydrogen and multicarbonated compounds. They can produce acetic acid, but they have methanogenic bacteria as competitors for hydrogen.
5) Oxidation of reduced compounds (alcohols, butyric and propionic acids) to carbon dioxide and acetate by sulphate-reducing bacteria (SRB) or nitrate-reducing bacteria (NRB), with the presence of sulphates and nitrates.

6) Acetate oxidation to carbon dioxide by sulphate-reducing bacteria (SRB) or nitrate-reducing bacteria (NRB).
7) Oxidation of hydrogen (or formate) by sulphate-reducing bacteria (SRB) or nitrate-reducing bacteria (NRB).
8) Conversion of acetic acid by acetoclastic methanogenic bacteria (AMB). Main AMB are *Methanosarcina* and *Methanothrix*. Both have a slow growth rate (doubling times of around 24 hours) and both are affected by the presence of hydrogen. Thus the maintenance of a syntropic metabolism of organic compounds is necessary.
9) Methanogenic respiration of carbon dioxide by hydrogenofil methanogenic bacteria (HMB). These bacteria are much faster than AMB, with doubling times of 4–6 hours.

Table 1.1. Free energy values for some key anaerobic digestion bioreactions.

	Reaction	**ΔG_0, kJ**
Oxidations (electron-donating reactions)		
1 Propionate → Acetate	$CH_3CH_2COO^- + 3H_2O \rightarrow$ $CH_3COO^- + H^+ + HCO_3^- + 3H_2$	+76.1
2 Butyrate → Acetate	$CH_3CH_2CH_2COO^- + 2H_2O \rightarrow$ $2CH_3COO^- + H^+ + 2H_2$	+48.1
3 Ethanol → Acetate	$CH_3CH_2OH + H_2O \rightarrow$ $CH_3COO^- + H^+ + 2H_2$	+9.6
4 Lactate → Acetate	$CHCHOHCOO^- + 2H_2O \rightarrow$ $CH_3COO^- + HCO_3^- + H^+ + 2H_2$	−4.2
5 Lactate → Propionate	$3CHCHOHCOO^- \rightarrow$ $2CH_3CH_2COO^- + CH_3COO^- + H^+ + HCO_3^-$	−165
6 Lactate → Butyrate	$2CHCHOHCOO^- + 2H_2O \rightarrow$ $CH_3CH_2CH_2COO^- + 2HCO_3^- + 2H_2$	−56
7 Acetate → Methane	$CH_3COO^- + H_2O \rightarrow HCO_3^- + CH_4$	−31.0
8 Glucose → Acetate	$C_6H_{12}O_6 + 4H_2O \rightarrow$ $2CH_3COO^- + 2HCO_3^- + 4H^+ + 4H_2$	−206
9 Glucose → Ethanol	$C_6H_{12}O_6 + 2H_2O \rightarrow$ $2CH_3CH_2OH + 2HCO_3^- + 2H^+$	−226
10 Glucose → Lactate	$C_6H_{12}O_6 \rightarrow 2CHCHOHCOO^- + 2H^+$	−198
11 Glucose → Propionate	$C_6H_{12}O_6 + 2H_2 \rightarrow$ $2CH_3CH_2COO^- + 2H_2O + 2H^+$	−358

Table 1.1. *(continued).*

Respirative (electron-accepting reactions)		
12 HCO3- → Acetate	$2HCO_3^- + 4H_2 + H^+ \rightarrow CH_3COO^- + 4H_2O$	−104.6
13 HCO3- → Methane	$HCO_3^- + 4H_2 + H^+ \rightarrow CH_4 + 3H_2O$	−135.6
14 Sulphate → Sulphide	$SO_4^{2-} + 4H_2 + H^+ \rightarrow HS^- + 4H_2O$ $CH_3COO^- + SO_4^{2-} + H^+ \rightarrow 2HCO_3^- + H_2S$	−151.9 −59.9
15 Nitrate → Ammonia	$NO_3^- + 4H_2 + 2H^+ \rightarrow NH_4^+ + 3H_2O$ $CH_3COO^- + NO^- + H^+ + H_2O \rightarrow$ $2HCO_3^- + NH_4^+$	−599.6 −511.4
16 Nitrate → Nitrogen gas	$2NO_3^- + 5H_2 + 2H^+ \rightarrow N_2 + 6H_2O$	−1120.5

1.2. THERMODYNAMICS OF REACTIONS

Thermodynamics is a very important aspect in understanding anaerobic metabolism. Obviously, an overall reaction with a positive value of the net free energy change (ΔG) is not possible. As can be seen in Table 1.1, where the free energy changes are given for some fermentation reactions, in the absence of nitrate and sulphate (the usual case in most anaerobic digesters of OFMSW) the only respirative reactions are those of bicarbonate to methane or acetate. Notably, the most negative value is presented by the reduction of glucose to propionate. However, propionate conversion to acetate and hydrogen is the most difficult reaction (Azbar et al., 2001). Hydrogen is a very important intermediate in the anaerobic digestion process. In many reactions in which H2 is a product, it is necessary for hydrogen to have a low partial pressure for the reaction to proceed. The presence of hydrogen scavengers (HMB) is absolutely necessary to assure the thermodynamic feasibility and thus the conversion of all the reactions producing this compound. For instance, consider the following case, which exemplifies the hydrogen transfer intraspecies (production plus consumption of hydrogen to achieve a proper development of the different kinds of species) involving the anaerobic oxidation of butyrate to acetate (reaction 2 of Table 1.1). This reaction is thermodynamically not possible, but it can take place if acetate and hydrogen are removed from the medium. Acetate can be consumed by AMB (step 8 and reaction 7 of Table 1.1), whereas hydrogen is consumed by HMB (step 9 and reaction 12 of Table 1.1). The overall reaction can then be formulated as follows:

$$2\text{Butyrate} + 2H_2O + H^+ \rightarrow 5CH_4 + 3CO_2 \qquad (\Delta G = -163.4 \text{ kJ})$$

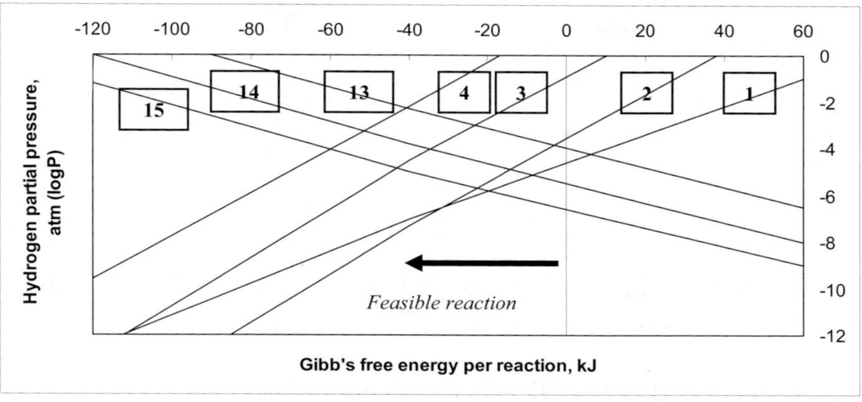

Figure 1.2. Role of hydrogen partial pressure in the thermodynamic feasibility of different reactions involved in the anaerobic digestion process (Pohland, 1992). **(1)** Propionic acid oxidation to acetic acid; **(2)** Butyric acid oxidation to acetic acid; **(3)** Ethanol oxidation to acetic acid; **(4)** Lactic acid oxidation to acetic acid; **(13)** Acetogenic respiration of bicarbonate; **(14)** Methanogenic respiration of bicarbonate; **(15)** Respiration of sulphate to sulphide.

Removal of hydrogen can also be achieved if inorganic oxidants (such as Fe^{3+} salts) are present in the medium.

Hydrogen partial pressure plays an important role in anaerobic degradation. Thermodynamically ΔG is related to the present activities of the species involved in reaction. In fact,

$$\Delta G = \Delta G_0 + \text{RT ln [`Equilibrium-constant-like expression']}$$

In the equilibrium-constant-like expression, the present activities of the reaction products, each raised to a power equal to its stoichiometric coefficient in the chemical reaction, appear in the numerator, and the present activities of the reactants, each raised to the appropriate power, appear in the denominator. For gas components, their partial pressure in the reaction medium is used as a measure of their activity.

If ΔG of oxidations 1–4 and of respirative reactions 13–15 (Table 1.1) is plotted against the logarithm of the partial hydrogen pressure in a logarithmic graph, a set of linear plots is obtained, as shown in Figure 1.2. These plots correspond to a given set of species concentration (acetic acid 25 mM; Propionic, butyric acids and methanol, 10 mM; bicarbonate, 10 mM; methane, 0.7 atm.; Pohland, 1992). For reactions 1–4 (Table 1.1) a negative slope is obtained, whereas for reactions 13–15 the slope is positive. As a consequence, the feasible region, that is the one with negative ΔG values, lies between a narrow range of low pH_2 values. Thus hydrogen partial pressure is an indicator of the health of the anaerobic digestion process.

1.2.1. Biodegradation of OFMSW

As described, complete degradation of OFMSW to methane requires the combined action of a series of microorganisms. Thus, when OFMSW comes into a digester, it contains a large fraction of suspended solids and complex soluble matter. Among the main components of biodegradable matter, namely, carbohydrates, lipids and proteins, carbohydrates are known to be easily and rapidly converted via hydrolysis to simple sugars and subsequently fermented to volatile fatty acids (VFAs) (Cohen, 1982; Miron et al., 2000). Lipids are hydrolysed to long chain fatty acids, and then oxidised to acetate or propionate if the hydrogen partial pressure is low enough to thermodynamically allow the conversion (see Table 1.1 or discussion above). As explained, hydrogen inhibits the oxidations, because the reaction is thermodynamically unfavourable under standard conditions. The reaction only takes place when the hydrogen partial pressure is kept low by the presence of HMB. It has been reported that the presence of methanogens enhances the hydrolysis of lipids, unlike the case of the hydrolysis of carbohydrates and proteins (see Figure 1.1). It has even been stated that lipid hydrolysis hardly occurs without methanogenic bacteria (Palenzuela-Rollon, 1999; Miron et al., 2000; Sanders et al., 2001). Finally, proteins are hydrolysed to amino acids (step 1) which are further degraded to VFAs, for instance, through anaerobic oxidation linked to hydrogen production (step 2). Hydrolysis is a very important step in anaerobic digestion and is thoroughly discussed in Section 2.3.1.

A major danger for overall anaerobic conversions is presented when microorganism populations are not balanced. Methane-formers have a growth rate much slower than acidogenic bacteria; the problem is that acid-formers outgrow methane-formers and acidic conditions may be prevalent in the digester, which would slow down the activity of methanogenic bacteria and eventually totally inhibit their activity. Digester failure must be avoided, as start-up is quite a slow process. Adequate control of alkalinity and VFA concentration can avoid such a situation.

1.3. ENVIRONMENTAL FACTORS CONTROLLING THE AD PROCESS

1.3.1. Nutrients

There are many substances – organic and inorganic – which are necessary for anaerobic digestion processes to run optimally. Not considering the obvious presence of organic carbon to be degraded, there is the requirement for phosphorus and nitrogen – generically called nutrients – sulphur, vitamins and some traces of minerals (Fe, Ni, Mg, Ca, Na, Ba, Tu, Mo, Se and Co), which are generically called micronutrients. It is also important also to note that the presence of micronutrients in small quantities can stimulate growth. However, if a certain threshold level is surpassed their presence can be inhibitory.

Considering the low yield of biomass production of an anaerobic process, the nutrients and micronutrient content of a waste is usually enough for digestion to proceed. However, it is necessary to specially check the availability of nutrients, as there are cases of deficits of these compounds. As nutrients and micronutrients are necessary to synthesize cellular matter, the quantitative requirement is a function of the operating conditions of the digester. Taking the following empirical formula $C_5H_9O_3N$ (Speece and McCarty, 1964) as representative of composition of proteinic matter, the ratio COD/N is 11.4. Similarly it can be deduced that for cellular matter the ratio P/N is between 1/5 and 1/7. Moreover, taking into account yields for cellular matter production (Mosey, 1983) and the digester load, values between 400/7 (high load) and 1000/7 (low load) have been reported for COD/N ratio (Henze and Harremoës, 1983). Consequently, an average ratio COD/N/P of around 600/7/1 can be recommended for a substrate to be anaerobically digested. OFMSW usually contains nutrients and micronutrients in high enough quantities for microorganism growth not to be limited. For instance, see Table 1.2 where a typical composition regarding nutrients is presented.

Table 1.2. Typical composition of OFMSW of different sources found in literature.

Country	TS (%w/w)	TVS (%TS)	TCOD/ TVS	C (%TS)	N (%TS)	P (%TS)	Reference
Belgium	30–35	—	—	33	1.15	—	Pauss et al. (1984)
Italy	20	88	1.2–1.6	48	3.2	0.4	Cecchi et al. (1986)
Spain	22.5	845	—	49	1.61	0.44	Marimon (1982)
Switzerland	—	—	—	39.9	0.48	0.12	Glauser et al. (1987)
Sweden	59	—	—		1.1	0.5	Szikristz et al. (1988)
The Netherlands	38.7	53	2	—	7.6	—	Gijzen et al. (1987)

1.3.2. Toxic substances

There are substances that at a given concentration inhibit bacterial activity, especially methanogenic bacteria. The problem of inhibitory or toxic substances is far from simple. There is no general agreement on threshold limits and, as mentioned before, at low concentrations, some toxicants can become stimulants for anaerobic digestion. Thus reported toxicity values in literature are sometimes misleading. The reason is that inhibitory capability is dependent not only on the concentration of the substance, but also on the environmental conditions; that is, pH, temperature, and the concentration of other substances, which can have synergies or antagonisms with a given toxicant.

There are several common substances that can affect the anaerobic digestion process and which are considered toxic or inhibitory at a given threshold level. VFAs, pH, free ammonia and hydrogen sulphur are the most frequent. Others can be salinity or some xenobiotics. Problems of this kind when digesting OFMSW are due to the excess of VFAs. Other toxic compounds are rare in this environment if source separation is carried out.

As pointed out above, VFAs are intermediary compounds of the anaerobic degradation of organic matter. The undissociated species have been reported as more toxic because they can more easily diffuse to the inner parts of the cell (Andrews, 1969; Pohland and Martin, 1969). Consequently, pH together with the alkalinity level exerts a definite effect on VFA toxicity, and the threshold level will depend on these parameters. Among VFAs, propionic and butyric have been described as the most inhibitory. In accordance with Boone and Xum (1987) propionic acid concentrations over 3000 mg/L are definitively toxic and cause digestion failure.

Ammonium, which is necessary as a nutrient, at some concentrations inhibits methanogenesis. Similarly, pH has also a definite effect on the threshold levels. The reason is the same as the case of VFAs, in which the toxic species is the undissociated one, in this case free ammonia. Solubility and temperature also have a large effect on ammonia toxicity. For instance pig manure, with very high ammonia-N concentrations (up to 4 g/L or greater), cannot be directly digested under thermophilic conditions (around 55 ºC).

In fact, the problem of ammonia inhibition has been addressed by many authors in the literature (Angelidaki and Ahring, 1993; Hansen et al., 1998; and from a more practical approach, De Baere et al. (1984)). More details directly related to OFMSW can be found in a comprehensive review of the literature done by Kayhanian (1999) to evaluate ammonia inhibition in anaerobic digesters. The review includes the fundamentals of biochemical pathways of nitrogenous compounds, mechanisms of ammonia inhibition and the role of free ammonia. With reference to the digestion of OFMSW in dry systems at thermophilic temperature, long-term experimental studies at the pilot scale revealed that ammonia inhibition occurred at ammonia concentrations of 1200 mg/L. To overcome this problem two methods have been suggested: (a) dilution of digester content with some adequate wastewater and (b) adjustment of feedstock C/N ratio. Both methods rely on the decrease of N concentration and are described in Kayhanian (1999). It is important to stress that it is very difficult to set threshold values for complex substrates. Moreover, the toxicity is even dependent on the type of methanogen: CN^- anion is toxic for MAB (step 8 described above) and not for MHB (step 9) (Stronach et al., 1986).

Hydrogen sulphide presents a similar behaviour as ammonia and its toxicity is also very much dependent on the environmental conditions of the system, namely pH and alkalinity. The literature describes threshold values from 200 mg/L to 1500 mg/L. In addition to the environmental factors, these differences can also be attributed to bacterial acclimation. Moreover, iron ion can also influence these values, as it can remove sulphur anion by precipitation of FeS. This precipitation can be a method to overcome toxicity problems derived from H_2S presence.

As mentioned above, the presence of metal ions in anaerobic environments can act as either stimulant or inhibitor. As reference values, concentrations over 1 mg/L for heavy metals or 5–8 g/L for metals of Group II can be toxic, in accordance with an E.P.A. report (E.P.A., 1974). These values are to be considered with care as they are dependent on environmental factors.

1.3.3. Xenobiotics

Xenobiotics are released in large quantities due to human activities and can be a problem for the anaerobic digestion of some residual streams. Typical examples are solvents, frequently present in industrial wastewaters, which can be very toxic for methanogenic bacteria. Together with these solvents, which can also include alcohols, organic acids, esters, ketones, etc., other inorganic substances and also small quantities of industrial products such as pesticides, dyes, drugs, and other synthetic compounds can become inhibitors or toxic to anaerobic environments. In addition other compounds used in the synthesis of industrial substances, especially those containing halogens, sulphur, hydroxyl, amines, nitrocompounds, ether, ketones and other functional groups should also be included in this classification. By way of example, Table 1.3 presents a list of the effect–concentration of organic chemicals (some surfactants, pesticides and phenols) on the production of gas in anaerobic digestion (Madsen and Rasmussen, 1996).

Table 1.3. Toxicity of organic chemicals in the anaerobic gas production test.

Chemical	Effect–concentration (mg/L)	
	EC20	EC50
Surfactants		
AES	6.8	11
LAS	18	59
AEO	71	190
APG	8.8	67
C_{10} EGE	167	—
C_{12} EGE	94	—
ADMBAC	< 5.0	6.7
Esterquat	> 400	> 400
Pesticides		
Esfenvalerate	> 200	> 200
Prochloraz	> 10	> 400
Phenols		
2,4-DCP	43	73
2,4,5-DCP	< 5.0	11
4-Nitrphenol	14	20
2,4-Dinitrophenol	6.4	8.9

However, it must be taken into account that some xenobiotics can be biodegraded in certain conditions, including the genetic manipulation of microrganisms or not.

As mentioned before, OFMSW does not present this kind of toxicity, especially if it comes from a source separation.

1.3.4. Temperature effects

Anaerobic digestion can function over a large range of temperatures: from psychrophilic temperatures at around 10 °C to some extreme thermophilic temperatures over 70 °C (see, for instance, Ahring (1994) or Scherer et al. (2000)).

Temperature very much influences anaerobic reactions, both from the kinetic and the thermodynamic point of view. In particular, methanogenesis is strongly influenced by this parameter, degradation rates and yields increasing, as usual, with temperature. Figure 1.3 presents a scheme in which the rate of the AD process is represented in front of the temperature. Thus, over this general increase, two optimal ranges with maximum activity have been identified: mesophilic (around 35 °C) and thermophilic (around 55 °C). Most of the digestion of solid wastes, and in particular OFMSW, is carried out only in these two temperature ranges. Figure 1.4 is taken from De Baere (2000) and shows a diagram in which the cumulative installed capacity of treatment of both types of anaerobic digestion plant for OFMSW is presented. As can be seen, the installed capacity is approximately around 50% for each temperature range. Looking carefully at this graph, it seems that the thermophilic range is increasingly preferred. One reason for this could be that thermophilic temperatures offer better yields and, consequently, higher biogas production. However, this surplus of energy should be balanced by the increased need of feed heating. In many cases this increased energy demand is the same as the energy excess. Among other factors, commented on below, substrate concentration, yields and kinetics should be carefully considered before taking any decision.

Although it is true that the kinetics are faster and the yields larger in thermophilic temperatures, some imbalances may occur due to the complexity of the several steps involved in the process. Thus acidogenesis can produce more acids than methanogenesis can convert at a higher temperature. This problem can affect the digestion of wastes with a high biodegradability, as was observed by Bernal et al. (1992) with market wastes, if the process is not carried out in separate reactors; that is, one for hydrolysis and fermentation and another for methanogenesis (for two-phase reactors, see Chapter 5, Section 5.2). In the literature there are references to successful operations at both temperature ranges. Cecchi et al. (1991) found thermophilic temperatures to be optimal for digestion of mechanically selected OFMSW in a pilot-plant study (one- or two-phase reactors).

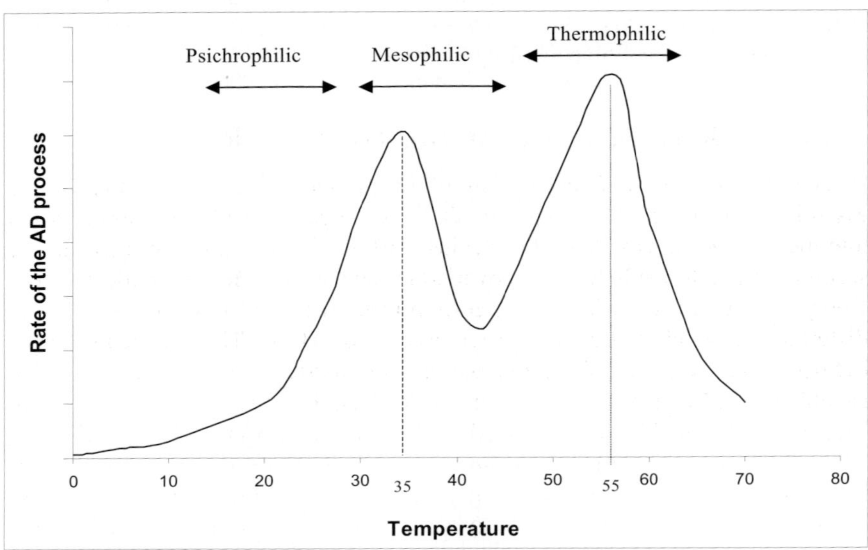

Figure 1.3. Temperature ranges for anaerobic digestion. Optima are for mesophilic around 30–35 °C and for thermophilic around 55–60 °C.

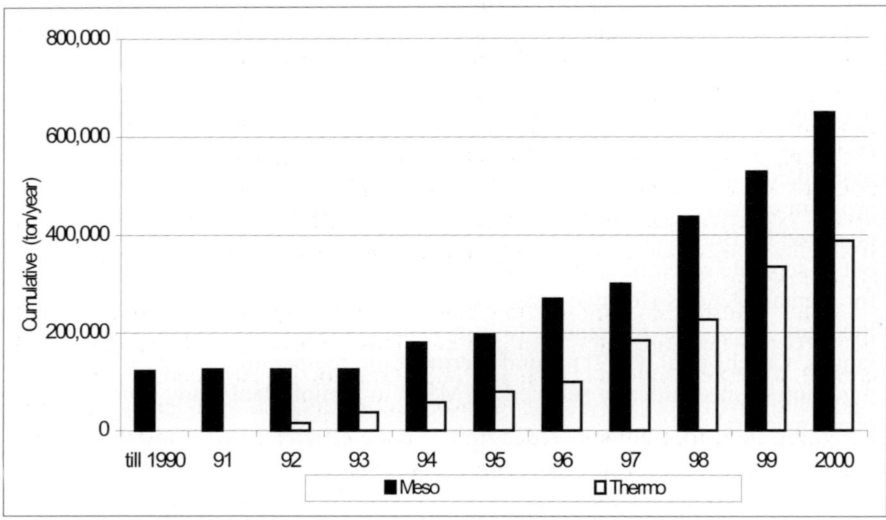

Figure 1.4. Installed capacity on both industrial ranges of temperature (mesophilic and thermophilic) for anaerobic digestion of OFMSW (De Baere, 2000).

To conclude, it can be said that although biogas production yields and bioreaction kinetics can be more favourable at thermophilic temperatures, optimal conditions depend on the type of substrate – concentration and biodegradability – and the type of system used.

1.4. CONTROL OF ANAEROBIC DIGESTERS

In accordance with the description given in Section 1.1, biomethanization essentially occurs in two steps. In the first, organic matter is converted to intermediates such as volatile fatty acids, carbon dioxide and hydrogen. In the second, these intermediates are converted to methane by methanogenic bacteria. One problem with the anaerobic digestion process is that it is very sensitive to disturbances which result in digester organic overload. This overload can be defined as an excess of biodegradable organic matter for the active population capable of digesting it. Thus digester overload can be caused by a real excess of organic biodegradable feed, as well as for any circumstance that produces a decrease in the active microorganism concentration (temperature decrease, toxic substances introduced in the digester, flow rate increase, etc.). These disturbances mainly affect methanogenic bacteria, whereas acidogenic bacteria, which are much more tolerant, continue to work, producing more acids. These acids, in turn, inhibit methane-formers. At the same time, fermentation products which are normally not intermediates can also be formed as an attempt to metabolise the accumulated hydrogen or formate. This imbalance, if not corrected, can finally result in a digester failure. Table 1.4 points out most of the disturbances that cause digester overload and the final effect on the digester.

The control of anaerobic digesters is very important to overcome these problems, especially when the process operates under severe conditions, that is, with high strength substrates and short hydraulic retention times. From a general point of view, a control system of an anaerobic digestion process may have several objectives such as disturbance detection, calculation of indirect variables from direct measurement, filtering noise from measurements, setting up of control strategies, etc.

Many researchers, through the study of the response of the microbial population when subjected to different kinds of overload, have developed control strategies, understanding the importance of digester control. In many cases the control of anaerobic digestion constitutes a mandatory step for designing industrial units in order first to cope efficiently with instability and, second, to operate the system at optimal conditions.

An efficient automatic control system needs to know the state of the anaerobic digester process. This state is described by many process variables, such as concentrations, flow rates, temperatures, pressures, levels, specific measurements such as pH or VFAs, etc. All these variables are related to process stability, but some of them are more sensitive to disturbances and may give an earlier indication of a process imbalance. On the other hand, some of these variables are more easily measured than others. The ultimate aim is, of course, to maintain the process at an optimum stable state. In order to measure the stability of the process several parameters have been described in the literature (Moletta et al., 1994). Table 1.5 summarizes these parameters in accordance with the nature of the phase where they are measured.

On the other hand, to operate a control system in a reliable way it is very important to have appropriate instrumentation and this is still a major difficulty in biological processes due to fouling problems.

Table 1.4. Possible disturbances of anaerobic digesters and predictable results.

Disturbance	Arising problem	Final effect if not digester total failure
Flow rate increase	Washout of microorganisms. Methanogens are the most affected, given its doubling time	Reduction in: Methane % in gas pH Methane production rate Alkalinity Increase in: VFAs Acids different from acetic acid
Feed concentration increase (overloading)	Imbalances mainly affecting methanogenic bacteria and resulting in an accumulation of VFAs	
Introduction of toxic substances		
Temperature fluctuations Oxygen exposure		

In addition to the measurement, which is of high importance, an appropriate control algorithm is necessary. In this sense, adaptive control schemes can be advantageously used. Thus an improved performance can be achieved if the parameters of the controller are continuously adapted to follow the changes in the operative conditions (Steyer et al., 1999). However, robust control strategies appear to be a very promising option and significant improvements over conventional proportional–integral (PI) controllers are the most appropriate way to tackle the insensitivity of some phenomena such as parameter variation when severe operational conditions prevail.

Table 1.5. Parameters that can be used for automatic control in anaerobic digesters in accordance with Moletta et al. (1994) and Molina et al. (1998).

Liquid phase	Gas phase
VFA	Gas production rate
Ph	Methane or CO_2 production rate
Alkalinity	CO content in biogas
ORP	H_2 content in biogas
COD	
Liquid level	
Others: polarographic sensors, dissolved hydrogen	

Digesters operating with OFMSW usually do not present problems related to the use of short retention times, as is the case for liquids in immobilized biomass reactors. In this sense, a large part of the control is left to the role of the operators. This role is, in any case, quite important. The ability to see, to smell, just to feel based on accumulated experience, is crucial in a biological process. Computer-based control helps operators in assimilating the sometimes large set of information available for many variables, but the good sense of an experienced operator is of the utmost importance in industrial digesters.

1.5. CONCLUSIONS AND PERSPECTIVES

Fundamental as well as applied research is quite relevant for the advances that have to be registered in the field of anaerobic digestion of solid wastes and, in particular, for that of the organic fraction of municipal solid wastes.

Aspects such as the degradation of chlorinated compounds need to be examined in greater depth, owing to the high potential that anaerobic degradation offers in this area (Verstraete et al., 2000). Another field in which AD needs to advance is that of the treatments to enhance the extent of bioconversion, in order to produce more biogas and, at the same time, a reduced amount of solids in the digester effluent. This is relevant because a large offer in compost products can be foreseen in the near future. These treatments, discussed in depth in Chapter 8, are of great interest right now in what concerns the digestion of sewage sludge, as witness the large number of papers being published in this area. A deeper insight into the mechanisms of hydrolysis attack so as to favour kinetics would also help the progress of this biotechnology. Finally it should be mentioned that scientific activity in this area is quite high, and a specialized conference now exists, organized by IWA and devoted to the AD of solid wastes. This offers a good indication of the progress that will be produced in the near future.

1.6. REFERENCES

Ahring, B.K. (1994). Status on science and application of thermophilic anaerobic digestion. *Water Science and Technology* **30**(12), 241–249.

Andrews, J.F. (1969). Dynamic model of the anaerobic digestion process. Journal Sanitary Engineering Division, *ASCE J.* **95**, SA1, 95–116.

Angelidaki, I. and Ahring, B. (1993). Thermophilic anaerobic digestion of livestock waste: the effect of ammonia. *Appl. Microbiol. Biotechnol.* **38**(4), 560–564.

Azbar, N., Ursillo, P. and Speece, R. (2001). Effects of process configuration and substrate complexity on the performance of anaerobic processes. *Water Research* **35**(3), 817–829.

Bernal, O., Llabrés, P., Cecchi F. and Mata Alvarez, J. (1992). A comparative study of the thermophilic biomethanization of putrescible organic wastes. *Odpadní vody / Wastewaters* **1**(1), 197–206.

Boone, D.R. and Xum, L. (1987). Effects of pH, temperature and nutrients on propionate degradation by a methanogenic enrichment culture. *Applied and Environmental Microbiology*, July 1987, 1589–1592.

Cecchi, F., Traverso, P.G. and Cescon, P. (1986). Anaerobic digestion of organic fraction of municipal solid waste-digester performance. *The Science of Total Environment* **56**, 183–197.

Cecchi, F., Pavan, P., Mata Alvarez, J., Bassetti, A. and Cozzolino, C. (1991). Anaerobic digestion of municipal solid waste: Thermophilic versus mesophilic performance at high solids. *Waste Management & Research* **9**, 305–315.

Cohen, A. (1982). Optimization of anaerobic digestion of soluble carbohydrate containing wastewaters by phase separation, Ph.D. Thesis, University of Amsterdam, Amsterdam, The Netherlands.

De Baere, L., Devocht, M., Assche, P. and Verstraete, W. (1984). Influence of high sodium chloride and ammonium chloride salts levels on methanogenic association. *Water Research* **18**(5), 543–548.

De Baere, L. (2000). State-of-the-art of anaerobic digestion of solid waste in Europe. *Water Science and Technology* **41**(3), 283–290.

E.P.A. (1974). Process design manual for sludge treatment and disposal. EPA 635/1-74 006. Technology Transfer, Washington D.C.

Gijzen, H.J., Lubberding, H.J., Verhagen, F.J., Zwart, K.B. and Vogels, G.D. (1987). Application of rumen microorganisms for enhanced anaerobic degradation of solid organic waste materials. *Biological Wastes* **22**, 81–95.

Glauser, M., Aragno, M. and Gandolla, M. (1987) Anaerobic digestion of urban wastes: sewage sludge and organic fraction of garbage. In: *Bioenvironmental Systems* Vol. 3 (ed. D.L. Wise), pp. 143–225. CRC Press, Boca Raton, Florida.

Golueke, C.G. and McGauhen, P.H. (1970). Comprehensive studies of solid waste management. First and Second Annual Reports, Bureau of Solid Waste Management, Washington D.C.

Golueke, C.G. (1971). Comprehensive studies of solid waste management. Third Annual Report, U.S. E.P.A., Washington D.C.

Gujer, W. and Zehnder, A.J.B. (1983). Conversion processes in anaerobic digestion. *Water Science and Technology* **15**(8–9), 127–167.

Hansen, K.H., Angelidaki, I. and Ahring, B.K. (1998). Anaerobic digestion of swine manure – inhibition by ammonia. *Water Research* **32**(1), 5–12.

Henze, M. and Harremoës, P. (1983). Anaerobic treatment of wastewater in fixed film reactors – A literature review. *Water Science and Technology* **15**(8–9), 1–101.

Kayhanian, M. (1999). Ammonia inhibition in high-solids biogasification – An overview and practical solutions. *Environmental Technology* **20**(4), 355–365.

Kispert, R.G., Sadek, S.E. and Wise, D.L. (1975). An economic analysis of fuel gas production from solid waste. *Res. Recov. and Conserv.* **1**, 95.

Madsen, T. and Rasmussen, H.B. (1996). A method for screening the potential toxcity of organic chemicals to methanogenic gas production. *Water Science and Technology* **33**(6), 213–220.

Marimon, S.R. (1982). Los residuos sólidos urbanos. Análisis de un servicio municipal. Servicios de los Estudios en Barcelona del Banco Urquijo.

Merkel, W., Werner, M., Ulrich, S. and Karlheinz, K. (1999). Population dynamics in anaerobic wastewater reactors: modelling and in situ characterization. *Water Research* **33**(10), 2392–2402.

Miron, Y., Zeeman, G., Vanlier, J.B. and Lettinga, G. (2000). The role of sludge retention time in the hydrolysis and acidification of lipids, carbohydrates and proteins during digestion of primary sludge in CSTR systems. *Water Research* **34**(5), 1705–1713.

Moletta, R., Escoffier, Y., Ehlinger, F., Coudert, J.P. and Leyris, J.P. (1994). On-line automatic control system for monitoring an anaerobic fluidized-bed reactor: response to organic overload. *Water Science and Technology*, **30**(12), 11–20.

Molina, F. J., Leon, C., Arnaiz, M. C. and Lebrato, J. (1998). Application of fuzzy logic for on-line control of a laboratory-scale anaerobic reactor. *Environmental Engineering*, **4**, 23–31

Mosey, F.E. (1983). Mathematical modelling the anaerobic digestion process: Regu1atory mechanisms for the formation of short-chain volatile acids from glucose. *Water Science and Technology* **15**(8–9), 209–232.

Palenzuela-Rollon, A. (1999). Anaerobic digestion of fish wastewater with special emphasis on hydrolysis of suspended solids. Ph.D. thesis, Agricultural University, Wageningen.

Pauss, A., Nyns, E.J. and Naveau, H. (1984). Production of methane by anaerobic digestion of domestic refuse. EEC Conference on Anaerobic and Carbohydrate Hydrolysis of Waste, 8–10 May, 1984, Luxembourg.

Pohland, F.G. and Martin, J.C. (1969). Dynamic model of the anaerobic digestion process – A discussion. *Journal Sanitary Engineering Division, ASCE*, **95**, SA6, 1197–1202.

Pohland F.G. (1992). In design of anaerobic processes for the treatment of industrial and municipal wastes (ed. J.F. Malina and F.G. Pohland). In *Water Quality Management*, Vol. 7. Technomic Publishing Co. Lancaster PA.

Sanders, W.T.M., Geerink, M., Zeeman, G. and Lettinga, G. (2001). Anaerobic hydrolysis kinetics of particulate substrates. *Water Science and Technology* **41**(3), 17–24.

Scherer, P.A., Vollmer, G.-R., Fakhouri, T. and Martensen, S. (2000). Development of a methanogenic process to degrade exhaustively the organic fraction of municipal "grey waste" under thermophilic and hyperthermophilic conditions. *Water Science and Technology* **41**(3), 83–91.

Siegrist, H., Renggli, D. and Gujer, W. (1993). Mathematical modelling of anaerobic mesophilic sewage treatment. *Water Science and Technology* **27**(2), 25–36.

Speece R.E. and McCarty, P.L. (1964). Nutrient requirements and biological solids accumulation in anaerobic digestion. *Adv. Wat. Pollut. Res.* **2**, 305–322.

Steyer, J.P., Buffiere, P., Rolland, D. and Moletta, R. (1999). Advanced Control of anaerobic digestion processes through disturbances monitoring. *Water Research* **33**(9), 2059–2068.

Stronach, S.M., Rudd, T. and Lester, J.N. (1986). Anaerobic digestion processes in industrial wastewater treatment. Springer-Verlag, pp. 21–38.

Szikristz, G., Frostell, B., Normann, J. and Bergstrom, R. (1988) Pilot scale anaerobic digestion of municipal solid waste after a novel pretreatment. In: *Anaerobic digestion 1988* (ed. E.R. Hall & P.N. Hobson), pp. 375–382. Pergamon Press, Oxford.

Verstraete, W., Van Lier, J., Pohland, F., Tilche, A., Mata-Alvarez, J., Ahring, B., Hawkes, D., Cecchi, F., Moletta, R. and Noike, T. (2000). Developments at the Second International Symposium on Anaerobic Digestion of Solid Waste (Barcelona, 15–19 June 1999). *Bioresource Technology* **73**, 287–289.

2

Reactor sizing, process kinetics and modelling of anaerobic digestion of complex wastes

J.L. García-Heras

2.1. INTRODUCTION

Processes of biological treatment of wastewater and wastes have been traditionally devised from a basic knowledge of biochemistry, physico-chemistry and engineering. Once implemented at full scale, processes were usually improved by means of trial and error. Understanding of processes advanced by studying cause–effect relationships in experiments where system conditions were changed. Thus lagoons, anaerobic digesters, activated sludge plants and biofilm reactors were established. However, in cases where certain serious performance problems affected full-scale facilities, the lack of a sound

Edited by J. Mata-Alvarez. ISBN: 1 900222 14 0

explanation was quite clear. The need for a deeper knowledge of systems behaviour was even more evident when the new process configurations and their design became more complex. Two-phase separation, thermophilic digesters and biofilm reactors are examples in the anaerobic treatment of wastewater and sludge. Examples of aerobic treatment systems include phosphorus removal and nitrification/ denitrification.

Therefore, engineers and scientists required sets of equations that somehow quantified the phenomena involved in the treatment process so that they could consider cause/effect relationships when results from full-scale facilities were lacking. In full-scale plants an absence of results arose from the technical impossibility of making the required tests, the high cost of the experiments and the excessive risk of changes in the facilities. Sometimes there were no full-scale installations at the time. The situation was usually solved by pilot or lab-scale tests and the setting up of a group of simple equations assumed valid for steady state. It was called the static or steady-state model of the process. This approach was useful for plant design and reactor sizing for a specific range of system conditions. The situation described still persists in many cases, although there has been enormous progress in every aspect mentioned.

Anaerobic processes, in particular, consist of simultaneous transformations, involving many compounds that can be both the reactants and the products of reactions. These systems involve a set of chemical compounds that are degraded by different groups of bacteria, resulting in a pool of microorganisms living in their feed substrates. Moreover, some of the latter act as inhibitors of microbial activity in some of the reaction steps. In addition certain components can not be measured, so must be considered as unknown from an experimental point of view. All this confirms the repeated assertion that the anaerobic process is a complex system. For the above-mentioned purposes, some sort of quantification of the involved phenomena is required; which, in turn, clearly demands a wider and more complex model.

A more complex model of anaerobic systems includes more variables and coefficients because more microorganisms, reactions and substrates are involved. This makes it much more difficult (and sometimes impossible) to obtain equations for a design at steady state because of the non-linearity of the kinetic expressions. For design purposes it is also important to be able to characterise the feedstock for different complex wastes. This characterisation must reflect the break down of the influent waste composition. These reasons suggest a dynamic model for this process. There are, however, other perhaps more important reasons, e.g., the system may be submitted to both influent and operational disturbances whose effect on reactor performance in the transient state must be identified; or, maybe, automatic control of the reactor is desired.

The requirements demanded of a dynamic model by design engineers can be different from those demanded by people pursuing plant operation or process research. In either case, a model that is large and complex will inevitably be required as the starting point. However, the model finally adopted will be different in each case.

Among the wide scope of modelling structures and methodologies that exist, the one chosen here (the mechanistic) is that based on the physico-chemical and biological principles with single values for the variables (deterministic). This produces a set of equations representing the system (functional model).

The basic fundamentals in developing such a model are considered in various sections of this book. First, a conceptual model must be elaborated by defining the microorganisms, processes and substrates composing the anaerobic process. A set of variables defining the system and a set of coefficients must be established, by which means the kinetic equations of the chosen processes can be expressed. A multivariable system is thus defined.

2.2. MICROBIAL REACTION KINETICS

When cells of a particular bacterium are put into contact with an excess of appropriate substrate in a batch reactor, under the proper environmental conditions, unrestrained microbial growth occurs. Substrate is channelled to cell synthesis as building blocks and to energy as a consequence of oxidation reactions. Several phases of microbial behaviour can be distinguished in the cited test, among which the exponential phase, the declining growth phase and the endogenous phase are noteworthy. The substrate and microorganism concentrations change as the batch experiment proceeds. In other words, the bacteria undergo the mentioned phases over time (Monod, 1949).

The growth phases through which this microbial culture passes are a consequence of its interaction with the environment within the test vessel. In a continuous flow reactor the cells can be maintained in a particular growth phase for a long time, depending on the operational conditions imposed on the reactor (Benefield and Randall, 1980).

2.2.1. Kinetics of basic processes

Most of the reactions taking place in the above-cited test are catalysed by microorganisms; hence the kinetics of the overall system is governed by the kinetics of the bacterial activity. If an understanding in depth and an accurate prediction of the anaerobic process behaviour is intended, microbial growth, substrate utilisation and product production must be precisely known. In any attempt to model the process a clear quantification of these phenomena has to be established in terms of stoichiometry and kinetics.

Biomass growth rate

When the energy, carbon source and electron acceptor requirements for growth are fulfilled in the batch culture, above, the increase of biomass concentration with regard to time is proportional to the biomass concentration itself (X) (Stanier et al., 1986). The proportionality coefficient (μ) is called the 'specific growth rate'.

$$\left(\frac{dX}{dt}\right)_{growth} = \mu \cdot X \tag{2.1}$$

This formula expresses the exponential growth phase providing there is an excess of substrate in the culture. Thus μ is a constant.

Substrate utilisation rate

The microorganism growth rate in a substrate, the concentration of which is S, is proportional to the substrate utilisation rate; Y being the coefficient,

$$\left(\frac{dX}{dt}\right)_{growth} = -Y \cdot \left(\frac{dS}{dt}\right)_{utilization} \tag{2.2}$$

Y is known as the 'biomass (or growth) yield coefficient'. It expresses the mass of microbial cells grown on a particular substrate. The substrate utilisation rate is proportional to the biomass concentration (Pirt, 1975) and is expressed as

$$\left(\frac{dS}{dt}\right)_{utilization} = -K \cdot X \tag{2.3}$$

with K being referred to as the 'specific substrate utilisation rate'. Thus a relationship between K and μ can be established, thereby allowing substrate utilisation to be evaluated for various values of growth rate

$$Y = \mu / K \tag{2.4}$$

The coefficient K remains constant in an environment in which the substrate is in excess.

Biomass decay rate

Substrate is used as cell material for synthesis as well as a source of energy for growth and maintenance. The role of the latter becomes apparent when substrate is scarce or the bacterial population grows too old. Three basic approaches have been published. Herbert (1958) proposed that the energy for maintenance was obtained by endogenous metabolism using the cell material for the oxidation reactions, promoting biomass loss due to endogenous respiration. McCarty (1966) introduced the concept of bacterial decay whereby bacteria eventually die as a consequence of difficult environmental conditions or old age. He considered it unnecessary to take maintenance energy into account. Pirt (1975) suggested that the energy for maintenance was obtained by oxidation of the external substrate. These discussions were concerned mainly with aerobic processes.

The theories of Herbert, McCarty and Pirt lead to a common acceptance in literature of the biomass decay concept from a practical engineering point of view. The living bacteria in a culture have a limited life span, after which they die. Their death, or lysis, produces substances that are incorporated into the biodegradation cycle. The kinetics of this decay process is usually expressed as being proportional to the biomass concentration by means of the decay coefficient K_d:

$$\left(\frac{dX}{dt}\right)_{decay} = -K_d \cdot X \tag{2.5}$$

Net biomass variation

The net increase of biomass concentration in the test can thus be expressed as growth rate plus decay rate.

$$\left(\frac{dX}{dt}\right)_{net} = \left(\frac{dX}{dt}\right)_{growth} + \left(\frac{dX}{dt}\right)_{decay} \tag{2.6}$$

$$\left(\frac{dX}{dt}\right)_{net} = -Y\left(\frac{dS}{dt}\right)_{utilization} - K_d \cdot X \tag{2.7}$$

It should be noted that several authors (Sherrard and Schroeder, 1973; Ribbons, 1970) use the concept of a 'net specific growth rate coefficient' (μ_{net}) to express the proportionality coefficient between $(dX/dt)_{net}$ and X in equation 2.7. They refer to the relationship between μ_{net} and K in equation 2.4 as the 'observed yield coefficient' (Y_{obs}). This concept will not be used in this chapter although it is consistent and compatible with the approach maintained here.

2.2.2. Usual kinetic expressions for microorganism growth

In the previous paragraphs a single culture – single substrate was described. The anaerobic treatment of complex substrates, however, is a multivariable system where a mixed culture of microorganisms promotes several simultaneous transformations. In most of them there are products of reaction that act as reactant or substrate for other reactions. For those reactions in which microorganism growth is involved the reaction kinetics is usually expressed as the rate of biomass growth or substrate utilisation. As has already been outlined above, the microbial growth rate can be expressed by equation 2.1. In it the specific growth rate coefficient (μ) takes into account the dependence of the reaction rate on the instantaneous and initial substrate concentration, and on the inhibition due to various internal compounds.

Monod

So as also to keep equation 2.1 valid for the declining growth phase in the aforementioned batch test, Monod suggested that μ was not a constant but a function of substrate concentration (S) which then becomes a limiting factor for culture growth. He established the expression

$$\mu = \mu_{max} \frac{S}{K_S + S} \tag{2.8}$$

where μ_{max} is the maximum specific growth rate. The 'half-saturation constant' (K_S) expresses the value of substrate concentration for which μ takes the value of μ_{max} / 2. For very high values of S (excess of substrate) the Monod kinetics becomes a zero order in S,

$$\mu = \mu_{max} \tag{2.9}$$

For values of S much smaller than K_S (limiting substrate conditions) the Monod expression leads to a first-order kinetics in S:

$$\mu = \frac{\mu_{max}}{K_S} \cdot S \tag{2.10}$$

Consequently, μ affects the growth kinetics by a different weight depending on the value of S in the reactor, relative to Ks.

Every description based on microbial growth can be similarly made for substrate uptake simply by using equations 2.3 and 2.4 to correlate both coefficients. Thus, the specific substrate utilisation rate coefficient (K) varies under substrate limiting conditions:

$$K = K_{max} \frac{S}{Ks + S} \tag{2.11}$$

K_{max} plays a similar role to μ_{max} in the following expressions.

Contois

The specific growth rate coefficient (μ) for Contois (1959) depends on the biomass concentration (X):

$$\mu = \mu_{max} \frac{S}{K \cdot X + S} \tag{2.12}$$

This expression is a variation of the Monod kinetics and takes into account that mass transfer limitation can lead μ to vary with microbial population density.

Chen and Hashimoto

The Chen and Hashimoto (1978) model is an application of that of Contois to the anaerobic digestion processes. These authors, however, include the influence of the initial concentration of substrate (S_o) in the kinetic expression, in order to express mass transfer limitation:

$$\mu = \mu_{max} \frac{S}{K \cdot S_o + (1 - K)S} \tag{2.13}$$

Co-substrate utilisation

When two substrates (S_1 and S_2) are transformed into one product by means of a group of bacteria, both of them can act as a limiting substrate. So two Monod factors with two half-saturation constants should be used:

$$\mu = \mu_{max} \frac{S_1}{K_{S1} + S_1} \cdot \frac{S_2}{K_{S2} + S_2} \tag{2.14}$$

2.2.3. Inhibition

Microbial growth rate so far has been described as dependent only on substrate and microbial population. However, it also depends on other factors, one of the most relevant being inhibition. Some compounds make the reaction velocity decrease in the cell processes by inhibiting the catalytic activity of enzymes. In modelling, from an engineering point of view, this effect can be expressed by a variation of the specific growth rate in equation 2.1, or the bacterial decay coefficient in equation 2.5.

In a microbial reaction, inhibition can be caused by substances either entering with the influent substrate or being produced by the anaerobic process itself. Regarding the latter, a simple classification could be:

- Substrate inhibition, where the substrate provokes enzymatic inhibition.
- Product inhibition, which is caused by products that are final or intermediate in the chain of simultaneous biochemical reactions, such as H^+ (pH), H_2, NH_3, H_2S and volatile or long chain fatty acids.

Correction of the specific growth rate coefficient (μ)

Three main ways of quantifying inhibition appear in the literature. All three can be expressed by a change in the two coefficients of the Monod kinetics (equation 2.8), which arranged as

$$\mu = \mu_{max} \frac{1}{1 + \frac{K_s}{S}} \tag{2.15}$$

clarifies the interpretation of the expressions that follow.

- *Competitive inhibition*

Competitive inhibition occurs with compounds that are similar to the substrate. These inhibitors compete with substrate for the active enzyme sites. When there is substrate in excess the inhibition decreases. Therefore the action of a competitive inhibitor substance can be modelled by increasing the half-saturation constant in the Monod equation (Costello et al., 1991) without any change in the μ_{max} value:

$$\mu = \mu_{max} \frac{1}{1 + \frac{K_S}{S}\left(1 + \frac{I}{K_i}\right)} \qquad (2.16)$$

I, being the inhibitor concentration, and K_i the inhibiting coefficient.

- *Non-competitive inhibition*

A non-competitive inhibitor fixes to those enzyme sites on which substrate never fixes. Thus, the enzyme decreases its activity independently of the substrate concentration. Hence, K_S remains constant but the value of the maximum specific growth rate coefficient decreases (Lehninger et al., 1982) according to the expression:

$$\mu = \mu_{max} \frac{1}{(1 + \frac{K_s}{S})(1 + \frac{I}{K_i})} \qquad (2.17)$$

- *Incompetitive inhibition*

In this type of inhibition the inhibitor only fixes on the complex enzyme-substrate. Its effect can be reflected by the decrease in the maximum specific growth rate (μ_{max}) and the half saturation constant (K_s) in the Monod equation (Desjardins and Lessard, 1992). The coefficient μ is affected as follows:

$$\mu = \mu_{max} \frac{1}{1 + \frac{K_s}{S} + \frac{I}{K_i}} \qquad (2.18)$$

A transformation of equation 2.18 shows that both μ_{max} and K_S decrease by the factor $K_i/(K_i{+}I)$. When the inhibiting substance is the substrate itself, equation 2.18 conforms to the Haldane (1930) expression:

$$\mu = \mu_{max} \frac{1}{1 + \frac{K_s}{S} + \frac{S}{K_i}} \qquad (2.19)$$

The presence of several incompetitive inhibitor substances can be modelled by incorporating additional terms (I/K_i) in the denominator of equation 2.18.

Correction of microbial decay coefficient

In some cases, such as the AD of swine manure, the previous descriptions of inhibition cannot accurately predict process behaviour. Several authors (Hill et al., 1983) suggest that the inhibitor compound affects the bacterial decay rate by increasing the coefficient K_d.

2.2.4. Influence of temperature

Reactor temperature affects not only the reaction velocities of physico-chemical processes but also (and principally) the biochemical conversion rates.

The average value of the temperature over a long period fixes the bacterial population, thus defining two major groups of microorganisms with their inherent kinetic coefficients. These are thermophilic (45–65 °C) and psychro-/mesophilic (10–40 °C). However, it is more common to classify them into thermophilic (45–65 °C) and mesophilic (25–40 °C). When there is a rapid change of temperature from one range to the other, there is no population shift as the groups are not compatible.

In each group of microorganisms a variation of the reactor temperature within its range changes the reaction velocity. The classic expressions for the variation of reaction rate (r) with temperature T are based on the Arrhenius equation (Andrews, 1989; Dinopolou et al., 1988) which becomes more manageable when expressed as

$$r_T = r_o \cdot e^{C(T - T_o)} \tag{2.20}$$

where

r_o = reaction rate at temperature (of reference) T_o

r_T = reaction rate at temperature T

C = A coefficient that takes into account activation energy and temperature; and that in the usual range of temperatures in anaerobic digestion can be considered as a constant.

The reaction rate r_T is expressed by equation 2.1 for microorganism growth, by equation 2.3 for substrate utilisation, and by equation 2.5 for decay. The variation of reaction rate expressed in paragraphs 2.2.2 and 2.2.3 can be considered as if the changes in temperature alter the value of certain biochemical coefficients. The maximum specific growth rate (μ_{max}) and decay coefficient (K_d) are usually considered to be the most affected by temperature change. Hydrolysis coefficients also vary with temperature.

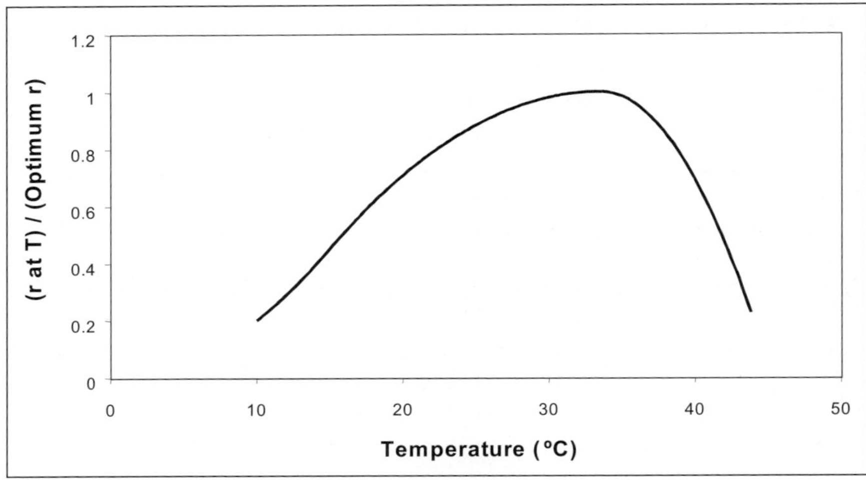

Figure 2.1. The theoretical ratio (reaction rate at temperature T)/(optimum reaction rate), as a function of temperature, in the mesophilic range.

Equation 2.20 shows a value that always increases or decreases with temperature (according to the value of C), which is not always accurate. Actually the theoretical curve of reaction rate versus temperature (Figure 2.1) follows a pattern with a maximum at the value of the optimum temperature (Brock, 1978). Other approaches, therefore, must be established, consistent with the curve in Figure 2.1. Some authors suggest using two different Arrhenius expressions: one increasing until the optimum temperature is reached and another decreasing for temperatures higher than the optimum. It is not a continuous curve but it agrees more with experimental results. Other authors (Angelidaki et al., 1993) suggest a different linear expression for either side of the optimum temperature.

$$r_T = r_{op} - K(T_{op} - T) \qquad \text{for } T<T_{op} \qquad (2.21)$$

$$r_T = r_{op} \frac{T_{max} - T}{T_{max} - T_{op}} \qquad \text{for } T>T_{op} \qquad (2.22)$$

where

T_{max}: Temperature at which there is no microbial growth

T_{op}: Temperature of optimum growth

K: Coefficient

r_{op}: Optimum reaction rate (at T_{op})

The decay coefficient (K_d), however, decreases with temperature. An Arrhenius expression (equation 2.20) with the appropriate constant value can express this influence.

2.3. PROCESS KINETICS IN ANAEROBIC DIGESTION

Anaerobic Digestion (AD) of organic wastes is a complex system involving several bacterial groups living in syntrophic association. The process takes place by virtue of many simultaneous reactions catalysed by the microorganisms. The compounds present exhibit different oxidation states until being finally converted into carbon dioxide and methane.

The simplified diagram in Figure 2.2 shows the growth of the microorganisms (from the substrate, whose degradation they are responsible for) forming reaction products and undergoing cell decay. This description has been expressed for a long time by simple diagrams (Figure 2.3) that show three elementary processes involving several reactants, microorganisms and products (McCarty, 1964; Andrews, 1969; McInerney and Bryant, 1981). Other authors, however, have produced more complex diagrams in an attempt to take into account the main processes involved in the overall anaerobic digestion system. Figure 2.4 (Gujer and Zehnder, 1983) summarises clearly their basic approaches. Figure 2.5 (García-Heras et al., 1999; Pohland, 1992) presents a similar reaction scheme, but in a matrix form which is appropriate to the further inclusion of a summary of process stoichiometry (Siegrist et al., 1993).

The main biochemical processes in AD have been classified, here, into four groups, i.e. hydrolysis, fermentation of monomers, acetogenesis/ hydrogenotrophic methanogenesis and aceticlastic methanogenesis. There are, however, two other processes that are not discussed because they are not relevant to the waste being dealt with, here. One is the action of homoacetogenic bacteria, which produce acetate from bicarbonate and hydrogen (acetogenic respiration of bicarbonate). The other is the nitrate and sulphate reduction to ammonium and sulphide by nitrate and sulphate-reducing bacteria, respectively. This process oxidises reduced organic products, acetate and hydrogen at the expense of nitrate and sulphate, which act as electron acceptors (Pohland, 1992).

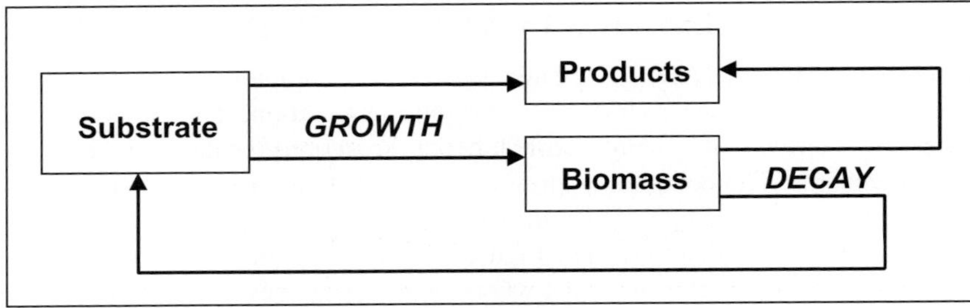

Figure 2.2. A simple scheme for microorganism growth and decay.

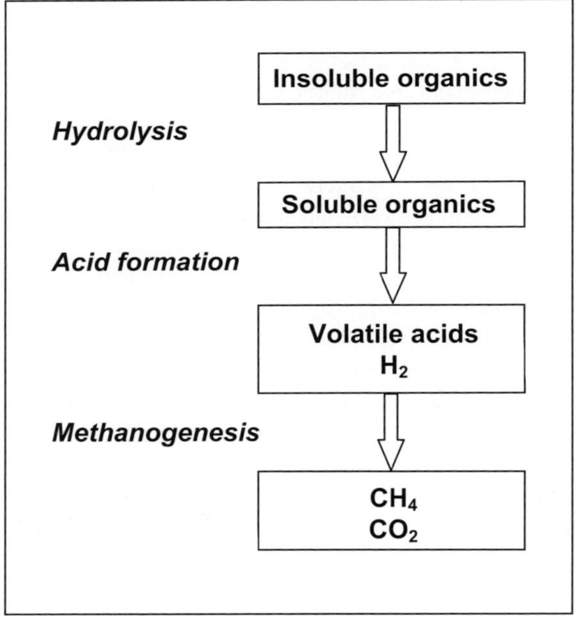

Figure 2.3. Anaerobic digestion described by three processes: hydrolysis, acidogenesis and methanogenesis.

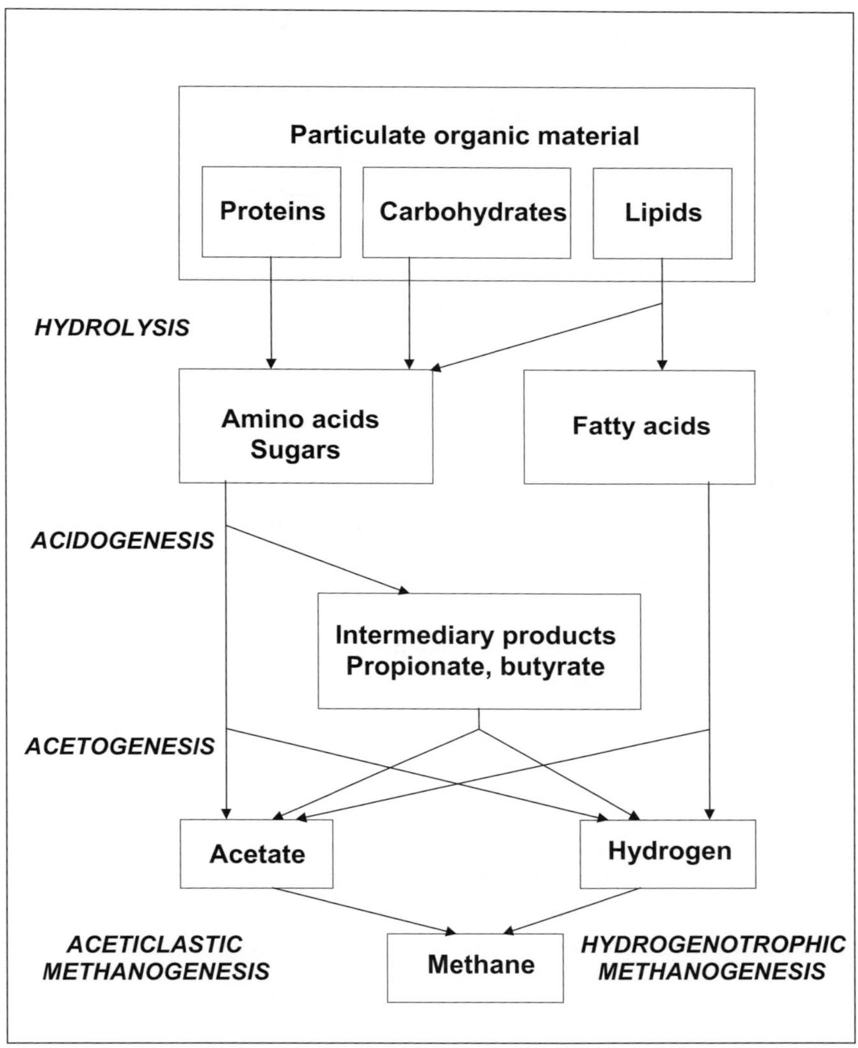

Figure 2.4. A complex diagram for anaerobic digestion considering five simultaneous processes (Gujer and Zehnder, 1983).

Process ↓	Organic polymers	Organic monomers	Reduced organics	NO_3^- $SO_4^=$	NH_4^+ H_2S	Acetic acid	Carbon dioxide	H_2	CH_4	*Microorganisms* ↓
Hydrolysis of organic polymers										
Fermentation of organic monomers										Fermentative bacteria
Oxidation of reduced organics										Obligate hydrogen producing bacteria (OHPA)
Acetogenic respiration of bicarbonate										Homoacetogen bacteria
Oxidation of reduced organics by SRB and NRB										Sulfate-reducing (SRB), nitrate-reducing (NRB) bacteria
Oxidation of acetate by SRB and NRB										SRB and NRB
Oxidation of Hydrogen by SRB and NRB										SRB and NRB
Aceticlastic methanogenesis										Aceticlastic methanogenic bacteria
Hydrogenotrophic methanogenesis										Hydrogenotrophic methanogenic bacteria

Organic polymers : Carbohydrates, Lipides, Proteins
Organic monomers : Sugars, organic acids, aminoacids
Reduced organics : Volatile fatty acids (propionic, butiric, valeric)

Figure 2.5. A complex diagram for anaerobic digestion in a matrix form (García-Heras et al., 1999; Pohland, 1992). Each arrow indicates the reactants and products (first row) of a process (first column) catalysed by a group of bacteria (last column).

2.3.1. Hydrolysis of organic polymeric material

The first step in anaerobic biodegradation is the conversion of the complex waste (particulate and soluble polymers) into soluble products by enzymatic hydrolysis. These soluble compounds can then be used as a substrate by the acidogenic/fermentative bacteria. The microorganisms involved in AD produce the extracellular enzymes necessary as catalysts in hydrolysis.

Complex organic wastes such as organic industrial wastes, the organic fraction of municipal solid wastes, and sludge from the treatment of both industrial organic wastewater and sewage, can all undergo hydrolysis. In the case of wastage of activated sludge or other biological treatment systems, microbial cells are considered simply as a substrate for the anaerobic process. They undergo a lysis process in which the cell membrane breaks down, so releasing organic matter to the bulk substrate (Gossett and Belser, 1982; Pavlostathis and Gossett, 1986).

Although hydrolysis of proteins and lipids have not been so widely studied as that of carbohydrates (particularly cellulose), the former represent a very significant proportion of the type of wastes dealt with here (McInerney, 1988). As a whole, the products from this first step are organic monomers, i.e., amino acids, long chain fatty acids and sugars.

Hydrolysis proceeds by adsorption of exocellular enzymes on the particulate substrate together with their reaction with the soluble substrate (Sanders et al., 1999; Noike et al., 1985), and also by attachment of the enzyme producing bacteria to the substrate organic particles (Vavilin et al., 1996; Hobson, 1987).

Process kinetics

Three main kinetic expressions can be used for the particulate substrate utilisation rate. First-order kinetics in substrate describes hydrolysis as a process unaffected by the microorganism concentration. Equation 2.23 thus includes all the above mentioned mechanisms (Eastman and Ferguson, 1981):

$$r_{XS} = -K \cdot X_s \tag{2.23}$$

Equation 2.24 describes hydrolysis as the attachment of bacteria on the particulate substrate and its consequent colonisation (Vavilin et al., 1996):

$$r_{XS} = -K \frac{X_s}{K_{XS} + X_s} \cdot \frac{X_B}{K_{XB} + X_B} \tag{2.24}$$

If the effect of microbial density is considered, the reaction rate in equation 2.25 can be established (Henze, 1995). It shows saturation when substrate concentration is high with regard to biomass. It then becomes kinetics of first-order in X_B. When biomass concentration is large, regarding substrate, the expression becomes first-order in X_s. Equation 2.25 shows the Contois concept of mass transfer (1959) exhibiting the same form as equation 2.12.

$$r_{XS} = -K \frac{Xs / X_B}{K_X + X_s / X_B} X_B \tag{2.25}$$

where

X_S is the substrate concentration

X_B is the microbial concentration

K is the maximum specific hydrolysis rate

K_{XS}, K_{XB} and K_X are half-saturation constants.

Hydrolysis can be inhibited by the accumulation of amino acids and sugars (Sanders et al., 1999) due to the hindrance of enzymatic production and activity. The kinetics may be expressed as non-competitive inhibition. Hydrolysis does not imply microbial growth, hence the effect of inhibition must be expressed by a factor in the reaction velocity (r_{XS}). The reaction velocity in equations 2.23, 2.24 and 2.25 is therefore corrected by

$$\frac{1}{1 + \frac{I}{K_i}} \tag{2.26}$$

I, being the concentration of sugars or amino acids, and K_i the coefficient of inhibition for these compounds.

Other possible inhibitors are un-ionized VFA (Brummeler et al., 1991; De Baere et al., 1985). The adverse effect of pH is associated with the presence of the un-ionized form of VFA; so it is not easy to separate their influence (Ratledge, 1994). A non-competitive inhibition is suggested, which is expressed by exposing equations 2.23, 2.24 and 2.25 to the factor in equation 2.26, where I is the un-ionized VFA concentration. Thus the effect of pH is included (Llabres-Luengo and Mata-Alvarez, 1988).

2.3.2. Fermentation of organic monomers (acidogenesis)

In this stage the fermentative bacteria degrade the organic monomers of sugars and amino acids (following Stickland reactions), producing volatile fatty acids (VFA), acetate, H_2 and CO_2 (Figures 2.4 and 2.5). Propionic, butyric and valeric acids produced in this step are referred to here as VFA. Ammonia is also produced by the degradation of amino acids.

Process kinetics

Microbial growth is generally expressed by the Monod expression (equations 2.1 and 2.8), where:

S is the concentration of sugars or amino acids
X is the concentration of biomass degrading sugars or amino acids.

So two kinetic expressions appear, one for sugar utilising bacteria, and another for amino acid utilising bacteria. However, a reasonable simplification is to assume only one type of microorganism for both (Gujer and Zehnder, 1983).

Fermentation of sugars is regulated by the concentration of Hydrogen. Since a non-competitive inhibition is observed (Mosey, 1983), a correction factor for the kinetics must be applied as shown in equation 2.17, where:

S is the concentration of sugars or amino acids
I is the concentration of dissolved hydrogen
K_i is the hydrogen inhibition constant.

Volatile fatty acids inhibit acidogenesis following an incompetitive pattern (equation 2.18) (Moletta et al., 1986; Hill and Barth, 1977), although Dinopolou et al. (1988) observed that a non-competitive pattern (equation 2.17) fits the experiments better. In these expressions:

S is the concentration of sugars or amino acids
I is the VFA concentration
K_i is the VFA inhibition coefficient.

2.3.3. Acetogenesis/hydrogenotrophic methanogenesis

Both long chain fatty acids (LCFA) and volatile fatty acids (VFA) are degraded by the obligate hydrogen producing acetogens (OHPA), generating acetate, carbon dioxide and hydrogen (Figure 2.4). Those organic acids having more than 5 atoms of carbon are considered here as LCFA (Angelidaki et al., 1998).

Fatty acids act as electron donors in producing CO_2 during their degradation, as well as electron acceptors in transforming H^+ into H_2. Carbon dioxide, on the other hand, is an electron acceptor, being converted to methane together with hydrogen (an electron donor) by the hydrogenotrophic methanogenic bacteria. The latter microorganisms grow in a syntrophic co-culture, together with the acetogenic OHPA bacteria. So acetogenesis must be considered jointly with hydrogenotrophic methanogenesis (Harper and Pohland, 1986; Boone, 1989).

Process kinetics

Two groups of bacteria act as acetogens. One utilises LCFA and the other VFA. Each has a specific growth rate expressed by the Monod kinetics (equations 2.1 and 2.8). X is the microbial concentration and S the concentration of LCFA or VFA. The values of μ_{max} and K_S are considered to be different for both groups of compounds as is shown, later, in Table 2.1.

The specific growth rate of the hydrogenotrophic methanogenic bacteria is expressed by the double Monod equation for a co-substrate (equation 2.14), where:

S_1 is the concentration of dissolved hydrogen

S_2 is the concentration of dissolved CO_2

X is the concentration of hydrogenotrophic methanogenic bacteria.

The syntrophic hydrogenotrophic methanogenesis seems to be inhibited mainly by acetate although other fatty acids also affect the process. A non-competitive expression is suggested (equation 2.17) (Moletta et al., 1986). Several authors (Andrews and Graef, 1971; Carr and O'Donnell, 1977) prefer an incompetitive expression (equation 2.18) but with reference to methanogenic bacteria in general. In them:

S is the concentration of dissolved hydrogen

I is the concentration of acetate or other fatty acids

K_i is the inhibition constant.

The decrease in activity of the hydrogen-utilising methanogens leads to H_2 accumulation.

Acetogenesis shows a double inhibition. On the one hand, hydrogen acts in a non-competitive way (equation 2.17) (Mosey, 1983) and on the other, acetate and other weak acids inhibit the process by competitive inhibition (equation 2.16) (Costello et al., 1991). In the equations:

S is the concentration of fatty acids (LCFA or VFA)
I is the hydrogen and acetate concentration, respectively
K_i is the hydrogen and acetate inhibition constant, respectively.

2.3.4. Aceticlastic methanogenesis

Methane is the only reaction product that is not a reactant in the whole process and can, therefore, be considered as an end product. Two processes generate it. hydrogenotrophic methanogenesis produces methane from the use of H_2 and CO_2 by the hydrogen-consuming bacteria in a syntrophic co-culture with the OHPA bacteria. The methanogenic aceticlastic bacteria, on the other hand, grow on acetate as substrate so producing methane and carbon dioxide. This second mechanism accounts for most of the CH_4 produced in the overall process.

Process kinetics

Aceticlastic methanogenic bacteria have a specific growth rate (μ) also expressed by equations 2.1 and 2.8, S being the acetate concentration.

Although aceticlastic methanogenesis has a simple reaction pattern, many internal compounds, some of which are toxic due to the pH value, inhibit it.

The inhibiting action of the un-ionized VFA can be expressed by an incompetitive expression (equation 2.18) (Buhr and Andrews, 1977; Dalla Torre and Stephanopoulos, 1986) where I is the un-ionized VFA and S the acetate concentration. Other authors (Hill, 1983) suggest increasing the decay rate coefficient by VFA as a way of modelling it. The possible inhibition by un-ionized LCFA is expressed by a non-competitive equation (equation 2.17) by others (Hwu et al., 1996).

Free ammonia (NH_3) also inhibits the process (Angelidaki et al., 1993). It is modelled as a non-competitive mechanism (equation 2.17) (Siegrist et al., 1993). Other researchers (Hill and Barth, 1977) have presented the combined action of free ammonia inhibition together with un-ionized VFA by expanding equation 2.18 applied to the specific growth rate of aceticlastic methanogens:

$$\mu = \mu_{max} \frac{1}{1 + \frac{K_S}{S} + \frac{I_1}{K_{i1}} + \frac{I_2}{K_{i2}}} \qquad (2.27)$$

I_1 and I_2 being the acetate and ammonia concentration respectively, and K_{i1} and K_{i2} their inhibition coefficients.

Influence of pH can be considered as included in the mentioned approaches provided that the concentrations of un-ionized acids as well as that of free ammonia are those variables managed in the previous inhibition expressions. Otherwise, the effect of pH should be included, in which case H^+ concentration would influence the inhibition (Rozzi, 1984; Costello et al., 1991).

2.3.5. Physico-chemical processes

The four biochemical processes defined above are accompanied by non-enzyme catalysed physico-chemical phenomena. They have reaction rates that can be correlated in the chemical equilibrium state.

Biogas stripping–absorption

The gas transfer phenomena in an anaerobic digester of CSTR type, occurs as a consequence of the production of soluble gas components in the bulk liquid which are immediately stripped to gas bubbles and simultaneously absorbed in a process of dynamic equilibrium. The biogas produced consists of three main components: carbon dioxide, methane and hydrogen.

For each gas component this process has been traditionally explained by means of the stationary liquid film theory described in reactor engineering literature (Levenspiel, 1981; Coulson and Richardson, 1984). The concept of overall gas transfer (GT) can be expressed by the K_La coefficient which takes into account the liquid film thickness as well as the gas–liquid surface of contact (Arrua et al., 1990; Ho et al., 1987). The mass of gas transferred from liquid to gas phase is expressed by the variation of the dissolved gas concentration as:

$$\left(\frac{dS}{dt}\right)_{GT} = K_L a(S - S_s) \qquad (2.28)$$

which, according to Henry's law, in the equilibrium or stationary state becomes

$$\left(\frac{dS}{dt}\right)_{GT} = K_L a\left(S - \frac{P_p}{H}\right) \qquad (2.29)$$

where

S is the concentration of dissolved gas in the bulk liquid (Kg/m^3)

S_s is the concentration of dissolved gas at the boundary layer interface in equilibrium with the gas phase (Kg/m^3)

K_La is the overall gas transfer coefficient (d^{-1})

P_p is the gas partial pressure (atm)

H is Henry's constant ($atm{\cdot}m^3/Kg$).

For modelling purposes, the variation in S is the sum of equation 2.29 together with the variation in S due to the biochemical reactions in which S is involved. An equation similar to equation 2.29 can be written to express the variation in the amount of each biogas component in the gas phase within the reactor. Each biogas component is thus defined by two variables: one representing that part which is dissolved in the liquid, and the other, that part which is in the gas state (Garcia-Heras et al., 1999). The biogas pressure in the reactor is the sum of the three partial gas pressures. These correspond to three state variables in the reactor, i.e., the volumetric fraction of each gas component in the gas phase.

Buffer systems and chemical equilibrium

The chemical transformations taking place in the liquid phase of an anaerobic reactor give rise to a variety of weak acids both organic and inorganic. Hence the H^+ concentration is a consequence of the whole system performance. It is also well known how relevant a role pH plays in anaerobic digesters. Prediction of pH is therefore a very important issue.

The pH value is strongly affected by the buffer capacity of the system. A useful simplification to compute pH is to take into account only a limited number of weak acids (Angelidaki et al., 1993). Bicarbonate, volatile fatty acids and ammonia are the main process controllers in terms of buffer capacity. Each buffer system predominates at a different pH range; therefore it is interesting to consider all the three, so as to be able to predict the process behaviour for different new advanced reactor configurations. A pH range of 5.5–8 can be covered by this approach, in which ammonia–ammonium dominates at high values, bicarbonate – carbon dioxide at the medium values and VFA at the smallest pHs in the range.

Each buffer solution involves a two-way reaction in which reaction rates, in dynamic equilibrium, can be defined. These are related by the ionisation constant (K_A) in a state of chemical equilibrium.

Volatile fatty acids follow a pattern similar to that of acetic acid, which dissociates as follows:

$$CH_3COOH \rightleftarrows CH_3COO^- + H^+ \tag{2.30}$$

The reaction rate to the left is: $K_1 \left[CH_3COO^-\right]\left[H^+\right]$ (2.31)

And to the right: $K_2\left[CH_3COOH\right]$ (2.32)

In equilibrium: $K_2 / K_1 = K_{A,CH3COOH}$ (2.33)

For pH values at which anaerobic digesters usually work, the presence of ion carbonate is neglected; hence, only the bicarbonate – carbon dioxide reaction is considered relevant. The dissolved CO_2 is converted to H_2CO_3 fast enough for the latter to be considered as the actual compound in the solution (Sperandio and Paul, 1997). Thus gas transfer modelling can be linked with the following equations:

$$H_2CO_3 \rightleftarrows HCO_3^- + H^+ \tag{2.34}$$

one reaction rate is: $K_3\left[HCO_3^-\right]\left[H^+\right]$ (2.35)

and the other: $K_4\left[H_2CO_3\right]$ (2.36)

At equilibrium: $K_4 / K_3 = K_{A,\,H2CO3}$ (2.37)

Ammonium appears in solution as NH_4^+ and dissociates, producing dissolved ammonia (NH_3) which is stripped to the gas

$$NH_4^+ \rightleftarrows NH_3 + H^+ \quad (2.38)$$

one reaction rate is: $K_5[NH_3][H^+]$ (2.39)

and the other: $K_6[NH_4^+]$ (2.40)

At equilibrium: $K_6 / K_5 = K_{A,NH4^+}$ (2.41)

So, in this approach, the pH value is related chiefly to three state variables: concentration of acetate, bicarbonate and ammonium.

2.3.6. Typical coefficient values

One of the most important tasks needed to obtain an accurate quantification of the process is to achieve reliable values of the kinetic and stoichiometric coefficients. This extensive area of study is referred to as parameter estimation. In the present chapter values taken from current literature are shown. Wide intervals between values can be observed, because different authors use different substrates and define different experimental conditions. Despite this, some indicative figures are presented here.

Table 2.1 refers to several authors from whom the above-cited figures have been drawn. It does not include all the literature dealt with, but considers the papers that review the data in the most detail. Those values found in the papers referred to in Table 2.1 were estimated in tests within the mesophilic range (around 35 ºC). The estimation of K_La in anaerobic processes, however, is less known (Suescun et al., 1998).

Table 2.1. Summary of literature references for numerical values of kinetic coefficients in the processes constituting anaerobic digestion.

Authors	P1	P2	P3		P4	P5
			LCFA	VFA		
Angelidaki et al. (1998)	×	×	×	×	×	×
Eastmann and Fergusson (1981)	×					
Gujer and Zehnder (1983)	×	×	×	×	×	
Harper and Pohland (1986)					×	×
Heyes and Hall (1983)				×		
Lawrence and McCarty (1969)				×	×	
Noike et al. (1985)		×			×	
Novak and Carlson (1970)			×	×		
O'Rourke (1968)	×		×	×		
Pavlostathis and Giraldo-Gomex (1991)	×	×	×	×	×	×
Romli et al. (1995)				×	×	×
Van den Berg (1977)					×	

P1 Hydrolysis
P2 Fermentation of sugars and amino acids
P3 Anaerobic oxidation
P4 Aceticlastic methanogenesis
P5 Hydrogenotrophic methanogenesis

Hydrolysis

Hydrolysis is considered by most of the authors as a limiting step in anaerobic digestion. It is one of the key factors involved in performance quantification and prediction of AD of complex wastes. The equations expressing hydrolysis in the above paragraphs include half-saturation constants (K_{XS}, K_{XB}, K_X), and maximum rates (K) when saturation curves reach a value equal to 1. Experimental estimation of these parameters is still a developing field of research. The coefficient of the first-order equation (K) is the most commonly found in the literature, where the respective values are:

Carbohydrates	$K = 0.5–2\ (d^{-1})$
Lipids	$K = 0.1–0.7\ (d^{-1})$
Proteins	$K = 0.25–0.8\ (d^{-1})$

Complex wastes can be characterised, on the one hand, by their composition of lipids, proteins and carbohydrates. On the other, the degree of polymeric complexity and particle size must be considered (Hobson, 1987). Therefore estimation of the coefficient is a difficult issue.

Fermentation of monomers

Fermentation of organic monomers is referenced in the literature by considering either sugars or amino acids as substrate. Numerical values of the model coefficients can be obtained by assuming both groups of compounds to be the overall substrate. These are in the ranges:

$\mu_{max} = 3–9\ (d^{-1})$
$K_{max} = 24–120$ (g COD / g COD·d)
$K_S = 300–1400$ (mg/l)
$Y = 0.1–0.06$ (gVSS / g COD)
$K_d = 0.02–0.3\ (d^{-1})$

However, these are typical values. Experimental evaluation is needed for complex wastes because the composition of hydrolysis products very much depends on the way in which hydrolysis proceeds.

Acetogenesis

Acetogenesis, considered as anaerobic oxidation of long chain (LCFA) and short chain (VFA) fatty acids, is widely dealt with in the bibliography. Common values for both groups of compounds are the following:

	LCFA	VFA
μ_{max} (d^{-1})	0.1– 0.5	0.3–1.3
K_{max} (g COD / g COD·d)	2–20	5–20
K_S (mg COD / l)	100–4000	100–4000
Y (g VSS / g COD)	0.04–0.1	0.02–0.07
K_d (d^{-1})	0.01	0.01–0.04

Methanogenesis

Both aceticlastic and hydrogenotrophic methanogenesis have more specific values for the model coefficients since microorganisms use more simple (or pure) substrates. The former consumes acetate, and the latter, hydrogen and carbon dioxide. The coefficient values are:

	Aceticlastic Methanogenesis	Hydrogenotrophic Methanogenesis
μ_{max}(d^{-1})	0.1–0.4	1–4
K_{max}(g COD / g COD·d)	2–7	25–35
K_S (mg COD / l)	50–600	0.01–0.1
Y (g VSS / g COD)	0.02–0.05	0.04–0.1
K_d (d^{-1})	0.02–0.04	0.01–0.04

2.4. REACTOR DESIGN

In a biological treatment system, design starts from consideration of the type and configuration of the treatment plant. Then the evaluation of the reactor size is carried out. In addition to economic factors, the choice of plant type depends not only on the waste to be treated, but also on the final use to be given to the treated waste (Hobson and Wheatley, 1993).

For feedstock with low suspended solids the anaerobic treatments usually consist of retained biomass systems. When there is a high content of solids in the feed, as in the case of complex wastes, completely stirred reactors (CSTRs) are used (Mata-Alvarez, 1987). A description of types of reactor can be found, elsewhere, in this book. The most common digesters of this type are installed in sewage wastewater treatment plants where sewage sludge is in some cases co-digested together with other organic wastes, such as OFMSW. In these CSTRs there is no biomass recycling, so the concept of solid (or cell) retention time (SRT) has no meaning.

2.4.1. Steady-state models

To assess the size of an anaerobic digester of complex organic wastes, design criteria must be chosen. One method involves forecasting reactor performance as a function of design parameters. This can be achieved either from graphs based on previous experience, or from simple models of anaerobic digestion.

The most popular models for design are those steady-state models that describe the AD performance for constant values of both influent and operational conditions. They usually take a very simple definition of substrate, define a kinetics for substrate uptake, and then create a mass balance in steady state to produce the final equation. As they are applied to a single substrate and, hence, to only one microbial group, the ability of the model to predict the process performance is rather limited due to some lack of specificity in describing the phenomena involved in the overall process.

First-order model

The basic equation is:

$$dS/dt = -K \cdot S \tag{2.42}$$

where K is the first-order kinetic constant, and S here represents the concentration of all biodegradable compounds (i.e., organic polymers, amino acids, sugars and fatty acids) in the reactor.

For a CSTR digester operating at steady state, a mass balance in substrate, that takes into account equation 2.42, yields the value of S at steady state as a function of the hydraulic retention time (HRT):

$$S = S_0 \frac{1}{1 + K \cdot HRT} \tag{2.43}$$

As S is a difficult parameter to measure, another approach can be used. If B denotes the accumulated specific methane production in the particular reactor for each unit of volatile solids (VS) added, and B_0 is the ultimate methane yield (Chen and Hashimoto, 1978) in the same units, the concentration of biodegradable VS in the fermenter (S) will be directly related to gas production, in accordance with:

$$B_0 \cdot (S_0 - S) = B \cdot S_0 \tag{2.44}$$

Combining equations 2.43 and 2.44

$$B = B_0 \frac{K \cdot HRT}{1 + K \cdot HRT} \tag{2.45}$$

In this case B (specific methane production) is obtained as a function of HRT at steady state.

Although this is not a sophisticated model, it can provide a single and useful kinetic constant (K) that has applicability when dealing with complex systems such as those involved in the fermentation of refuse (Pfeffer, 1974). In addition, the model has been used in the anaerobic digestion of cornstover (Gaddy, 1981) and pre-treated straw (Baccay and Hashimoto, 1984). Other studies using first-order kinetics are those that deal with the anaerobic digestion of sewage sludge and pre-treated straw (Pavlostathis and Gossett, 1985, 1986), and those that deal with the anaerobic digestion of brewery by-products (Keenan and Kormi, 1977).

Monod model

Lawrence and McCarty (1970) developed a kinetic model at steady state for Activated Sludge by focusing on microbial growth in the system. It can be applied to an anaerobic process as well (Lawrence and McCarty, 1969). The procedure considers microbial growth in a substrate-limiting environment (equations 2.1 and 2.8). The net variation of the biomass concentration (X) is obtained by combining equations 2.7, 2.3 and 2.11. Then, by establishing a mass balance of X in the digester and assuming a sterile feed ($X_0=0$), the value of substrate concentration (S) is obtained in steady state:

$$S=\frac{K_S \cdot (1+K_d \cdot HRT)}{HRT \cdot (Y \cdot K_{max}-K_d)-1} \tag{2.46}$$

By using equation 2.44, the specific methane production (B) is obtained as a function of HRT

$$B=B_0\left(1-\frac{S}{S_0}\right) \tag{2.47}$$

S being the value derived from equation 2.46.

If the K_d value is neglected, equations 2.46 and 2.47 can be simplified, such that:

$$S=\frac{K_S}{HRT \cdot Y \cdot K_{max}-1} \tag{2.48}$$

$$B=B_0\left(1-\frac{K_S/S_0}{HRT \cdot Y \cdot K_{max}-1}\right) \tag{2.49}$$

This model is rigorously applicable to soluble substrates. Lawrence (1971) extended the application of Monod kinetics to municipal sewage sludge. In spite of the fact that this more complex waste made calculations somewhat more difficult, the estimated constants gave reasonably good results when methanogenesis was assumed as the limiting step. The Monod model has been applied to several wastes, as was pointed out by Chin (1981), and has been the basis of most anaerobic digestion models, such as the structured model developed by Bryers (1985).

Model for process limited by mass transfer

Chen and Hashimoto (1978) proposed the growth kinetics in equation 2.13 including in it the initial substrate concentration S_0. Following the same procedure as in the two previous paragraphs and neglecting decay (K_d=0) for a CSTR digester, a mass balance in microorganism concentration leads to the following equation for methane produced per mass of VS added:

$$B = B_0 \left(1 - \frac{K}{HRT \cdot Y \cdot K_{max} - 1 - K} \right) \tag{2.50}$$

In addition to their own data, these authors tested their model with O'Rourke's (1968) sewage sludge data, Pfeffer's (1974) municipal refuse data, and the animal manure data of several other authors. Many others have used this model. These include Samson and Leduy (1986), who studied the anaerobic digestion of algal biomass, and Lema et al. (1987) who modelled the biomethanization of landfill leachates.

2.4.2. Reactor sizing

AD of complex wastes may be employed with the objective of optimising organic matter removal and/or methane production. The degree of organic matter reduction required is a function of the final use of the treated waste, and must be in accordance with the link that follows in the waste treatment chain. Methane production, however, may be regarded as an alternative source of energy or a means by which energy can be saved in the whole plant. Both objectives condition the design criterion. In addition, both performance parameters are also a function of the waste composition as expressed by the model coefficients. So either an AD model or accumulated experience from pilot (or full-scale) tests on different wastes is needed for design.

Variables and sizing criteria

In complete mix digesters (CSTR) the main design parameter is volume. The effective reactor volume is affected by scum and grits accumulation due to the presence of inorganic matter or possibly to poor mixing performance. Consequently, a larger total volume is required. Precise details on this can be found in engineering handbooks (Water Environment Federation, 1995). Effective volume is the sizing parameter which is considered here.

The tank effective volume (V) is related to the hydraulic retention time (HRT) and the feed flow rate (Q) by:

$$V = HRT \cdot Q \tag{2.51}$$

In addition, the volume is also related to the feed organic matter concentration (S_0) and to the organic loading rate (OLR), also referred to as hydraulic loading rate:

$$OLR = \frac{Q \cdot S_0}{V} = \frac{S_0}{HRT} \quad (2.52)$$

The effective digester volume required depends on the sizing criterion chosen. Two criteria are usually considered. One fixes the hydraulic retention time, and the other fixes the organic loading rate; although both criteria are related.

HRT criterion

In practice the HRT value affects digester efficiency with respect to organic matter removal and to specific gas production (Figures 2.6 and 2.7). This efficiency also depends on feedstock composition and reactor temperature.

The proportion of carbohydrates, lipids and proteins in the raw waste implies a different overall biodegradability, which is exhibited as different values of removal efficiency and biogas production (Parkin et al., 1986). Primary sludge is more anaerobically biodegradable than wastage of activated sludge, as the latter contains many more bacterial cells than the former. This behaviour is observed in many digesters fed by other mixed wastes. It can be explained by the different values of the kinetic and stoichiometric coefficients (mentioned in previous sections) of the compounds present in the feedstock.

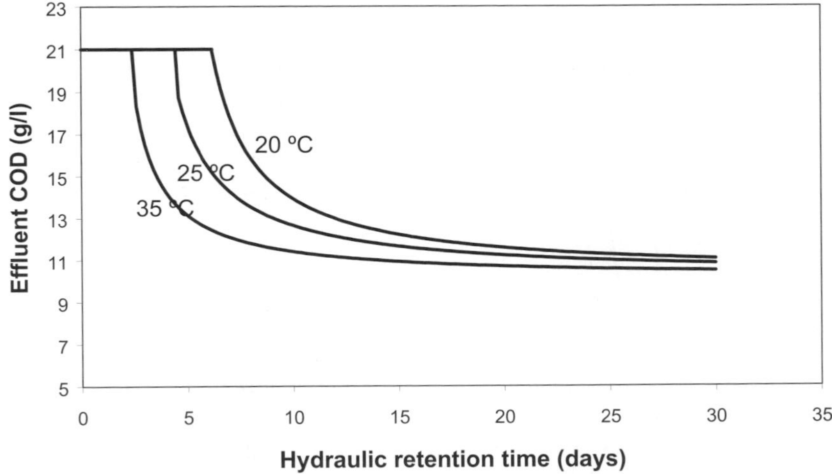

Figure 2.6. The variation of organic matter concentration in the effluent of an anaerobic digester, versus hydraulic retention time, as a function of temperature (WEF and ASCE 1992).

Digester performance is influenced by the operation temperature in the digester for the same type of waste. There are three ranges of temperature, i.e., psychrophilic, mesophilic and thermophilic (around 15, 35 and 55 ºC respectively). They involve two large groups of bacteria (15–35 ºC; and 55 ºC). These exhibit different kinetic behaviour and, hence, different reactor efficiencies. Within the same temperature range and, therefore, for the same group of microorganisms, temperature affects the digester efficiency as kinetic coefficients vary. Figure 2.6 is an example of the variation of reactor organic matter concentration for the psychro–mesophilic microbial group in a digester fed by sewage sludge. Figure 2.7 shows the variation of the specific gas production (WEF and ASCE, 1992).

Figures 2.6 and 2.7 were obtained empirically but show a clear agreement with equations 2.46 and 2.47 obtained at steady state. Their graphic representation agrees fully with these graphs. The residual effluent COD in Figure 2.6 corresponds to the non-biodegradable fraction of the influent substrate. The presence of 3 kinetic and 1 stoichiometric coefficients shows the influence of substrate composition and temperature on digester performance.

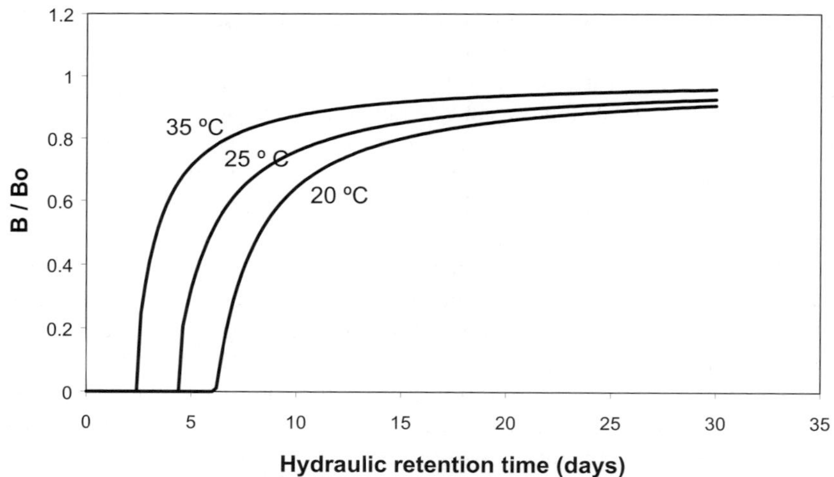

Figure 2.7. The ratio of specific gas production to maximum specific gas production (B/Bo) versus hydraulic retention time, as a function of temperature (WEF and ASCE, 1992).

Equation 2.46 is used to define a minimum hydraulic retention time (critical HRT), below which the digester does not degrade the substrate at all and, hence, does not produce any biogas. This is illustrated in Figures 2.6 and 2.7. The value of critical HRT is obtained simply by making S equal to S_0:

$$HRT_c = \frac{1}{\frac{S_o \cdot Y \cdot K_{max}}{K_S + S_o} - K_d} \tag{2.53}$$

Consistently, in equation 2.47, B equals zero for the critical HRT.

So the effective digester volume can be calculated from equation 2.51, for a particular feed flow rate, provided an HRT value has been fixed by the designer. This value must be greater then the critical HRT (equation 2.53). The HRT value is selected in order to achieve certain organic matter removal and certain specific methane production as shown in Figures 2.6 and 2.7 or equations 2.46 and 2.47.

OLR criterion

The organic loading rate (OLR) criterion is the same as the HRT criterion for a constant feedstock concentration (S_o) (equation 2.52); i.e., fixing OLR establishes the value of HRT, and hence the volume. Thus an increase in the OLR design value can be made with the intention of decreasing HRT, thereby affecting design, as in the previous paragraph.

A different concept, however, is involved when an increase in the design value of OLR is imposed in the interest of increasing S_o at a constant HRT value. This affects the digester performance in two ways: physically, owing to higher feedstock viscosity; and biochemically. On the one hand, the physical change in viscosity of the digester content can negatively affect mixing efficiency for CSTR. It is a relevant engineering issue for complex wastes and is dealt with elsewhere (Water Pollution Control Federation, 1987). For this type of feedstock in a single stage anaerobic digester, the upper limit in concentration is around 12–15% of total solid, and is referred to as 'wet' condition (De Baere et al., 1985). On the other hand, biochemically speaking, the increase in OLR obtained by raising the feedstock concentration at a constant HRT can be compared with the change of food:microorganism ratio (F/M) in the activated sludge process (expressed as $S_o/(HRT \cdot BH)$, BH being the biomass concentration). There, the proportion S_o/BH can increase when BH is controlled physically by sludge recycling (Benefield and Randall, 1980). The phenomena of mass transfer, and variation in hydrolysis efficiency, can both be considered as factors, which have a significant influence when digester performance responds to an increase in OLR, given a constant value of HRT.

The reactor size can be calculated using equation 2.52 with the selected OLR value. Then, either HRT is fixed, resulting in S_0; or S_0 is fixed, resulting in HRT. The influence of OLR as a design criterion, however, is not widely described in the literature (Henze, 1995).

2.5. DYNAMIC MODELLING

In the 1970s some waste treatment engineers considered the use of dynamic models as somewhat academic and without significant practical application. In modern process engineering, however, they are increasingly in use to deal with the design and control of complex multivariate systems. Fortunately, the field of wastewater treatment has recently adopted this change of direction (Ekama and Marais, 1984; Gujer et al., 1999; IWA Task Group, 2000; Dochain et al., 1999).

With regard to anaerobic digestion, the reasons for dynamic modelling are quite clear, as follows. The steady-state models for AD of complex wastes described in previous paragraphs consider only one substrate, one type of bacteria and hence one group of coefficients. There are also references in the literature to many other particular steady-state equations. Anaerobic treatment processes, however, behaves in a much more complicated manner. As has been already seen through this book, the system involves many components, many processes and microorganisms, each with its kinetic and stoichiometric coefficients. Moreover, several processes affect each other in terms of reaction components. These considerations illustrate why steady-state model predictions have a limited application. Inherently they involve very few components; the interaction between them is not considered; and complex substrates are poorly characterised in terms of the model variables.

In addition, two more aspects are relevant in this regard. One is that disturbances imposed on the treatment plant, such as influent variations in flow rate or composition, usually influence process performance. Therefore it is convenient to have some means of prediction of the system's transient state. The other aspect is the increasing trend towards an integrated treatment of wastewater, sludge and organic wastes. Thus the AD of complex wastes is sometimes preceded by pre-treatment or followed by post-treatment; the liquid sidestreams produced in the digesters are recycled to the water train; there are physical and hydraulic constraints in reactors; and, finally, there is a continuous improvement in the configuration of reactors. This leads to a tendency to link partial sub-models together to produce an overall model with good predictive capacity.

The latter consideration might seem to suggest that simple models are required for this purpose. However, if an accurate prediction is pursued, the simplification only comes after partial dynamic models have been developed, checked and validated. This continues to offer a broad field of research and application.

2.5.1. Model complexity

The simplest classification of models (Jeepson, 1996) distinguishes those that are independent of time (steady-state models) from those that predict performance as a function of time (dynamic). Models can also be classified according to their structure: 'black box' are those based only on empirical relationships between real input–output; 'mechanistic' are those based upon the biochemical and physical laws ruling the process; and 'grey box' are those that are a mixture of both. Models can also be divided into those that consider inputs and outputs as probability distributions (stochastic) and those that disregard any uncertainty in data (deterministic). This distinction is not always clear, however, owing to the inherently statistical nature of real measurements. Alternatively, models which consider the form of mathematical equations can be defined by using qualitative expressions (qualitative) and by using deterministic functions (functional). Throughout this chapter attention has been drawn to models that are dynamic, mechanistic, deterministic and functional at the same time.

Anaerobic digestion systems are governed by manipulating the operational variables according to the operating pattern, e.g., feed frequency and flow rate, feed concentration, temperature, recycling loops and so on. The system proceeds to a situation defined by a group of variables resulting from the process performance itself and that defines the state of the system (state variables). They are usually the concentration of process components in the liquid or gas phase. The system is thus defined. If a model tries to describe the process, a third set of parameters is required, together with the states and manipulated variables: the vector of stoichiometric and kinetic coefficients. So the equations constituting the model are set up using operational variables, state variables and coefficients. The former two are functions of time, whereas the later generally have constant values.

Therefore a multivariable system, such as AD, will be reflected by a complex model. Its complexity, however, should be appropriate to the purpose of the model. Operation/control, design and research demand an increasing degree of model complexity. This relates to the fact that the quantity of data does not necessarily reflect the value of the information contained. The description of the system is derived either from data or from process information. Olsson and Newell (1999) refer to this as 'data description' and 'process description'. The former is obtained by gathering real experimental data and the latter by studying the model of the process. The first description requires more data than the second does, but the second can produce more valuable information than the first, providing that the model is good enough.

The predictive capacity of a model used for design is restricted to the circumstances under which it has been verified in tests under controlled conditions. So the complexity demanded of the model is less than when the intention is to predict wider specific conditions, as in the case of process research. Even less model complexity is required for plant control because the number of controllable variables is much smaller than the number of state variables in the model. The model must be reduced so as to improve its ability to predict those aspects of plant performance capable of responding to the appropriate controllers.

2.5.2. Model development

Whatever the model's purpose is, a set of steps must be followed to obtain a dynamic model of the process, and to validate it. Figure 2.8 (Olsson and Newell, 1999) shows a clear picture of the steps involved. The upper three tasks deal with building the model, whilst the remaining three involve an experimental check on the accuracy of its predictions.

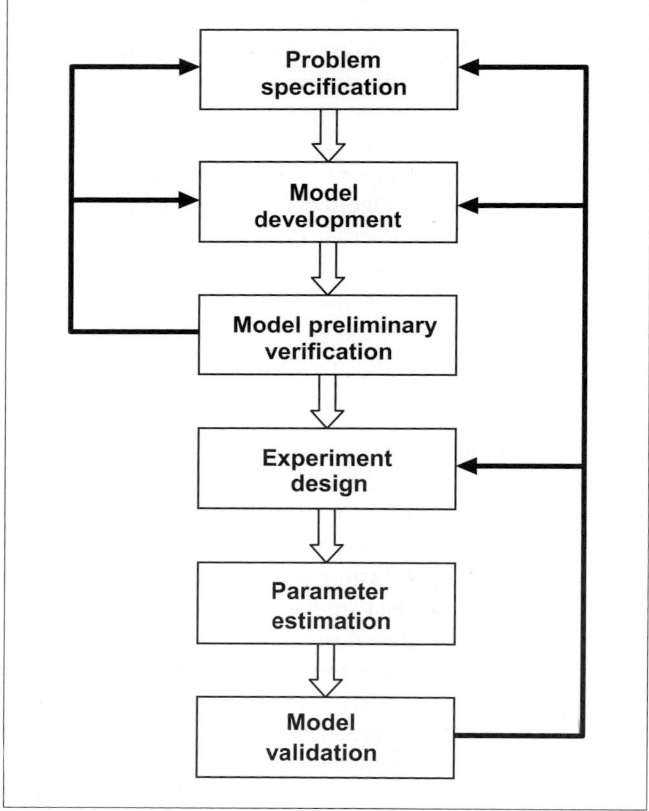

Figure 2.8. The steps to develop a dynamic model of anaerobic digestion (Olsson and Newell, 1999).

Problem specification

This first step can be compared to a chain with three links. Firstly, the purpose of the model must be clearly defined. Is its use intended for research, operation/control or design? Consistent with this, a working interval must be specified, in which the values of the operational variables will fall for the above-defined use (operation space). Then, the required degree of accuracy and reproducibility of the values of the chosen state variables (predicted by the model within the operation space) must be decided. The third link, as a consequence of all this, is the model size and complexity necessary to accomplish the allocated role.

Model building

Firstly, a conceptual model is needed to establish the equations that represent the process. It is derived from knowledge of biochemistry and physics. This work defines the processes involved and therefore the state variables. Then, the kinetic expression for every process is established, taking into account the set of coefficients of the model. The variation of each state variable over time, owing to the transformations in the reactor, must be expressed by considering all the processes in which each variable is involved. Finally, in the case of a continuous flow reactor, the derivative of each variable is calculated as the sum of the reaction term plus the transport term. Thus the system of differential equations is established (Roels, 1982).

A practical way to build such a type of model is to put processes, state variables and reaction kinetics into a matrix where the stoichiometry of the reactions is provided (Figure 2.5). This method has proved very useful in dealing with biological aerobic wastewater treatment systems for some years (IWA Task Group, 2000).

Preliminary verification

An initial analysis of identifiablity of the model should be conducted here. The question is whether there is – at least theoretically – only one result for the parameter estimates, or whether the model structure makes it impossible (Beck, 1989). An initial study of parameter sensitivity must be also carried out. In this step, the model is provided with a set of theoretical numerical values for its coefficients. By carrying out simulations in extreme conditions for the operational and influent variables, the soundness and consistency of predictions can be evaluated, in order to define the interval in which the model behaves as expected. If this does not fall within the interval in which the model is intended to be used, a return to the first and second steps is required (Figure 2.8).

2.5.3. Model validation

Once a reasonable model has been elaborated, an evaluation of the accuracy of its predictions is demanded. In other words, the degree of agreement between the values obtained from the model and the actual measured values provides the degree of validity of the model. If the evaluation is not satisfactory, a return to previous steps is necessary.

Experimental design

In this step, results of the studies on parameter sensitivity, practical identifiability and error propagation are exploited (Bastin and Dochain, 1990; Larrea et al., 1992). The optimum set of tests producing the best data for model fitting and validation is chosen at this point (Walpole and Myers, 1989; Schenk, 1968). These experiments must excite the system during dynamic tests.

Parameter estimation

The model is fitted to the experimental data by adjusting the parameters (kinetic and stoichiometric coefficients). This work is done by minimising the function of differences (residuals) between real measurement and model prediction for every state variable as a set. This is commonly known as linear and non-linear regression (Johansson, 1993; Draper and Smith, 1981). The theory of Process Identification provides a methodology for this difficult step (Ljung, 1987; Stearns and David, 1988).

The first relevant issue in this step of parameter estimation is waste characterisation. The waste composition has to be expressed in terms of the model state variables. As some of them cannot be measured in the laboratory, specific tests must be devised to identify them. The second issue concerns the estimation of kinetic parameters of the model. This crucial task has been widely tackled by many authors and is still a subject which continues to be researched in anaerobic digestion (Henze and Gujer, 1995).

Model validation

Finally, the model with the numerical values for its coefficients already estimated (generally known as the 'calibrated' model), must be tested so as to check its predictive accuracy and precision. Experimental data are compared with simulation results. This evaluation is based on the theory of State Estimation.

The type and number of experimental data used for model validation lead to two approaches. Integrated fitting and validation uses the same set of data for both purposes. Separated technique uses some data for fitting and others for validation. In both cases the residuals of validation data and the residuals of fitting data are compared by specific statistics (Box et al., 1978; Cooper, 1969). As both have advantages and disadvantages, users must choose the technique according to their specific needs.

2.5.4. Simulators

The software produced by implementing a model of a process is referred to as a simulator. Such a computer program includes equations of a model preliminarily verified, and the mathematical methods required to solve them. Thus the software predicts the process performance under certain operational conditions. This programme still requires numerical values of the set of coefficients of the model. Once they have been provided in a particular case, the software can be referred to as the process simulator because its predictions are consistent with the real data. Implementation of the model in a computer programme is needed for those tasks relating to parameter estimation and model validation. Therefore, the above observations on dynamic models of anaerobic digestion are, still more appropriately applicable to process simulators.

For purposes of research or where a broader understanding of the process (in general) is sought, it is usually more convenient for researchers to develop their own simulator, either from their own model or from the models of others. A platform such as Matlab-Simulink, aided externally by Excel, can be very efficient, and should be well-within the capabilities of normal computing facilities. Programming sophisticated user-interfaces makes access to the internal simulator content quite inflexible, and is very time consuming.

Engineers and practitioners, on the other hand, need simulators that are easy to use both as design tools and as a means of operation and control. In these cases a commercial simulator is the best option. There are an increasing number and variety of simulators of wastewater treatment plants (WWTPs) on the market today. The user must, therefore, define clearly what his objectives are before searching for, and finally selecting, a product. Three main aspects need to be taken into consideration. Firstly, the purchaser needs to know what technical information is provided with the simulator. Next, the type of results generated and their presentation must be evaluated. Thus the model contained and the degree of its validation can be assessed. Perhaps some of these selection criteria might require a certain degree of help from research and development professionals. Finally, he needs to know what after-sales technical assistance can be expected. It is important, in addition, to remark that a sound knowledge of both process and model are essential in assessing the capability and power of a simulator to extract useful and reliable data. Inconsistent and even contradictory predictions are, otherwise, inevitable.

This is not the proper place to advise on commercial software. There is, however, very good and up-to-date information on simulators in the book by Olsson and Newell (1999). They deal mainly with software for WWTPs treating sewage but some of the packages include anaerobic digesters of sewage sludge. Although many of these programmes are good at predicting WWTPs as a whole, the content of the models included in AD is rather incomplete as far as matters described in this chapter are concerned. It is, thus, desirable to look for AD specific software that is based on sound and complete dynamic AD models. Among the many authors quoted in the present chapter that have produced complex models for AD, only a few have implemented them on a platform which is both easy to use and, at the same time, reliable in terms of prediction capacity (Siegrist et al., 1993; Vavilin et al., 1994; Angelidaki et al., 1998; Batstone et al., 2000). Much research work remains to be undertaken in this area.

2.6. PERSPECTIVES AND CONCLUSIONS

Anaerobic digestion of OFMSW can be seen as an attractive and, in some cases, more sustainable alternative to other systems. Co-digestion of the OFMSW and sludge from a WWTP that treats urban wastewater suggests an integrated wastewater treatment. Digesters produce a liquid sidestream after dewatering that (in many cases) is recycled to the WWTP head. In addition, a fraction of leachates from landfills receiving MSW is accepted in sewers. In other words, OFMSW can be considered, in many cases, as a part of an integrated urban waste treatment.

Design of anaerobic digesters of a CSTR type is, today, carried out usually by either the criterion of hydraulic retention time or by organic loading rate. Thus, the reactor volume is calculated for a steady state over the predetermined temperature range. The influence of the type of waste on the design criteria is assessed using data obtained from previous experience. Operation of this type of digester of sewage sludge is traditionally conducted by following established feed patterns with regard to flow rate and feed concentration. Little risk of poor performance is involved because many of them are oversized in urban WWTPs in the mesophilic range. Furthermore, methods for automatic control of digesters in these plants are presently based on rather simple algorithms.

Somewhat different is the case of other complex wastes such as animal manure, agricultural wastes, sludge from industrial WWTPs and OFMSW. Here the risks of reactor instability caused by the feed system, or inhibition due to internal compounds, makes design and operation more difficult. Then, traditional methods suffer from serious limitations. In addition, when designers of the anaerobic process adopt new configurations, or when associated treatments (pre- and post-treatments) are coupled to the digester, these limitations are more apparent. Under such circumstances, reactor design and sizing are not always safe, as no analogous cases exist from which experience can be drawn. More modern and advanced control algorithms are also required because phenomena, such as inhibition or instability, increase in significance and can not be tackled with such simple methods.

The above suggests an alternative approach, at least as the near future is concerned. A general methodology based on dynamic models of AD can provide suitable and powerful tools, although their use requires professionals with a high degree of expertise. The accuracy of model predictions depends on the reliability of model calibration and validation. Therefore a compromise must be established between complexity in models for research, and simplicity in models for design, operation and control. Complex models are needed to increase understanding of the process. Although this implies increased difficulty in model calibration, it will ensure more reliable simplified models to be obtained for use in practical engineering.

The use of simulators of AD of complex wastes for practical engineering, demands that users increase their knowledge of the fundamental principles involved in order to understand the performance of software packages better. They will find that familiarising themselves with the AD model included in the simulator is an excellent way to identify those variables involved in the design and operation, which are a consequence of the variables contained in the model. An understanding of the kinetic and stoichiometric coefficients completes the field of knowledge needed to interpret simulator results. The link between progress in research and the application of these findings in engineering is, perhaps, one of the most difficult and challenging points to address. This link is manifest in some cases but absent in many others. Incorporating the know-how obtained by experienced practitioners into the new methodologies provides the most realistic means by which this can be achieved.

Enormous progress in modelling activities for AD has been achieved through the work of many researchers in the field. However, several areas remain where more research is demanded. In particular, more progress in crucial areas, such as the modelling of hydrolysis in the AD of complex wastes, is required. A clearer definition of some of the inhibition phenomena due to reaction products would also be useful. In addition, new methods of characterising feedstock better are necessary. An improvement in the chemical and mathematical tests for parameter estimation of AD models is also required. Finally, activities leading to model simplification need to be promoted more; since simplified models provide the means by which research findings can be applied to actual problems.

2.7. REFERENCES

Andrews, J.F. (1969). Dynamic model of the anaerobic digestion process. *J. Sanitary Engineering Division*, February, 95–116.

Andrews, J.F. (1989). Dynamics, stability and control of the anaerobic digestion process. In: *Dynamics modelling and expert systems in wastewater engineering* (ed. G.G. Patry and D. Chapman). Lewis Publishers inc, Chelsea, MI, USA.

Andrews, J.F. and Graef S.P. (1971). Dynamic modeling and simulation of the anaerobic digestion process. *Adv. Chem Ser. (ASCE)* **105**, 126–162.

Angelidaki, I., Ellegaard, L. and Ahring, B.K. (1993). A mathematical model for dynamic simulation of anaerobic digestion of complex substrates: focusing on ammonia inhibition. *Biotech. Bioeng.* **42**, 159–166.

Angelidaki, I., Ellegaard, L. and Ahring B.K. (1998). *Matematisk model for dynamisk simulering of den anaerobe biogas proces*. Inst. Miljotecknology, DTU EFP-1383/96-0004.

Arrua L.A., McCoy B.J. and Smith J.M. (1990). Gas–liquid mass transfer in stirred tanks. *AIChE Journal* **36**(11), 1768–1773.

Baccay, R.A. and Hashimoto, A.G. (1984). Acidogenic and methanogenic fermentation of caustized straw. *Biotechnol. Bioengng* **17**, 885–891.

Bastin, G. and Dochain, D. (1990). *On-line estimation and adaptive control of bioreactors*, Elsevier Science Publishers, B.V., Amsterdam, The Netherlands.

Batstone, D., Keller, J., Newell, B. and Newland, M. (2000). Modelling anaerobic degradation of complex wastewater. I: Model development. *Bioresource Technol.* **75**, 67–74.

Beck, M.B. (1989). System identification and control. In: *Dynamic model and expert systems in wastewater engineering*. Lewis Publishers, Chelsea, Michigan.

Benefield L.D. and Randall C.W. (1980). *Biological process design for wastewater treatment*. Prentice-Hall, Englewood Cliffs, N.J.

Boone, D.R. (1989). Diffusion of the interspecies electron carriers H2 and formate in methanogenic ecosystems and its implications in the measurement of Km for H2 or formate uptake. *Applied and Environmental Microbiology*, **55**(7), 1735–1741.

Box, G.E.P., Hunter, W.G. and Hunter, J.S. (1978). *Statistics for experimenters*. J. Wiley.

Brock T.D. (1978). *Biology of microorganisms*. Prentice-Hall, New Jersey, WI.

Brummeler, E., Horbach, H.C.J.M. and Koster, I.W. (1991). Dry anaerobic batch digestion of the organic fraction of municipal solid waste. *J. Chem. Tech. Biotechnol.* **50**, 191–209.

Bryers, J.D. (1985). Structural modelling of the anaerobic digestion of biomass particulates. *Biotechnol. Bioengng* **28**, 638–649.

Buhr, H.O. and Andrews J.F. (1977). The thermophilic anaerobic digestion process. *Water Res.* **11**, 129–143.

Carr, A.D. and O´Donnell R.C. (1977). The dynamic behaviour of an anaerobic digester. *Prog. Water Technol.* **9**, 727–738.

Chen, Y. and A. Hashimoto (1978). Kinetics of methane fermentation. *Biotech. Bioengng Symp.* **8**, 269–282.

Chin, K.K. (1981). Anaerobic treatment kinetics of palm oil sludge. *Water Res.* **15**, 199–202.

Contois, D.E. (1959). Kinetics of bacterial growth: relationship between population density and specific growth rate of continuous cultures. *J. gen. Microbiol.* **21**, 40–50.

Cooper, B.E. (1969). *Statistics for experimentalists*. Pergamon Press.

Costello, D.J., Greenfield, P.F. and P.L. Lee (1991). Dynamic modelling of a single-stage high-rate anaerobic reactor – I. Model derivation. *Water Res.* **25**, 847–858.

Coulson, J.M. and Richardson, J.F. (1984). *Chemical engineering*. Volume III. Pergamon Press, Oxford.

Dalla Torre, A. and Stephanopoulos G. (1986). Mixed culture model of anaerobic digestion: application to the evaluation of startup procedures. *Biotechnol. Bioengng* **28**, 1106–1118.

De Baere, L., Verdonck, O. and Verstraete, W. (1985). High rate dry anaerobic composting process for the organic fraction of solid wastes. In: *Seventh symposium on biotechnology for fuels and chemicals, 14–17 May*, (ed. Scott, C.D.) pp. 321–330. Gatlinburg, Tennessee, USA, Wiley, London.

Desjardins, B. and Lessard, P. (1992). Modélisation du procédé de digestion anaérobie. *Sciences et techniques de l'eau* **25**(2), 119–136.

Dinopoulou, G., Sterritt, R.M. and J.N. Lester (1988). Anaerobic acidogenesis of a complex wastewater: II. Kinetics of growth, inhibition and product formation. *Biotechnol. Bioengng* **31**, 969–978.

Dochain, D., Van Rolleghem, P. and Henze, M. (1999). *Integrated wastewater management. European concerted action project.* Report 1995–98. ISBN 92-828-9219-0. European Communities, 2000.

Draper, N.R. and Smith, H. (1981). *Applied regression analysis*. New York: Wiley.

Eastman, J.A. and Ferguson, J.F. (1981). Solubilization of particulate organic carbon during the acid phase of anaerobic digestion. *J. Wat. Pollut. Control Fed.* **53**, 352–366.

Ekama, G.A. and Marais, G.v.R (1984). *Nature of municipal wastewaters. Theory, design and operation of nutrient removal activated sludge process.* Water Research Commission, Pretoria, Republic of South Africa.

Gaddy, J.L. (1981). Economic and kinetic studies of the biological production of farm energy and chemicals from biomass. *Annual progress report to SERI, Document SERI/TR-98020-1.*

García-Heras, J.L., García, S. and Carricas, J. (1999). Mechanistic model of sludge anaerobic digestion implemented in computer to explore the model structure and calibration methods. *3rd International Research Conference. Water reuse. GRUTTEE. Toulouse, November 9–10 1999.* Proceedings pp. 303–306.

Gossett, J.M. and Belser, R.L. (1982). Anaerobic digestion of waste activated sludge. *J. Environ. Eng. Div (ASCE)* **108**, 1101–1120.

Gujer, W. and Zehnder, A.J.B. (1983). Conversion processes in anaerobic digestion. *Wat. Sci. Tech.* **15**, 127–167, 1983.

Gujer, W., Henze, M., Takashi, M. and van Loosdrecht, M. (1999). Activated Sludge Model No. 3. *Wat. Sci. Technol.* **38**(1), 183–193.

Haldane, J.B.S.(1930). *Enzymes.* Longmans, London.

Harper, S.R. and Pohland, F.G. (1986). Recent developments in hydrogen management during anaerobic biological wastewater treatment. *Biotechnol. Bioeng.* **28**, 585–602.

Henze, M. (1995) Anaerobic wastewater. In: *Wastewater treatment*, Springer-Verlag, Berlin.

Henze, M. and Gujer, W. (ed.) (1995). Modelling and control of activated sludge processes. Selected proceedings of the IAWQ International Specialized Seminar on Modelling and Control of Activated Sludge Processes, Copenhagen, Denmark, 22–24 August 1994. *Water Sci. Technol.* **31**(2), 1–273.

Herbert, D., (1958). Recent Progress in Microbiology. *VII International Congress for Microbiology*, (ed. Tunevall, G., Almquist and Wiksell), Stockholm, pp. 381.

Heyes, R.H. and Hall, R.J. (1983). Kinetics of two subgroups of propionate-using organisms in anaerobic digestion. *Appl. Environ. Microbiol* **46**, 710–715.

Hill, D.T. (1983). Simplified Monod kinetics of methane fermentation of animal wasters. *Agric. Wastes* **5**, 1–16.

Hill, D.T. and C.L.Barth (1977). A dynamic model for simulation of animal waste digestion. *J.Water Pollut.Control Fed.* **49**, 2129–2143.

Hill, D.T., Tollner, E.W. and R.D. Holmberg (1983). The kinetics of inhibition in methane fermentation of swine manure. *Agric. Wastes* **5**, 105–123.

Ho C.S., Smith M.D. and Shanahan J.F. (1987). Carbon Dioxide Transfer in Biochemical Reactors. *Advances in Biochemical Engineering Biotechnology* **35**, 93–125.

Hobson, P.N. (1987). A model of some aspects of microbial degradation of particulate substrate. *J. Ferment. Technol.* **65**(4), 431–439.

Hobson, P.N. and Wheatley, A.D. (1993). *Anaerobic Digestion. Modern Theory and Practice*. Elsevier Science Publishers Ltd, Barking, U.K.

Hwu, C. S., Donlon, B. and Lettinga, G. (1996). Comparative toxicity of long chain fatty acid to anaerobic sludges from various origins. *Wat. Sci. Technol.* **34**(2), 351–358.

IWA Task Group on mathematical modelling for design and operation of biological wastewater treatment (2000). *Activated sludge models ASM1, ASM2, ASM2D and ASM3*. IWA Publishing, London.

Jeepson, U. (1996). *Modelling aspects of wastewater treatment processes*. LUTEDX/TEIE-1010/1-444. Lund Institute of Technology.

Johansson, R. (1993). *System modeling and identification*. Prentice Hall Intl. Editions.

Keenan, J.D. and Kormi, I. (1977). Methane fermentation of brewery by-products. *Biotechnol. Bioengng* **19**(6), 867–878.

Larrea, L., García-Heras, J.L., Ayesa, E. and Florez, J. (1992). Designing experiments to determine the coefficients of activated sludge models by identification algorithms. *Wat. Sci. Technol.* **25**(6), 149–165.

Lawrence, A.W. (1971). Application of process kinetics to design of anaerobic processes. *Advan. Chem. Ser.* **105**, 163–189.

Lawrence, A.W. and McCarty P.L. (1969). Kinetic of methane fermentation in anaerobic treatment. *J. Water Pollut. Contr. Fed.* **41**, R1–R17.

Lawrence, A.W., and McCarty P.L. (1970). Unified basis for biological treatment design and operation. *J. Sanitary Enginng División, ASCE* **96**, SA3, 757.

Lehninger A.L., Nelson D.L. and Cox M.M. (1982) *Principles of biochemistry*. Worth Publishers, New York.

Lema, J. M., Ibañez, E. and Canals, J. (1987). Anaerobic Treatment of Landfill Leachates Kinetics and Stoichiometry. *Environ. Technol. Lett.* **8**(11), 555.

Levenspiel, O. (1981). *Chemical reaction engineering*. John Wiley, New York.

Ljung, L. (1987). *System identification: Theory for the user*. Prentice-Hall.

Llabres-Luengo, P. and Mata-Alvarez, J. (1988). The hydrolytic step in a dry digestion system. *Biological Wastes* **23**, 25–37.

Mata-Alvarez, J. (1987). A dynamic simulation of a two-phase anaerobic digestion system for solid wastes. *Biotechnol. Bioengng* **30**, 844–851.

McCarty, P.L. (1964). Anaerobic waste treatment fundamentals. *Public Works* **107**, September.

McCarty, P.L. (1966). Kinetics of Waste Assimilation in Anaerobic Treatment. In: *Developments in industrial microbiology*, Vol. 7, American Institute of Biological Sciences, Washington, D.C., p. 144.

McInerney, M.J. (1988). Anaerobic hydrolysis and fermentation of fats and proteins. In: *Biology of anaerobic microorganisms* (ed. Zehnder, A.J.). John Wiley and Sons, New York.

McInerney, M.J. and Bryant, M.P. (1981). Basic principles of bioconversions in anaerobic digestion and methanogenesis. In: *Biomass conversion processes for energy and fuel*, Plenum Press, New York.

Moletta, R., Verrier, D. and G. Albagnac (1986). Dynamic modelling of anaerobic digestion. *Water Res.* **20,** 427–434.

Monod, J. (1949). The growth of bacterial cultures. *Ann. Rev. Microbiol.* **9**, 97.

Mosey, F.E. (1983). Mathematical modelling of the anaerobic digestion process: regulatory mechanisms for the formation of short-chain volatile acids from glucose. *Water Sci. Technol.* **15**, 209–232.

Noike, T., Endo, G., Chang, J-E., Yaguchi, J-I. and Matsumoto, J.-I. (1985). Characteristics of carbohydrate degradation and the rate-limiting step in anaerobic digestion. *Biotechnol., Bioengng* **27**, 1482–1489.

Novak, J. T. and Carlson, D.A. (1970). The kinetics of anaerobic long chain fatty acid degradation. *J. Wat. Poll. Con. Fed.* **42**(11), 1932–1943.

Olsson, G. and Newell, B. (1999). *Wastewater treatment systems. Modelling, diagnosis and control*. IWA Publishing.

O'Rourke, J.T. (1968). Kinetics of anaerobic waste treatment at reduced temperatures. Ph.D. Dissertation. Standford University.

Parkin, G.F. (1986). Fundamentals of anaerobic digestion of wastewater sludges. *J. Environmental Engineering* **112**(5), 867–921.

Pavlostathis, S.G. and Gossett, J.M. (1985). Modeling alkali consumption and digestibility improvement from alkaline treatment of wheat straw. *Biotechnol. Bioengng* **27**(3), 345–354.

Pavlostathis, S.G and Gossett, J.M. (1986). A kinetic model for anaerobic digestion of biological sludge. *Biotechnol. Bioengng* **28**(10), 1519–1530.

Pavlostathis, S.G. and Giraldo-Gomez, E. (1991). Kinetics of anaerobic treatment: a critical review. *Crit. Rev. Environ. Control* **21**(5–6), 411–490.

Pfeffer, J.T. (1974). Temperature effects on anaerobic fermentation of domestic refuse. *Biotechnol. Bioengng* **16**, 771–787.

Pirt, S.J. (1975). *Principles of microbe and cell cultivation*. Halsted Press, a division of John Wiley & Sons, Inc., New York.

Pohland F.G. (1992). Anaerobic treatment: fundamental concepts, applications and new horizons. In: *Design of anaerobic processes for the treatment of industrial and municipal wastes* (ed. Malina, J.F. and Pohland , F.G.). Technomic Pub. Co.

Ratledge, C. (1994). Biodegradation of oils, fats and fatty acids. *Biochemistry of Microbial Degredation*. Dordrecht, Kluwer Academic Publishers.

Ribbons, D.W. (1970) Quantitative relationships between growth media constituents and cellular yield and compositio. In: *Methods in microbiology*, Vol. 3A (ed. Norris, J.W. and Ribbons, D.W.). Academic, London.

Roels, J.A. (1982). Mathematical models and the design of biochemical reactors. *J.Chem, Tech. Biotechnol.* **32**, 59–72.

Romli M., Keller, J., Lee, P.L. and Greenfield, P.F. (1995). Model prediction and verification of a two-stage high-rate anaerobic wastewater treatment system subjected to shock loads. *Trans. I Chem E*, **73** (Part B): 151–154.

Rozzi, A. (1984). Modelling and control of anaerobic digestion processes: *Trans.Inst.M.C*, **6**, 153–159.

Samson, R. and Leduy, A. (1986). Detailed Study of Anaerobic Digestion of Spirulina-Maxima Algal Biomass. *Biotechnol. Bioengng* **28**(7), 1014–1023.

Sanders, W. T. M., Geerink, M., Zeeman, G. and Lettinga, G. (1999). Anaerobic hydrolysis kinetics of particulate substrates. *II ISAD-SW.* (ed. Mata-Alvarez J.). Barcelona, Spain. Pp. 25–32.

Schenk, H. 1968. *Theories of engineering experimentation*. New York: McGraw-Hill.

Sherrard, J.H., and Schroeder, E.D. (1973). Cell Yield and Growth Rate in Activated Sludge. *J. Wat. Poll. Control Fed.*, **45**, 1889.

Siegrist, H., Renggli, D. and Gujer, W. (1993). Mathematical modelling of anaerobic mesophilic sewage treatment. *Wat. Sci. Tech.* **27**(2), 25–36.

Sperandio, M. and Paul, E. (1997) Determination of Carbon Dioxide evolution rate using on-line gas analysis during dynamic biodegradation experiments. *Biotechnol. Bioengng* **53**(3), 243–252.

Stanier, R.Y., Ingraham, J.L., Wheelis, M.L. and P.R. Painter (1986). *The microbial world*, 5th edition. Prentice-Hall, Englewood Cliffs, N.J.

Stearns, S.D. and David, R.A. (1988). *System identification*. Prentice-Hall.

Suescun, J., Irizar J., Ostolaza, X. and Ayesa, E. (1998). Dissolved oxygen control and simultaneous estimation of oxygen uptake rate in activated-sludge plants. *Water Environment Res.* **70**(3), 316–322.

Van den Berg, L. (1977). Effect of temperature on growth and activity of a methanogenic culture utilising acetate. *Can J. Microbiol.* **23**, 898–902.

Vavilin, V.A., Vasiliev, V.B., Rytov and S.V., Ponomarey, A.V. (1994). Simulation model methane as a tool for effective biogas production during anaerobic conversion of complex organic. *Bioresource Technol.* **48**, 1–8.

Vavilin, V.A., Rytov, S.V. and Lokshina, L.Y. (1996). A description of hydrolysis kinetics in anaerobic degradation of particulate organic matter. *Bioresource Technol.* **56**, 229–237.

Walpole, R.E. and Myers, R.H. (1989). *Probability and statistics for engineers and scientists.* New York: Macmillan.

Water Environment Federation (1995). *Wastewater residuals stabilization. Manual of Practice FD-9.* WEF, Alexandria, VA.

Water Pollution Control Federation (1987). *Anaerobic sludge digestion. Manual of practice No.16.* WPCF, Alexandria, VA.

WEF and ASCE (1992). Design of municipal wastewater treatment plants. Volume II. Manual of Practice No. 8. WEF, Alexandria, VA.

3

Analysis and optimisation of the anaerobic digestion of the organic fraction of municipal solid waste

W.T.M. Sanders, A.H.M. Veeken, G. Zeeman and J.B. van Lier

3.1. INTRODUCTION

The anaerobic digestion process of complex organic waste can be described by four stages: hydrolysis, acidification, acidogenesis and methanogenesis (see also Chapter 2). The main intermediary and end products of the process are volatile fatty acids (VFAs) and biogas, respectively. Because the methanogenic bacteria are very sensitive to accumulation of VFAs and the corresponding drop in pH. The digestion of organic waste is a delicate balance between the rate of hydrolysis and the rate of methanogenesis (Veeken and Hamelers, 1999).

Edited by J. Mata-Alvarez. ISBN: 1 900222 14 0

3.2. OPTIMISATION OF YIELDS

The performance of a reactor digesting a complex substrate is a function of the biomass and enzyme activity and the physical characteristics of the substrate at the applied reactor conditions. These activities and physical characteristics manifest themselves through three measurable parameters, the methanogenic activity, biodegradability and the first-order hydrolysis constant (Figure 3.1). Analysis and optimisation of a digester can therefore be done through analysis and optimisation of these three parameters. Direct measurement of the enzyme activity in an activity assay does not give any valuable information for optimisation of the hydrolysis because an enzyme assay is conducted with a fairly easily degradable substrate. Usually the amount of enzymes is not rate limiting during the digestion of complex wastes (Hobson, 1987) but the amount of available adsorption sites and the accessibility of the substrate to the enzymes (Veeken and Hamelers, 1999).

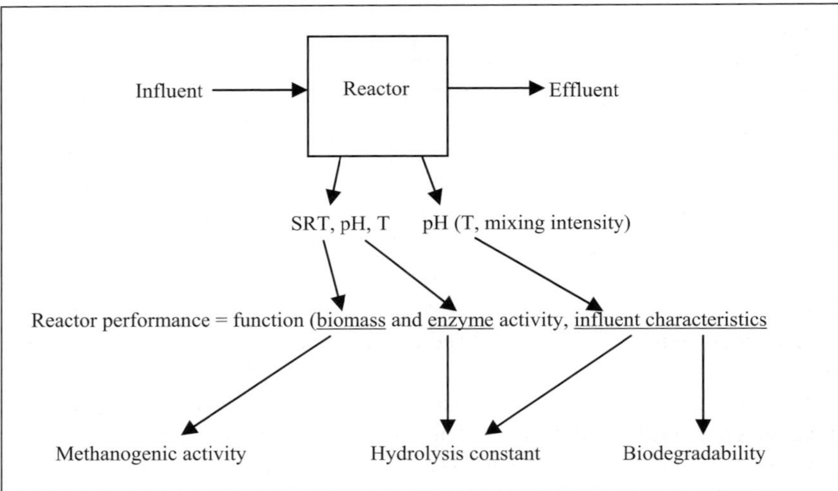

Figure 3.1. Schematic diagram of the relations between the reactor conditions (SRT, pH, T), performance of the reactor and measurable parameters (methanogenic activity, biodegradability, first-order hydrolysis constant).

3.3. METHANOGENIC ACTIVITY TESTS

3.3.1. Introduction

For the start up of the digestion, process optimisation or calamities it is important to have insight in the activity of the methanogenic bacteria in the reactor. During anaerobic digestion the methane can be formed from acetate and from hydrogen and carbon dioxide. Under the conditions in the digester the saturation of the acetate degrading capacity is usually 40–70%. The saturation of the hydrogen degrading capacity is only 1–3%. However, the growth rate of the acetate-converting methanogenic bacteria is lower than the hydrogen-degrading methanogenic bacteria. The acetate-degrading capacity (acetoclastic) is therefore considered to be more vital to the reactor performance than the hydrogen-degrading capacity (hydrogenotrophic).

In principle in a methanogenic activity test the maximal methanogenic activity is assessed by adding an excess amount of substrate to the sludge and after placing it at the desired conditions (pH, T) the substrate conversion rate is measured.

The acetoclastic methanogenic activity can be measured with acetate as a substrate. The hydrogenotrophic methanogenic activity has to be determined with hydrogen or propionic acid as a substrate. However, the use of hydrogen has too many practical disadvantages to serve as a substrate in a methanogenic activity test. Propionic acid can only serve as a substrate for the acetoclastic methanogens after it has been converted to acetate and hydrogen by the propionate-oxidising bacteria. This entails that propionic acid can only be used as a substrate when the propionic acid oxidation is not rate limiting. Moreover enough acetate and hydrogen has to be formed to supply both groups of methanogens with sufficient substrate to reach their maximum conversion rates. This last condition cannot, however, be achieved because the hydrogen concentration that leads to a maximal activity of the hydrogenotrophic methanogens will inhibit the propionic acid oxidation which subsequently lowers the hydrogen concentration. A reproducible activity test for the hydrogenotrophic methanogens does not therefore seem feasible.

This chapter will therefore focus only on an activity test for the acetoclastic methanogens.

3.3.2. Methods

Adding an excess amount of acetate together with nutrients, trace elements and a buffer to a sludge sample after which the methane production rate is measured under the desired conditions performs the methanogenic activity test. The test is conducted in glass bottles that can be closed with a screw cap. This cap usually contains a septum of some sort that makes it possible to sample the bottle without opening it. However, for the OFMSW the use of other types of bottle might be considered as the bottles are quite small and the neck of the bottle is too narrow to add the often large and chunky sludge from a digester treating the OFMSW. Irrespective of the type of bottle that is used the following items have to be considered.

The sludge concentration

The amount of sludge that is suitable for an activity test has to be balanced between a measurable methane production rate and the occurrence of diffusion limitation. If the sludge concentration is very low the sludge production rate will be very low which increases the contribution of the analytical error into the gas production rate. On the other hand too much sludge increases the effect of the diffusion rate into the gas production rate, as the conversion of acetate in the sludge will be higher than diffusion of acetate from the bulk to the sludge. Not all of the sludge will be saturated with acetate. For very active granular sludge a sludge concentration of 1.5–3 g/l is advised. However, for the less active sludge from the digestion of the OFMSW, as a rule of thumb, a sludge concentration of 5–10 g/l has to be considered.

The acetate concentration

For the amount of acetate that is to be used in the experiment the same considerations apply as for the sludge concentration. Moreover, because acetate could cause substrate inhibition, the acetate concentration must be below the inhibitory concentration. An acetate concentration of 2.5 g COD/l seems to be the golden mean in this case.

Buffering

Because the methane production rate is affected by the pH, which will change during the course of the test owing to production of CO_2, the activity test has to be conducted at a controlled pH. This is achieved by addition of a buffer salt. If the sludge is well-stabilised, bicarbonate can be used as a buffer. However, it must be kept in mind that part of the CO_2 will be removed when flushing the bottles with nitrogen. Moreover the CO_2 can interfere with several chemical equilibria. For the OFMSW, especially when it is less stabilised, it is advisable to use a P buffer instead of bicarbonate, because the CO_2 can, together with H_2 from the sludge, serve as a substrate for the hydrogenotrophic methanogens. In that way the acetoclastic methanogenic activity can be overestimated. The activity test is started at pH 6.8 and is allowed to increase to pH 7.7.

Nutrients and trace elements
To prevent deficiency of nutrients and trace elements during the test it is advisable to add a nutrient and trace element solution as given in Table 3.1.

Table 3.1. Nutrient and trace element stock solution.

Stock solution (10 times concentrated)	
NH_4Cl	2.8 g/l
K_2HPO_4	2.5 g/l
$MgSO_4.7H_2O$	1.0 g/l
CaCl2.$2H_2O$	0.1 g/l
Yeast extract	1.0 g/l
Trace element solution	10 ml/l
Trace element solution	
$FeCl_2.4H_2O$	2000 mg/l
H_3BO_3	50 mg/l
$ZnCl_2$	50 mg/l
$CuCl_2.2H_2O$	38 mg/l
$MnCl_2.4H_2O$	500 mg/l
$(NH_4)6Mo_7O_{24}.4H_2O$	50 mg/l
$AlCl_3.6H_2O$	90 mg/l
$CoCl_2.6H_2O$	2000 mg/l

Temperature
The activity test is performed at constant temperature such as 30 °C, 55 °C or the temperature at which the reactor is operated.

Stirring
To prevent diffusion limitation and floating layers it is desirable that the content of the bottles or reactors that are used in the test is mixed regularly. When glass bottles are used to conduct the methanogenic activity test it is not possible to stir the content. Shaking of the whole bottles, however, is very possible, for instance by means of a wrist action shaker. In the absence of any means of continuous shaking the bottles have to be shaken manually at least before gas measurements are done.

Analysis of methane production

During the course of the test methane production is determined regularly in order to calculate the methane production rate. To measure the methane production different methods can be used. The most commonly used methods are described here. The first is the headspace method. In this method the gas that is produced during the course of the test is allowed to accumulate in the headspace of the bottle. This headspace is sampled regularly with a chromatographic gas syringe. The sample is analysed with a gas chromatograph for methane. Although the headspace method is very quick and accurate it is rather costly as it requires a gas chromatograph. The second method is the method in which the produced methane is measured through the displacement of a sodium hydroxide solution from a Marriott flask.

3.3.3. Calculation

The methane production in a methanogenic activity test in a batch system can be described by (Veeken and Hamelers, 2000):

$$CH_4(t) = -(A_0/\mu_{max}).\exp(\mu_{max} \cdot t) \qquad (3.1)$$

where $CH_4(t)$ is the cumulative methane production at time t (in Nl, i.e. litres at 0 °C and 1 atm), A(0) the initial acetoclastic activity (mg COD.g^{-1} VS.d^{-1}), μ_{max} the maximum growth rate (in d^{-1}) and t time (in d). When assessing the methanogenic activity of a highly active sludge such as anaerobic granular sludge, the acetoclastic activity does not change significantly during the test. In that case, the methane production is linear in time and the activity can directly be calculated from the slope of the cumulative methane production as a function of time. This method is not, however, suitable for calculating the methanogenic activity of digested OFMSW or vegetable food and yard waste because the methane production is too low and we have to monitor the methane production over longer periods of time. Then, the methane production is not linear because of significant changes in biomass concentration in time (see Figure 3.2). The initial acetoclastic activity (A(0)) then has to be calculated using non-linear least-square curve fitting of the cumulative methane production to equation 3.1. The maximum growth rate (μ_{max}) of the methanogens is set at 0.2 day^{-1} but sensitivity analysis showed that a variation of μ_{max} between 0.1–0.5 day^{-1} only resulted in small changes in the initial activity.

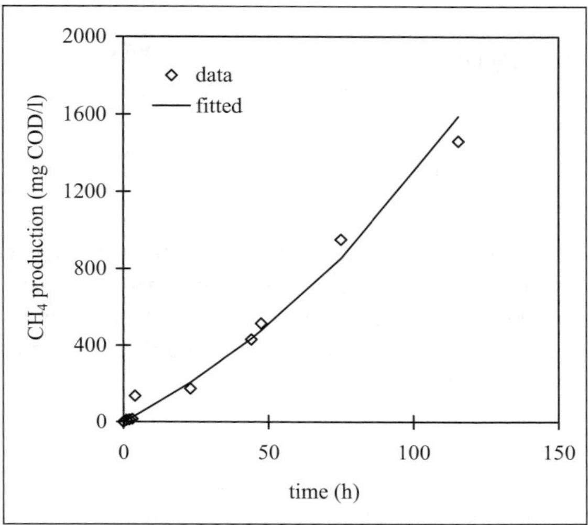

Figure 3.2. The cumulative methane production during a methanogenic activity test with digested vegetable food and yard waste.

3.4. THE FIRST-ORDER HYDROLYSIS CONSTANT

3.4.1. Introduction

Although first-order kinetics (see also 2.3.1) do not reflect the actual kinetics of the hydrolysis process (Hobson, 1987; Sanders et al., 2000) they are commonly used to describe the hydrolysis of particulate substrates during anaerobic digestion (Eastman and Ferguson, 1981; Pavlostathis and Giraldo-Gomez, 1991a,b). Provided the underlying microbiological and physical processes are well understood, the first-order kinetics represent a good tool for reactor design.

$$dX_{degr}/dt = -k_h . X_{degr} , \qquad (3.2)$$

where:
X_{degr}: concentration biodegradable substrate (kg/m^3),
t: time (days),
k_h: first-order hydrolysis constant (1/day).

The first-order kinetics originally were used to describe the degradation of particulate substrate (Eastman and Ferguson, 1981) but recently it is also used to describe the degradation of dissolved polymeric substrates (San Pedro, 1994; San Pedro et al., 1994).

Several researchers showed that the hydrolysis mechanism of particulate substrates is surface related (Hills and Nakano, 1984; Sanders et al., 2000). In the digestion of particulate substrates the amount of enzymes present is usually in excess relative to the available surface area (Hobson, 1987); the hydrolysis constant in fact is determined by the latter factor. Such surface limited kinetics can very well be described with a first-order relation (Vavilin et al., 1996; Veeken and Hamelers, 1999).

This section surveys the limitations of the first-order relation and the effects of digestion conditions on the first-order hydrolysis constant. Moreover the applicability of k_h values from literature will be evaluated, while guidelines for the experimental assessment of the hydrolysis constant will also be provided.

3.4.2. Methods

3.4.2.1. Experimental set ups for assessment of the hydrolysis constant

When evaluating literature it is clear that two different experimental set-ups are commonly used to investigate the hydrolysis of particulate substrates, i.e. batch (Veeken and Hamelers, 1999) or continuous (Miron et al., 2000) experiments. In the batch approach, the selected substrate is incubated at a specific temperature with or without an excess amount of methanogenic sludge. The continuous set-up uses completely stirred tank reactors (CSTR) operated at a specific temperature and at several hydraulic retention times (HRT). Analyses are made from the effluent once steady state has established.

To monitor the hydrolysis process during the digestion two methods are used:

1. Measurement of the amount of methane, and possibly also the concentration of soluble COD and volatile fatty acids (VFA) produced (Veeken and Hamelers, 1999).
2. Measurement of the concentration of the individual relevant components (Miron et al., 2000).

To be able to follow the degree of hydrolysis just by means of the methane production, the hydrolysis needs to be the rate-limiting step in the degradation process. This in fact implies that the methanogenic activity present during the test is sufficient to prevent accumulation of volatile fatty acids, and obviously it means that this method can only be applied to determine the hydrolysis constant under methanogenic conditions. Although for a batch experiment the required concentration of methanogenic inoculum can be calculated, measurement of the concentration of dissolved COD and volatile fatty acid in addition to the methane production is advised. In that case the degree of hydrolysis is calculated from the methane production, and the production of dissolved COD during the digestion.

The second method to follow the degree of hydrolysis, namely by measurement of the individual components, requires sampling of the reactor contents during the test for assessment of the suspended COD, proteins, carbohydrates or lipids in the system. This method is not always applicable, especially in tests that are conducted under methanogenic conditions, as with addition of inoculum sludge extra protein and carbohydrates are introduced. Because it is not possible to distinguish between proteins and carbohydrates originating from the biomass and the waste(water) this method is insufficiently accurate for waste(water) that contains only a small amount of hydrolysable proteins and carbohydrates.

If the hydrolysis constant is assessed in a batch experiment, two types of system can be used. The first is a large stirred batch reactor that is regularly sampled during the degradation process and will further be referred to as the 'classic' batch reactor. The classic batch reactor is best used when the degree of hydrolysis is determined via measurement of the methane production and of dissolved products acids in the reactor mixture. This reactor type is less suitable for the direct measurement of suspended COD or individual particulate components in the reactor mixture as possible floating or settling layers and lipids attached to the reactor wall make representative sampling difficult. In that case, a system where several small reactors or flasks represent one large reactor is more suited. At the start of the experiment, all flasks have the same contents and it is assumed that the progress of the hydrolysis is similar in all flasks. Each time the reactor is sampled one whole flask is sacrificed and its contents are analysed. This type of batch reactor will be further referred to as the 'multiple flask' reactor. The 'multiple flasks' procedure allows thorough homogenisation of the flasks' contents before sampling. Moreover, after emptying the flask can be rinsed with an organic solvent to remove the lipids that were attached to the wall.

In this section, the results of several investigations on the hydrolysis of complex waste(water) were used to evaluate the first-order kinetics for hydrolysis. Table 3.2 provides a list of the applied experimental set up and analyses that were used in the research in question.

Table 3.2. Experimental set up and performed analysis of the research used in this chapter.

Authors	**Experimental set-up**	**Analysis**	**Kh (re)calculated***
Veeken & Hamelers (1999)	Batch (classic)	CH_4 and VFA	No
Canalis (1999)	Batch (classic)	CH_4 and VFA	No
O'Rourke (1968)	CSTR	Grease-free LCFA Cellulose Organic-N	Yes
Boon (1994)	Batch (classic)	NH_4^+-N	Yes
Palenzuela-Rollón (1999)	Batch (multiple flask)	NH_4^+-N	No
Eastman & Ferguson (1981)	CSTR	Kjeldahl-N Cellulose Grease	Yes
Miron et al. (2000)	CSTR	Kjeldahl-N Carbohydrates Grease-free LCFA	Yes
Hills & Nakano (1984)	CSTR	CH_4	Yes
Sanders et al. (2000)	Batch (Classic)	CH_4 VFA Dissolved carbohydrates	No

* Hydrolysis constant was (re)calculated.

3.4.2.2. Calculations

From equation 3.2 the relation between the hydrolysis constant, digestion time and effluent concentration for a batch (equation 3.3) and CSTR (equation 3.4) can be derived.

$$X_{ss,effluent} = X_{ss,influent}(1-f_h) + f_h X_{ss,influent}\, e^{-k_h.t} \tag{3.3}$$

$$X_{ss,effluent} = \frac{X_{ss,influent}\, f_h}{1+\theta k_h} + X_{ss,influent}(1-f_h) \tag{3.4}$$

with

$X_{ss,effluent}$: concentration of total substrate in the effluent (biodegradable + non biodegradable part) (g/l),

$X_{ss,influent}$: concentration of total substrate in the influent (biodegradable + non biodegradable part) (g/l),

f_h : biodegradable fraction of substrate, $f_h \in [0;1]$,

k_h : first-order hydrolysis constant (d^{-1}),

θ : hydraulic retention time of reactor (d),

t : digestion time of batch (d).

From results of several investigations reported in literature (Table 3.2) the hydrolysis constant was (re)calculated to assess its dependency on the digestion conditions. For the estimation of the first-order hydrolysis constant from the two set-ups, equation 3.3 and 3.4 can be linearised which results in equation 3.5 and 3.6, respectively.

$$\ln\left(\frac{X_{ss,effluent}-X_{ss,influent}(1-f_h)}{X_{ss,influent}f_h}\right)=-k_h.t \tag{3.5}$$

$$\theta=(f_h.X_{ss,influent})\left(\frac{\theta}{X_{ss,influent}-X_{ss,effluent}}\right)-\frac{1}{k_h} \tag{3.6}$$

For CSTR systems (equation 3.6) ‘θ’ (y-axis) can be plotted against $\theta/(X_{ss,influent}-X_{ss,effluent})$. The hydrolysis constant then follows from the intercept of the line with the y-axis. The biodegradability of the substrate follows from the slope of the line (Eastman and Ferguson, 1981). In batch experiments, the hydrolysis constant is determined by plotting equation 3.5. The biodegradability cannot be derived from the graph, but has to be assessed directly from the experimental results. The biodegradability can be calculated from the maximum methane yield of the substrate (Veeken and Hamelers, 1999) or the final effluent concentration. The hydrolysis constant is calculated from the slope of the line.

A more direct and accurate method for assessing the hydrolysis constant and biodegradability from batch and continuous experiments is the non-linear least squares fit on the assessed effluent concentration. However, with the linear presentation of the results according to equations 3.5 and 3.6 more insight is gained in possible deviations from the first-order relation. Therefore, for the evaluation of the first-order kinetics the linearised equations are used in this section.

Additionally, it should be emphasised that the biodegradability in equations 3.3 to 3.6 refers to the biodegradability under the applied 'steady-state' conditions and may change with the imposed reactor conditions. The 'ultimate' biodegradability is discussed in section 3.5.

3.4.2.3. Possible errors

The effluent values for estimation of the hydrolysis constant obviously are important, especially in the batch digestion set-up. By determining the derivative of the hydrolysis function, the sensitivity of the hydrolysis constant to the effluent concentration can be assessed. The equations, $dk_h/dX_{ss,effluent}$, are presented in equations 3.7 and 3.8, for batch and CSTR systems, respectively. From Figure 3.3 it can be seen that the sensitivity of the calculated hydrolysis constant increases when the effluent concentration approaches the minimum value of $X_{ss,influent}(1\text{-}f_h)$. Since at very low biodegradable substrate levels a small error in the effluent concentration has a large impact on the value of the calculated hydrolysis constant it is advised to use only effluent values, obtained before 50% of the biodegradable substrate is converted.

$$\frac{dk_h}{dX_{ss,\,effluent}} = \left| \frac{1}{t} * \frac{-1}{X_{ss,\,effluent} - X_{ss,\,influent}(1-f_h)} \right| \tag{3.7}$$

$$\frac{dk_h}{dX_{ss,\,effluent}} = \left| \frac{1}{\theta} * \frac{X_{ss,\,effluent} \cdot f_h}{\left(X_{ss,\,effluent} - X_{ss,\,influent}(1-f_h)\right)^2} \right| \tag{3.8}$$

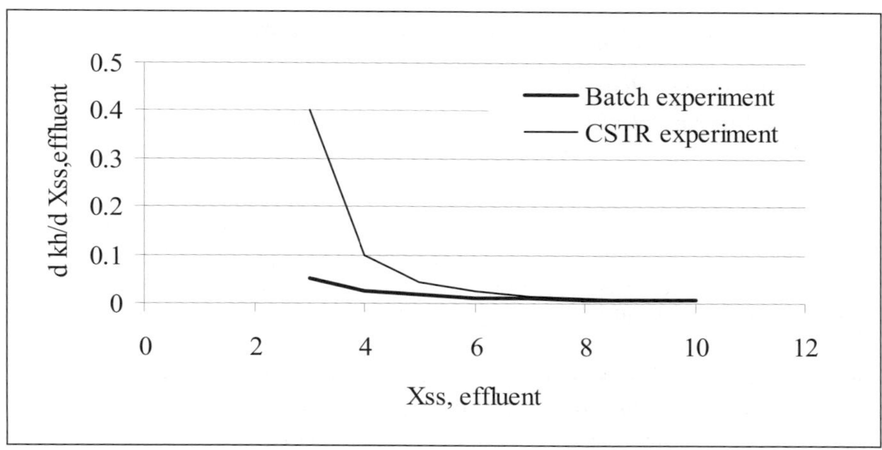

Figure 3.3. The sensitivity of the value of the first-order constant to the assessed effluent values. (Calculations have been made for $X_{ss,\ influent}$= 10 g/l, f_h=0.8, θ or t=20 days.)

Another concern for errors in the estimation of the hydrolysis constant can be found in the formation of new biomass during the digestion. The growth of biomass, in both analytical methods discussed in 3.4.2.1, leads to an underestimation of the hydrolysis constant. However, the biomass yield during anaerobic digestion is relatively low and it can be calculated that a biomass yield of 0.1–0.2 g/g only results in a 0.5–2.5% error in the value of the hydrolysis constant. The effect of biomass growth on the value of the hydrolysis constant can therefore be neglected.

When waste(water) contains sulphate or nitrate and the degree of hydrolysis is measured via methane production the hydrolysis obviously will also be underestimated due to consumption of volatile fatty acids for the formation of H_2S and N_2. Direct measurement of the individual components is advised for this type of waste(water).

3.4.3. The effect of temperature on hydrolysis constant and biodegradability

In general, the rate of all reactions varies with temperature in accordance with the Arrhenius equation:

$$k = A.e^{-\Delta G^*/RT} \tag{3.9}$$

where

k = kinetic rate constant, in this case the hydrolysis constant

A = the Arrhenius constant,

G* = the standard free energy of activation ($J.mol^{-1}$),
Typical standard free energies of activation are 15–70 kJ mol^{-1} (Chaplin and Bucke, 1990),

R = the gas law constant ($J.mol^{-1}.K^{-1}$),

T = the absolute temperature (K).

Veeken and Hamelers (1999) found an Arrhenius relation (R^2 = 0.984–0.999) between the first-order hydrolysis constant for several organic wastes and the digestion temperature in the range of 20–40 °C. They calculated an average standard free energy of activation of 46±14 kJ mol^{-1} for the organic waste components. This is within the range of 15–70 kJ mol^{-1}, which is the typical range for standard free energies of activation as given by Chaplin and Bucke (1990). Moreover the results showed no significant difference in the biodegradability of the organic waste components within that temperature range.

Canalis (1999) digested particulate starch at 15, 20, 25 and 30 °C. Most of the batch experiments were terminated before the starch hydrolysis had reached its asymptote but it was assumed that the biodegradability of the starch was 1.0 at all temperatures. The calculated hydrolysis constants were 1.24±0.57 d^{-1}, 1.24±0.29 d^{-1}, 0.13±0.05 d^{-1} and 0.03±0.01 d^{-1} for 30, 25, 20, 15 °C, respectively. From the results an Arrhenius relation (R^2 = 0.91) was derived between the hydrolysis constants and the temperature, but the calculated standard free energy of activation of 190 kJ mol^{-1} was beyond the range of 15–70 kJ mol^{-1}. A reason for this high calculated value cannot be given.

O'Rourke (1968) digested primary sludge in CSTRs at 15, 20, 25 and 35 °C and assessed the concentrations of grease, cellulose and protein in the influent and effluent. From his results, the first-order hydrolysis constant and biodegradability of the sludge at the different digestion conditions was calculated using equation 3.6. Table 3.3 summarises the results of our calculations and Figure 3.4 presents them graphically. Table 3.4 provides the pH values and whether acid or methanogenic conditions prevailed in the digesters.

It appears that only the hydrolysis of protein followed first-order kinetics over the whole temperature range. Cellulose was degraded according to first-order kinetics at 20–35 °C. At 15 °C the first-order constant could only be calculated for cellulose in the HRT range of 30–60 days. Below 30 days HRT the hydrolysis of cellulose was not in accordance with first-order kinetics. Cause for this could be a lower enzyme activity at HRT<30 days because of which the enzyme activity instead of the surface of the substrate became rate limiting.

The Arrhenius relation of the calculated hydrolysis constants for proteins and cellulose in Table 3.3 was poor. For protein a standard free energy of activation of 53 kJ/mol was calculated with $R^2 = 0.79$. For cellulose it was 173 kJ/mol with R^2 0.88. This again is above the range of 15–70 kJ mol^{-1} as was specified by Chaplin and Bucke (1990).

The amount of neutral lipid (total grease minus free long chain fatty acids) was only ~25% of the total amount of grease in the sewage sludge. The hydrolysis did not proceed according to first-order kinetics at any temperatures investigated. The assessed values of the biodegradability (f_h) presented in Table 3.3, appear to be hardly effected by the temperature.

Table 3.3. The calculated values of the first-order hydrolysis constant and of the biodegradability calculated with equation 3.5 from the results of the digestion of raw domestic sewage sludge as conducted by O'Rourke (1968), values between brackets are the calculated standard errors (in %).

	kh (d^{-1})				**f_h**			
	15 °C	**20 °C**	**25 °C**	**35 °C**	**15 °C**	**20 °C**	**25 °C**	**35 °C**
Cellulose	0.14[1]	0.13	0.38	12.0	0.92	0.97	0.95	0.92
					(3.6)[1]	(3.6)	(1.6)	(0.5)
Protein	0.14	0.39	0.33	0.67	0.37	0.35	0.41	0.40
					(2.5)	(1.3)	(2.2)	(2.7)

(1) Calculated for the range HRT 30–60 days.

Table 3.4. The prevalence of acidic or methanogenic conditions and measured values of the pH during the digestion experiments conducted by O'Rourke (1968) at different temperatures.

	15 °C	**20 °C**	**25 °C**	**35 °C**
Acid conditions	HRT 5–60 days	HRT 5–15 days	HRT 3.75–10 days	HRT 0–5 days
Methanogenic conditions		HRT 30–60 days	HRT 15–60 days	HRT 7.5–60 days
PH[(1)]	6.95±0.35	7.05±0.35	7.1±0.3	7.1±0.3

(1) Variations in pH showed little if any relation to the HRT.

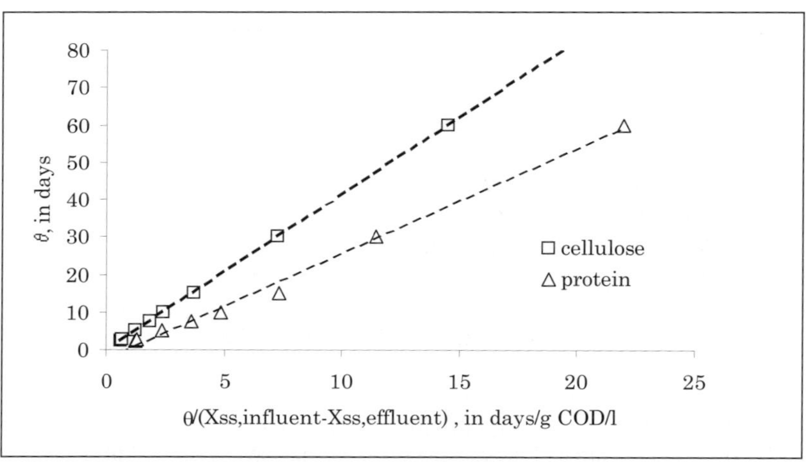

Figure 3.4. Presentation of the linearised equation (equation 3.6) of the hydrolysis of cellulose and protein in primary sludge at 35 °C as calculated from O'Rourke (1968). (Slope=$f_h.X_{ss,influent}$, y-intercept= $-k_h^{-1}$.).

3.4.4. The effect of pH on hydrolysis constant and biodegradability

The simplest relation between pH and activity of an enzyme is the 'bell shaped' curve (Figure 3.5) which has its optimum pH at:

$$pH_{optimum}=(pKa_1+pKa_2)/2 \tag{3.10}$$

where
pKa_n = the pKa of the nth dissociation form of the enzymes.

This relation is simplified as compared to the real situation as it is assumed that only one charged form of the enzyme is active and the enzyme is a single ionised species, when it could contain a mixture of different ionised groups. Despite its simplifications the 'bell-shaped' relation is commonly encountered in single enzyme substrate reactions (Fersht, 1999).

However, during anaerobic digestion it is, however, very likely that several enzymes all with different pH optima are present. Moreover, the 'bell-shaped' relation ignores ionisation of the substrate, products and enzyme–substrate complexes (Chaplin and Bucke, 1990). Obviously, the effect of the pH on the anaerobic digestion is much more complex. The net effect of pH on the hydrolysis rate is specified by the pH optima of the different enzymes present in the digester and the effect of pH on the charge/solubility of the substrate. The latter especially applies to the digestion of substrates that contain proteins.

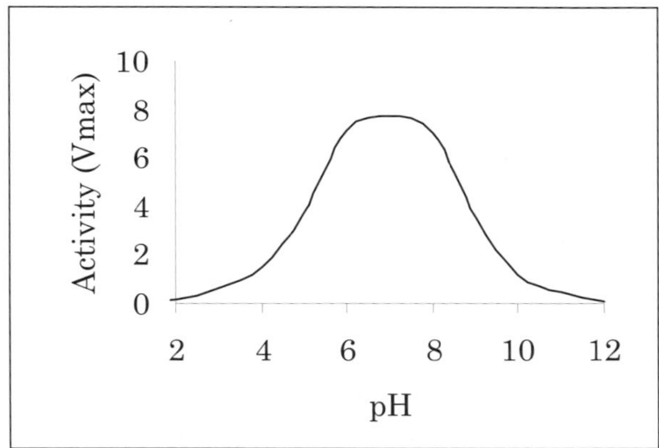

Figure 3.5. 'Bell-shaped' curve of the relation between enzyme activity and pH (Chaplin and Brucke, 1990).

Boon (1991) and Palenzuela-Rollón (1999) performed batch digestion experiments at controlled pH between 5 and 8 with primary sludge (35 °C) and fish processing wastewater (30 °C), respectively.

Figure 3.6 shows the assessed values of the first-order hydrolysis constant and biodegradability of protein COD and total COD in primary sludge in relation to the pH. It appears that the biodegradability (f_h) slightly increases with increasing pH, while the relation found between the hydrolysis constant and the pH resembles the 'bell-shaped' curve (Figure 3.5) with an optimum at pH 6.5 for both the total COD and the proteineous COD.

Concerning the pH effect on the biodegradability of protein during the digestion of fish processing wastewater at methanogenic and acidogenic conditions Palenzuela-Rollón (1999) assumed an equal biodegradability at all pHs investigated of 0.8±0.01. From the results of the batch experiment this degradability could only be established at pH 8 because those at pH 4–7 were terminated before the conversion of protein had reached its asymptote. The main protein in the wastewater was myosin which structure almost remains unaffected by pH and temperature (Engbersen and de Groot, 1995). This supports the assumption that also at pH 4–7 the biodegradability was 0.8. Palenzuela-Rollón concluded that there was no significant effect of the pH on the value of the hydrolysis constant between acid and methanogenic conditions, except at pH 8 (Figure 3.7).

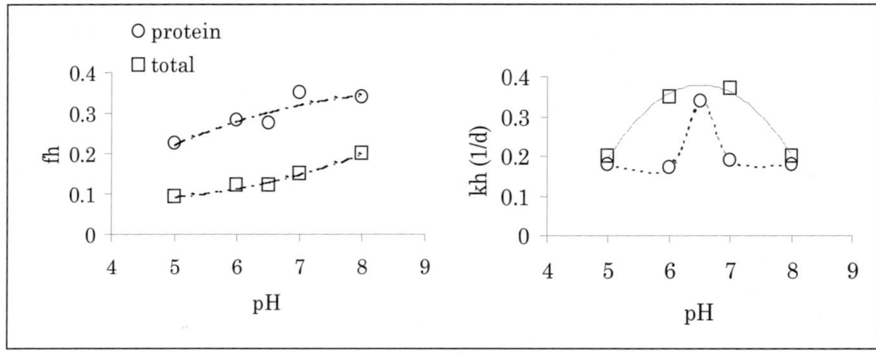

Figure 3.6. The assessed relationship between the pH and the biodegradability (left) and the first-order hydrolysis constant (right) of the proteins and total COD in primary sludge during a batch digestion at 35 °C. Calculated from Boon (1994).

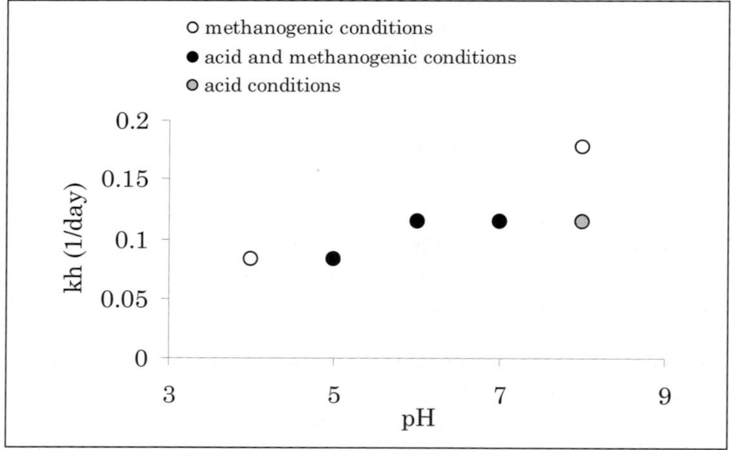

Figure 3.7. The assessed biodegradability and first-order hydrolysis constant of the proteins in fish-processing wastewater during a batch digestion experiments at 30 °C conducted at methanogenic and acidogenic conditions and different pH values (Palenzuela-Rollón, 1999).

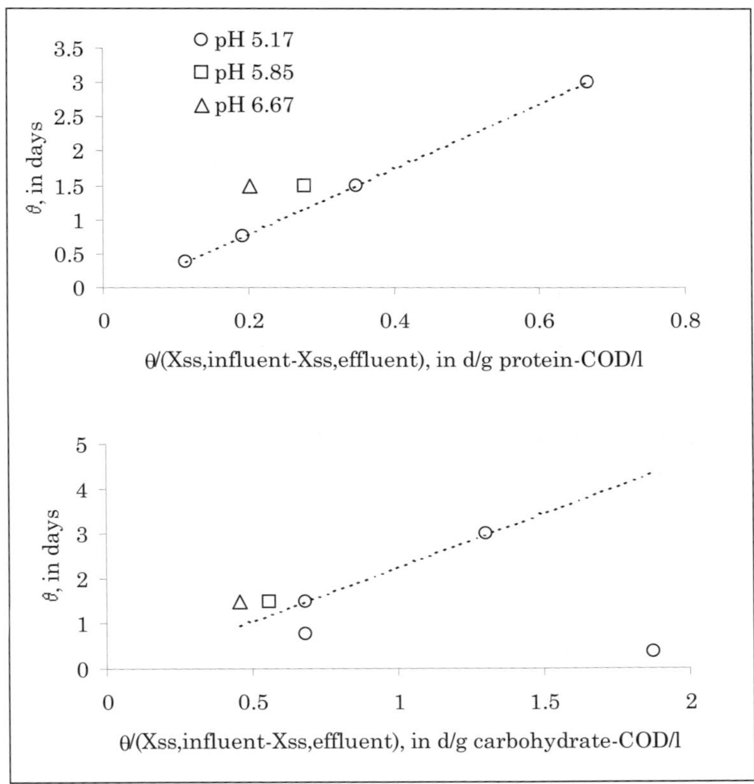

Figure 3.8. Graphical presentation of the linearisation equation (equation 3.6) concerning the hydrolysis of protein (upper graph) and carbohydrates (bottom graph) in primary sludge at 35 °C and under acid conditions as calculated from the results obtained by Eastman and Ferguson (1981) (slope=$f_h.X_{ss,influent}$, y-intercept=$-k_h^{-1}$).

Eastman and Ferguson (1981) conducted digestion experiments with primary sludge in CSTRs at 35 °C, at very short retention times (9–72 h) and pH 5.17±0.04. Although they only calculated the k_h and biodegradability on the basis of total COD, the hydrolysis of the separate components in the sludge also followed. At pH 5.17 nitrogenous components were degraded very fast and followed first-order kinetics (Figure 3.8). For carbohydrates only first-order kinetics can be used for HRT 1.5 and 3 days. At retention times below 24 h, a lower enzyme activity could have caused the hydrolysis kinetics to change from a surface-limited to an enzyme-limited situation. Greases remained completely un-degraded.

Calculation of the biodegradability and hydrolysis by using equation 3.6 results in a value for k_h of 6.2 and 7.0 1/d and for the biodegradability of 0.30 and 0.31 for the protein and carbohydrates, respectively. In addition to the reactors operated at pH 5.17 Eastman and Ferguson (1981) also conducted two experiments at pH 5.85 and pH 6.67 and a retention time of 36 h (1.5 days). Although one HRT obviously does not allow calculation of the k_h and biodegradability, the results clearly reveal that the biodegradability of both nitrogenous and carbohydrate compounds is positively effected by the increase of the pH from 5 to 7.

Miron et al. (2000) conducted digestion experiments with primary sludge in CSTRs at HRTs 3, 5, 8, 10 and 15 days at 25 °C. Three reactors acidified and the pH reduced to 4.8, namely those operated at a HRT of 3, 5 and 8 days. In two reactors, methanogenic conditions and pH = 6.5 established, namely the reactors at a HRT of 10 and 15 days. From the results Miron et al. (2000) concluded that the hydrolysis process of none of the single components present in the sewage sludge proceeded according to first-order kinetics.

3.4.5. Accumulation of hydrolysis intermediates

As discussed in section 3.4.2.1, both a batch set-up and completely stirred tank reactor set-up can be used to determine the hydrolysis constant for a certain type of waste(water). The main difference between these two methods is that in the batch set-up never a steady state situation is reached, whereas this is a pre-condition in the continuous set-up. In the batch set-up intermediates, such as VFA, LCFA and H_2, can accumulate in time, which implies changing process conditions during the assessment of the k_h value.

In a CSTR in steady state, possibly accumulated intermediates remain at a constant level. However, as the assessment of the hydrolysis constant in a continuous set-up requires at least three different HRTs, also here different levels of accumulated intermediates will prevail for the different HRTs. Eastman and Ferguson (1981) investigated the hydrolysis of lipids, proteins and carbohydrates during the acidification of primary sludge at different HRTs but at a constant pH (see also 3.4.4). Although likely different levels of accumulated intermediates prevailed at the different HRTs, a first-order relation was established for the hydrolysis of protein and carbohydrate hydrolysis. This indicates that for proteins and carbohydrates the accumulation of degradation intermediates can be neglected for the hydrolysis of these components. Moreover, it also means that the hydrolysis constant of proteins and carbohydrates under acid conditions can also be determined in a batch set-up, provided the pH is controlled (Palenzuella-Rollón, 1999).

With respect to the assessment of the hydrolysis constant for neutral lipids it can be concluded from the results in Sanders (2001) that accumulated intermediates such as H_2 and LCFA do not seem to effect the hydrolysis constant.

3.4.6. The effect of the particle size distribution on the first-order hydrolysis constant

For particulate substrates several researchers showed that the hydrolysis mechanism is surface related (Hills and Nakano 1984; Sanders et al., 2000). This means that in presence of excess amounts of enzymes the available surface to the hydrolysis is the rate-limiting factor (Hobson 1987; Sanders et al., 2000). The hydrolysis constant in the first-order kinetics is, however, based on the degradation rate of the total mass. Obviously, the mass:surface ratio is an important controlling parameter for the value of the hydrolysis constant and is strongly related to the particle size distribution of the substrate (Sanders et al., 2000). According to the results of experiments of Hills and Nakano (1984) with tomato solid waste and of Sanders et al. (2000) with particulate starch with different initial size distributions, the average particle diameter of the substrate (D_p) multiplied by the sphericity of the particles (Φ_s) (Hills and Nakano, 1984) is linearly related to the hydrolysis constant (Figure 3.9).

South et al. (1995) proposed an enzyme-adsorption based kinetic model (ABK model) to describe hydrolysis of solid lignocellulosic substrates in ethanol production. The model predicts that the rate of hydrolysis increased for increasing enzyme concentrations and for an increased amount of available biodegradable adsorption sites. The ABK model transforms into a first-order kinetic model when the hydrolytic enzyme concentration exceeds the amount of adsorption sites. As hydrolysis of OFMSW can generally be described by first-order kinetics, it can be assumed that the amount of hydrolytic enzymes is never rate-limiting for hydrolysis. This is also expected regarding the fast growth of hydrolytic and fermenting bacteria (Pavlosthatis and Giraldo-Gomez, 1991b). It was also observed that the hydrolysis rate of solid organic substrates increased with increasing biodegradability of the substrate. This can also be explained by the ABK model, as increased biodegradability results in an increase in adsorption sites for enzymes.

The results of Perot et al. (1988) show a positive relation between mixing intensity and the hydrolysis rate. Based on the results presented in Sanders (2001) the absence of methanogensis negatively effects the hydrolysis of neutral lipids, which likely can be attributed to changes in the particle size distribution of the lipids due to the lack of gas mixing.

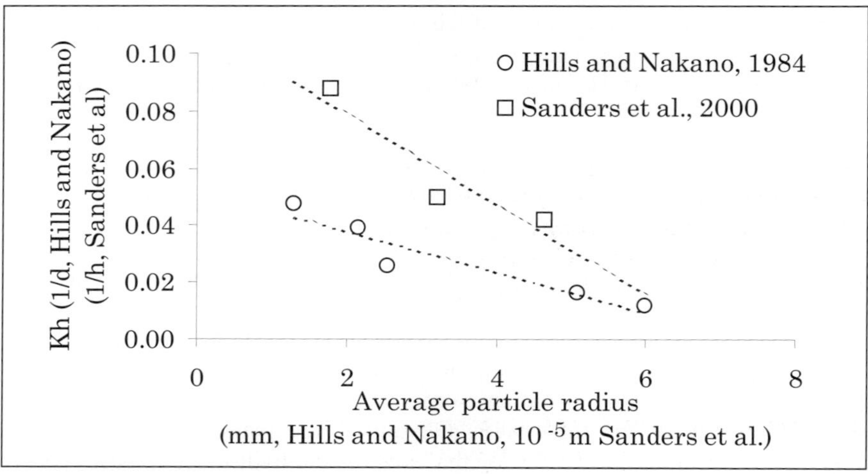

Figure 3.9. The relation between the average particle radius of a substrate and the first-order hydrolysis constant as assessed from Hills and Nakano (1984) and Sanders et al. (2000).

Lipids can also have an effect on the bioavailability of other components. Palenzuella-Rollón (1999) performed experiments in which fish processing wastewater was digested with different lipid concentrations in batch experiments. The results showed that increasing the lipid concentration to ≥4300 mg COD.l^{-1} decreased the hydrolysis constant for proteins significantly. Palenzuella-Rollón (1999) suggested that this may be due to physical hindrance by non-hydrolysed lipids and probably also LCFA, i.e. covering the surface of the proteins which results in a decrease of the amount of protein surface available to the hydrolysis.

3.4.7. Discussion and conclusion

Although Eastman and Ferguson (1981) proposed first-order kinetics for particulate complex heterogeneous substrates, these can only be applied when the rate limiting factor, in general the surface of the substrate, bioavailability or biodegradability does not alter during the conversion of the substrate. Especially for lipids the latter is not always the case. Reduction of the lipid–water interface due to coagulation might be the reason that lipid hydrolysis does not follow first-order kinetics under acidogenic conditions. First-order hydrolysis kinetics of lipids were only assessed in a multiple flask batch experiment at methanogenic conditions (Sanders, 2001). The deviation from first-order kinetics as found for lipid hydrolysis in CSTR experiments of O'Rourke (1968) and Miron et al. (2000) is probably related to problems encountered with sampling, namely scum layers and adsorption of lipids to the reactor wall. The results also indicate that gas mixing effects the particle size distribution, and with that the hydrolysis rate of lipids. This implies that a value assessed for the hydrolysis constant of neutral lipids in a multiple flask batch experiment might not be applicable to a full-scale application in a CSTR.

3.5. BIODEGRADABILITY

3.5.1. Introduction

As already mentioned in paragraph 3.4.2, two types of biodegradability can be discerned, the 'ultimate' biodegradability and the biodegradability at the applied reactor conditions. The ultimate biodegradability is reached at the most optimum reactor conditions.

Chandler et al. (1980) proposed that the biodegradability of particulate organic substrates could be predicted on the basis of its lignin content. Tong et al. (1990), however, showed that the biodegradability also depended on the structure of the lignocellulose complex. Cellulose is readily degradable but becomes less degradable or even refractory when incorporated into the lignocellulose complex.

The results of Veeken and Hamelers (1999) and Tong et al. (1990) showed that there is a positive relation between the hydrolysis rate and the biodegradability. This indicates that the accessibility of the substrate not only increases that hydrolysis rates but also the biodegradability.

The major part of OFMSW is composed of lignocellulosic organic matter besides small amounts of soluble compounds such as carbohydrates, fats and proteins (Ten Brummeler, 1993). Anaerobic digestion of the solid organic matter starts with the enzymatic degradation of the solid structure of the substrate as the organic polymers cannot be utilised directly by microorganisms. The rate and extent of enzymatic degradation depends on the physico-chemical properties of the substrate, the type of microorganisms involved and the environmental conditions (Eriksson et al., 1990). The environmental aspects such as temperature and pH were already discussed in paragraph 3.4. This paragraph focuses on the effects of physico-chemical structure on the enzymatic degradation of organic solid substrates. The following physico-chemical properties affect the rate and extent of degradation of solid organic substrates:

the surface area of the solids, which is directly related to the particle size (see 3.4.6);

the chemical composition of the substrate.

3.5.2. Chemical composition of the substrate

Chandler et al. (1980) proposed that the biodegradability of solid organic substrates could be predicted on basis of the lignin content. This is most probably true for substrates with are highly similar with respect to their physico-chemical structure as was also shown by Scharff and Veeken (1995) for landfill wastes of different ages. However, Tong et al. (1990) showed that the biodegradability strongly depended on the structure of the compound, i.e. solid organic compounds with the same lignin content showed a distinctly different hydrolysis rate and extent of degradation. Jimenez et al. (1990) showed that reducing the lignin content of vine shoots with sodium chlorite in acid medium increased the methane yield by 54%.

The rate and degradation of the lignocellulosic complex is related to the accessibility of enzymes to the sites of degradation. The accessibility depends on factors such as lignin content, the degree of crystallinity of cellulose, and the porosity of the cellulose fibre. The extent of degradation cannot be easily generalised as among others the morphology of the lignocellulosic complex is highly variable, different types of lignin are found and cellulose can be present in crystalline and amorphous forms (Eriksson et al., 1990; Gil and Pascoal Neto, 1999). Degradation of the lignocellulosic complex also depends on the type of microorganisms involved (e.g. white-rot fungi, brown-rot fungi, bacteria, actinomycetes) and the environmental conditions (e.g. oxygen content, temperature). It is assumed that lignin can only be degraded by fungi and is therefore limited to aerobic conditions (Eriksson et al, 1990). There are no publications that report the degradation of lignin under anaerobic conditions. Under aerobic conditions, several fungi are able to degrade lignin and thereby enhance the degradation of cellulose (Gilardi et al., 1995). Lignocellulosic wastes with high lignin contents will therefore give low methane yields and can better be treated by composting, an aerobic process. The chemistry and biochemistry of wood degradation are extensively studied but scarce for anaerobic degradation of solid organic wastes.

3.6. PERSPECTIVES AND CONCLUSIONS

When designing or optimising a reactor it is very tempting to use one of the many values for the first-order hydrolysis constant and biodegradability reported in literature. However, from the above results and discussions it is clear that the use of these literature values is only legitimate with a detailed knowledge of the substrate, process conditions and calculation procedure from which the hydrolysis constant was obtained. As it is often difficult to find a hydrolysis constant obtained for the correct temperature and pH, corrections for both effects will likely have to be made.

Corrections for temperature seem allowed as the hydrolysis increases with temperature according to the Arrhenius equation. For hydrolysis of total particulate COD and protein a standard free energy of activation of ~46 kJ/mol can be used. For the hydrolysis of carbohydrate components such as cellulose and starch the apparent standard free energy of activation is much higher, i.e. ~181 kJ/mol. Temperature corrections to the hydrolysis constant of neutral lipids cannot be made, because the changes in temperature affect the solubility and coagulation behaviour of the lipids. These changes in bioavailablity are not accounted for in the Arrhenius relation.

Obviously corrections to account for differences in pH are quite difficult, as the effect of pH depends on the characteristics of the substrate and of the different enzymes involved in the process. From the results obtained in this chapter no general relation between the first-order hydrolysis constant and the pH can be established. If the hydrolysis constant is assessed under methanogenic conditions the pH should be in the neutral range, namely pH 7±0.5. For acid conditions it is recommended that the hydrolysis constant be assessed at several pHs.

But even when digester conditions and the substrates seem similar, the hydrolysis rates can be different. Table 3.5 allows a comparison of the calculated hydrolysis of protein in primary sludge from data of O'Rourke (1968), Eastman and Ferguson (1981) and Boon (1994). It appears that the biodegradability of the proteins in raw domestic sewage remains rather constant, even at different pHs, but the hydrolysis constant is affected significantly even up to a factor of 30. This is likely due to differences in particle size distribution between the three primary sludges. However, as the particle size distribution was not measured in these researches a definite conclusion on this matter cannot be made. For substrates such as primary sludge it would be useful to also determine the particle size distribution when assessing the first-order hydrolysis constant as it would make the comparison of hydrolysis rates easier. However, owing to variations in the morphology of the lignocellulosic complex in substrates with a high lignin content, such as OFMSW, comparison of hydrolysis constants and biodegradability of these substrates is complicated even when the particle size distributions are known.

Table 3.5. Calculated first-order hydrolysis constant and biodegradability of the protein in raw domestic sewage sludge at pH 5 and 7, from data of O'Rourke (1968), Eastman and Ferguson (1981) and Boon (1994).

	k_h (d^{-1})	F_h	pH	T (°C)
Boon (1994)	0.2	0.32	7.0±0.2	35
O'Rourke (1968)	0.67	0.40	7.1±0.3	35
Boon, (1994)	0.2	0.23	5.0±0.2	35
Eastman and Ferguson (1981)	6.2	0.31	5.17±0.04	35

The fact that the hydrolysis rate is affected by particle size can also be used to decrease the solid retention time of the anaerobic reactor by additional pre-treatment of OFMSW. Increasing the porosity of the lignocellulosic complex can enhance the accessibility of sites susceptible for anaerobic degradation. In this way the anaerobic degradation is increased, thus decreasing the solid retention time of the anaerobic reactor. Chemical treatment by acid hydrolysis (Jimenez et al., 1990) and alkali lime cooking (Azzam and Nasr, 1993) and physical treatments by steam explosion and milling are possible ways to increase the degradation rate. As for particle size reduction, the additional costs of the pre-treatment step must be taken into account but also the treatment of wastewater with additional chemical reagents.

3.7. REFERENCES

Azzam A.M. and M.I. Nasr (1993). Physicothermochemical pretreatments of food processing waste for enhancing anaerobic digestion and biogas generation. *J. Environ. Sci. Health* **A28**, 1629–1649.

Boon, F. (1994) Influence of pH, high volatile fatty acid concentrations and partial hydrogen pressure on the hydrolysis. MSc. Thesis, Wageningen University, in Dutch.

Canalis, G. (1999). Hydrolysis of particulate starch under sub-optimal temperature conditions: Determining the effect of temperature on the rate of hydrolysis, enzyme production and enzyme activity. MSc. Thesis, Wageningen University.

Chandler, J.A., Jewell, W.J., Gossett, J.M., Soest, P.J. van, and Robertson, J.B. (1980). Predicting methane fermentation biodegradability. *Biotech. Bioengng Symposium* **10**, 93–107.

Chaplin, M.F. and Bucke, C. (1990). *Enzyme technology*. Cambridge University Press, Cambridge 1990.

Eastman, J.A. and Ferguson, J.F. (1981). Solubilization of particulate organic carbon during the acid phase of anaerobic digestion. *J. Wat. Poll. Control Fed.* **53**, 253–266.

Engbersen, J.F.J. and de Groot. Æ. (1995). *Introduction to bio-organic chemistry*. Wageningen Pers, Wageningen , in Dutch.

Eriksson K.E.L., Blanchette, R.A. and Ander, P. (1990). *Microbial and enzymatic degradation of wood and wood components*. Springer Series in Wood Science, Springer, Berlin.

Fersht, A. (1999). *Structure and mechanism in protein science: a guide to enzyme catalysis and protein folding*. W.H. Freeman and Company, NewYork.

Gil, A.M. and Pascoal Neto, C. (1999). Solid-state NMR studies of wood and other lignocellulosic materials. *Annual Reports of NMR Spectroscopy* **37**, 75–117.

Gilardi, G, Abis L. and Cass, A.E.G. (1995) Carbon-13 CP/MAS solid-state NMR and FT-IR spectroscopy of wood cell wall biodegradation. *Enzyme-and-Microbial-Technology* **17**(3), 268–275.

Hills, D. J. and Nakano, K. (1984). Effects of particle size on anaerobic digestion of tomato solid wastes. *Agricultural Wastes* **10**, 285–295.

Hobson, P.N. (1987). A model of some aspects of microbial degradation of particulate substrate, *J. Ferment. Technol.* **65**(4), 431–439.

Jimenez S., Cartegena, M.C. and Arce, A. (1990). Influence of lignin on the methanization of lignocellulosic wastes. *Biomass* **21**, 43–54.

Miron, Y., Zeeman, G., van Lier, J.B. and Lettinga, G. (2000). The role of sludge retention time in the hydrolysis and acidification of lipids, carbohydrates and proteins during digestion of primary sludge in CSTR systems. *Wat. Res.* **34**(5), 1705–1713.

O'Rourke, J.T. (1968). Kinetics of anaerobic waste treatment at reduced temperatures. PhD. Thesis, Stanford University.

Palenzuela-Rollón, A. (1999) Anaerobic digestion of fish processing wastewater with special emphasis on hydrolysis of suspended solids. PhD, Thesis, Wageningen University.

Pavlostathis, S.G. and Giraldo-Gomez, E. (1991a). Kinetics of anaerobic treatment. *Wat. Sci. Technol.* **24**(8), 35–59.

Pavlostathis, S.G. and Giraldo-Gomez, E. (1991b). Kinetics of anaerobic treatment: a critical review. *Critical Reviews in Environmental Control* **21**, 411–490.

Perot, C., Sergent, M., Richard, P., Phan Tan Luu, R. and Millot, N. (1988). The effects of pH, temperature and agitation speed on anaerobic sludge hydrolysis acidification. *Environ. Technol. Lett.* **9**, 741–752.

San Pedro, D.C. (1994) Study of hydrolysis of slowly biodegradable COD (SBCOD) under aerobic, anoxic and anaerobic conditions in activated sludge processes, Dissertation of the University of Tokyo, Department of Urban Engineering.

San Pedro, D.C., Mino, T. and Matsuo, T. (1994), Evalution of the rate of hydrolysis of slowly biodegradable COD (SBCOD) using starch as substrate under anaerobic, anoxic and aerobic conditions. *Wat. Sci. Technol.* **30**(11), 191–199.

Sanders, W.T.M., Geerink, M., Zeeman, G. and Lettinga, G. (2000). Anaerobic hydrolysis kinetics of particulate substrates. *Wat. Sci. Technol.* **41**(3), 17–24.

Sanders, W.T.M. (2001). Anaerobic hydrolysis during digestion of complex substrates. PhD thesis, Wageningen University.

Scharff H. and Veeken, A. (1995). Influence of Mechanical Pretreatment of MSW on Landfill Emissions. In: *Proceedings Sardinia 95. Fifth International Landfill Symposium*. CISA, Cagliari, Italy.

South, C.R., Hogsett, D.A.L., and Lynd, L.R. (1995). Modeling simultaneous saccharification and fermentation of lignocellulose to ethanol in batch and continuous reactors. *Enzyme-and-Microbial-Technology* **17**(9), 797–803.

Ten Brummeler, E. (1993). Dry anaerobic digestion of the organic fraction of municipal solid waste. Ph.D. Thesis. Wageningen, The Netherlands: Wageningen Agricultural University.

Tong, X., Smith, L.H., McCarthy, P.L. (1990) Methane fermentation of selected lignocellulosic materials. *Biomass* **21**, 239–255.

Vavilin, V.A., Rytov, S.V. and Lokshina, L.Ya. (1996). A description of the hydrolysis kinetics in anaerobic degradation of particulate organic matter. *Bio. Res. Tech.* **56**, 229–237.

Veeken, A. and Hamelers, B. (1999). Effect of temperature on the hydrolysis rate of selected biowaste components. *Bioresource Technol.* **69**(3), 249–255.

Veeken, A. and Hamelers, B. (2000). Effect of substrate-seed mixing and leachate recirculation on solid state digestion of biowaste. *Wat. Sci. Technol.* **41**(3), 255–262.

4

Anaerobic digestion of the organic fraction of municipal solid waste: a perspective

J. Mata-Alvarez

4.1. INTRODUCTION

Municipal Solid Waste (MSW), with a daily production in Europe of about 400,000 tons, represents one of the most important solid organic wastes generated by our society. Currently, in most western countries, separate collection of MSW has increased significantly. Biological treatments are the clearest alternative for the putrescent fraction coming from this separate collection. These treatment technologies can maximise recycling and recovery of waste components. Among biological treatments, anaerobic digestion (AD) is

Edited by J. Mata-Alvarez. ISBN: 1 900222 14 0

frequently the most cost-effective, owing to the high energy recovery linked to the process and its limited environmental impact, especially considering its limited greenhouse gases effect. Edelmann et al. (1999) compared, in both ecological and economical terms, different processes for treating biogenic wastes in plants with a treatment capacity of 10,000 tons/year of organic household wastes. After a series of measurements at compost plants they found that the methane emissions were greater than they had assumed and showed, with LCA (Life Cycle Analysis) tools, that anaerobic digestion had the advantage over composting, incineration or combination of digestion and composting, mainly because of AD's improved energy balance. They concluded that anaerobic processes will become much more important in the future for ecological reasons. In fact, the future of anaerobic digestion should be sought in the context of an overall sustainable waste management perspective.

In terms of global warming, which is often used as a reference value for ecological balance, anaerobic digestion scores much better than other options, as can be seen in Table 4.1, extracted from Baldasano and Soriano (2000). In addition, the total electrical energy produced exceeds the amount of energy used for the erection and operation of the plant. In fact, for a plant treating 15,000 tons/year of organic fraction of MSW (OFMSW) by composting, around 0.75 million kWh/year are needed, whereas for AD the net production is approximately 2.40 million kWh/year (Tilche and Malaspina, 1998). The exact yields depend on the quality of the OFMSW treated. Other similar yields referred to the treatment of 100 kg of OFMSW are schematically presented in Figure 4.1, calculated from our own experiments. Chapter 10 discusses these important aspects in detail and finds in favour of the environmental advantages of AD technology.

Anaerobic digestion of solid organic wastes is today an established technology. However, its development is relatively recent. In this chapter, we review the research on this process, from its origins in the laboratory to its present situation with more than 1,500,000 tons/y of installed capacity in industrial digesters.

Table 4.1. Contribution of the different greenhouse gases and different sources of methane emissions.

Source	Contribution (%)	Reference
Energy production	26	IPPC (1994)
	28	IPPC (1992)
Enteric fermentation	24	IPPC (1994)
	23	IPPC (1992)
Rice cultivation	17	IPPC (1994)
	21	IPPC (1992)
Wastes	7	IPPC (1994)
	17	IPPC (1992)
Landfill	11	IPPC (1994)
	—	IPPC (1992)
Biomass burning	8	IPPC (1994)
	11	IPPC (1992)
Urban wastewater	7	IPPC (1994)
	—	IPPC (1992)
CO_2	65	IPPC (1996)
	66	USEPA (1993)
CH_4	20	IPPC (1996)
	18	USEPA (1993)
CFCs	10	IPPC (1996)
	11	USEPA (1993)
N_2O	5	IPPC (1996)
	5	USEPA (1993)

IPPC: Intergovernmental Panel on Climate Change.

4.2. LABORATORY-SCALE RESEARCH

Many papers have been published dealing with the performance of different reactor configurations digesting organic solid wastes. Most of them focus on aspects of the anaerobic biodegradation of the putrescent fraction of municipal solid wastes.

Research activity on this field began around the second half of the 1970s. Of great help was the decision in June 1978 of the Council of European Communities to grant financial support for projects in the field of alternative energy sources, including energy from waste. Pioneering work on this was done by several authors such as Pauss et al. (1984) at the Catholic University of Louvain (Belgium) on hand-sorted organic fraction of MSW (HS-OFMSW); Marty et al. (1986) on a dry digestion study using a mechanically sorted organic fraction of MSW (MS-OFMSW) at the University of Aix Marseille (France); Le Roux and Warkeley (1978) and Le Roux et al. (1979) at the Warren Springs

Laboratory (United Kingdom), on a wet process with MS-MSW and Glauser et al. (1987), University of Neuchatel (Switzerland), on a comparison of mesophilic and thermophilic digestion and the effect of pretreatments on a mixture of sewage sludge and OFMSW. Also relevant are the studies carried out by Cecchi at the University of Venice, where an experimental pilot plant was set up in the municipality of Treviso (Italy). In collaboration with the University of Barcelona, these authors studied the performances of the anaerobic digestion processes of several substrates in mesophilic and thermophilic conditions (primary sewage sludge (PSS), primary and secondary sewage sludge (PSSS), PSSS mixed with SS-OFMSW or mixed MS-OFMSW). They looked at different aspects of the digestion such as the fate of the particulate matter fed to the digester (Traverso and Cecchi, 1988); the effects of external factors such as temperature and changes in the quality of the organic matter on digestion performance (Mata-Alvarez et al., 1990); the stability of the process in relation to process monitoring and recovery from digester failure conditions (Cecchi et al., 1987); or the behaviour of the digester when treating different mixtures of PSS and SS-OFMSW. All these activities, as well as others not mentioned here, gave rise several demonstration plants, which were the seed for the industrial eclosion in the 1990s.

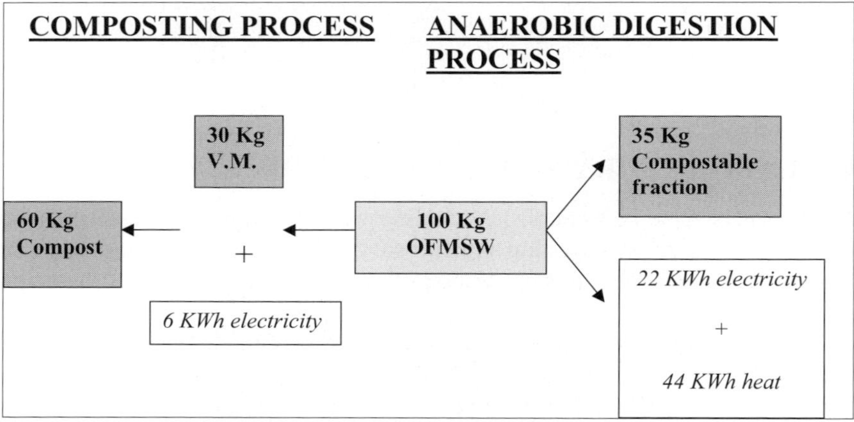

Figure 4.1. Yields of composting and anaerobic digestion technology for the treatment of 100 kg of OMFSW. (VM: Vegetal Matter.)

4.3. DEMONSTRATION AND DEVELOPMENT PLANTS: THE NEXT STEP

The first demonstration plants were built in Europe at the beginning of the 1980s. Among them it is worth mentioning the one constructed at Broni (Italy) with a digester of 2000 m^3. The plant was built with EEC economic support (De Baere and Verstraete, 1984) and a mixture of sewage sludge and OFMSW was used as digester feed. Other plants which were built at this time were those at La Buisse (France), built by VALORGA and that of ARBIOS S.A. which, together with the University of Ghent (Belgium), developed the DRANCO technology to treat the organic fraction of household refuse (see Figure 4.2). Both VALORGA and DRANCO, belong to what has been called dry technology owing to the high solids contents of the substrate fed to the digesters (around 35% Total Solids, TS). The opposite of the 'dry' technologies, the 'wet' ones, are those with a substrate fed to digesters with a TS contents below 15%, which means a TS content in the digester somewhere below 8%. There are also semi-dry technologies, which are between the two just described. A discussion of these technologies, pointing out advantages and drawbacks can be found in Chapter 5.

Figure 4.2. Pilot plant in Ghent, Belgium: the beginning of the Dranco process.

In West Germany, Klein and Rump (1987) designed a pilot plant to handle 5 tons of waste per day using the so-called KWU–Fresenius process. Similarly to the previously cited dry fermentation processes, total solids feed was around 40%. One of the first two-phase anaerobic digestion systems to treat the organic fraction of municipal solid waste was studied in Europe by Hofenk et al. (1984). The experiments carried out at pilot-plant level consisted of two stages of batch digesters (60 m^3 liquefactioning reactors for the first step and a single 10 m^3 UASB methane reactor for the second step). A sorting plant produced the substrate used in the experiments. These plants and many others not cited here marked the beginning of biomethanization technology, which brought a large number of commercial plants into being in the year 2000.

4.4. COMMERCIAL PLANTS

The first commercial plants began nearly at the same time as the demonstration plants. In France the first important digestion plant was built in 1988 at Amiens by VALORGA. This plant, with a total reactor volume of 2400 m^3, has been in operation ever since, with OFMSW coming from a front-end sorting plant. Figure 4.3 shows a picture of this plant. In Finland the DN-BIOPROCESSING LTD developed a fermentation technology called WABIO-Process (Pipping and Valo, 1988). The first WABIO-process was a treatment system for the organic fraction of MSW, sewage sludge and other organic wastes, which were mixed and homogenised in the compost unit. As a result of this step, the temperature of the organic mixture increased and was then fed to the biogas reactor. The WABIO-process has evolved a wet fermentation system to treat OFMSW. These plants are an example of the beginnings of anaerobic digestion on a commercial scale. Possibly, other types of plants not mentioned here have also contributed to the development of biomethanization. Today, in industrial terms, anaerobic digestion of solid waste can be seen as a mature technology (Riggle, 1998). Over the past ten years, for the treatment of OFMSW, it has evolved positively from an overall capacity of 122,000 tons/year in 1990 to more than 1,000,000 in the year 2000, according to the paper by De Baere (2000). Compared with the installed capacity for composting plants, this is not a large figure, but it must be remembered that most aerobic plants were constructed before anaerobic digestion was considered a fully established technology. De Baere's study identified a total of 53 plants (with a capacity larger than 3000 tons/year). Around 60% of the plants operate at the mesophilic range (40% thermophilic). Capacity has increased at a rate of some 30,000 ton/year during the period 1990–95 and around 150,000 during the past five years. An increase of around 200,000 tons/year by 2002 is envisaged.

Figure 4.3. Industrial plant in Amiens, France, in operation since 1988.

Yields from the biomethanization process are very dependent on the particular situation of each plant. Of course the main factor affecting this yield is the kind of substrate used (see section 4.6). In the case of the OFMSW it is interesting to see how different values are obtained over a long period of time, using the same process but different sorting procedures (Saint-Joly et al., 2000).

But there are also some other influences to be considered such as the desired end destination of the product and the type of process selected.

In the past few years, a remarkable interest has come about in digesting 'grey wastes' or 'residual refuse', i.e. what remains after source separation. Options for this fraction are landfilling or incineration. However, anaerobic digestion offers several advantages, such as: (a) greater flexibility, (b) the possibility of additional material recovery (up to 25%), and (c) a more efficient and ecological energy recovery: the low-calorific organic fraction is digested, the high-calorific fraction is treated thermally and the non-energy fractions can be recovered and reused (DeBaere and Boelens, 1999). It is expected that this 'residual refuse' will open up a new 'market' for AD. Table 4.2 shows some of the present market technologies, classified as dry and wet systems. Many others exist, but only the ones with a larger installed capacity have been reflected in this table.

Table 4.2. Some of present market AD technologies.

	Technology	Owner
DRY TECHNOLOGIES	BRV	LINDE
	DRANCO	OWS
	KOMPOGAS	KOMPOGAS
	VALORGA	BABCOCK BORSIG POWER
WET TECHNOLOGIES	BIOSTAB	ROS ROCA
	BTA	BTA
	KCA	LINDE
	WAASA	ALCYON/CYTEC

4.5. PRESENT RESEARCH ON ANAEROBIC DIGESTION

Despite the increasing number of full-scale plants, research activity continues on different aspects of anaerobic digestion of OFMSW, from the fundamentals, including modelling of the process (which is a very important tool for design) to process aspects, which include digester performance, inhibition problems, temperature influence, two-phase systems, etc. For instance, Pavan et al. (2000a) using a pilot-scale stirred reactor, considered the application of the semi-dry single-phase thermophilic anaerobic digestion process for different mixtures of mechanically sorted and source-sorted organic fractions of municipal solid waste, which have different degrees of biodegradability. During the study, and in order to obtain a stable operation, the hydraulic retention time (HRT) was increased when the percentage of source-sorted OFMSW increased. The main conclusion of the study was that when digesting highly biodegradable wastes, such as source-sorted OFMSW or fruit and vegetable wastes, it is advisable to use a two-phase anaerobic digestion system. This option allows much higher loads in the digester. In another detailed study, including kinetic aspects, these authors found optimal operating conditions for both the hydrolytic reactor (operated at mesophilic and thermophilic temperatures) and the methanisation reactor (thermophilic temperature). Again the main conclusion was that the best solution to digest highly biodegradable wastes was a two-phase digestion system (Pavan et al., 2000b). Somehow a similar conclusion was reached by Vieitez and Ghosh (1999) with high-solids wastes.

Scherer et al. (2000), using a simple two-phase system, processing 'grey waste' as feed, achieved 80% degradation of the volatile solids (VS). Temperatures in the reactors were 65 ºC in the first stage (HRT = 4.3d) and 55 ºC in the second stage (HRT = 14.2 d). Degradation rates obtained by the biogas yield (up to 797 L/kg VS fed) revealed up to 98% of the theoretical possible yield.

Ghosh et al. (2000) treating RDF (refuse-derived fuel) in bench-scale digesters, compared a conventional high-rate reactor with a two-phase configuration and studied the influence of temperature and particle size. Thermophilic digestion at 55 ºC only increased the methane yield by 7% over digestion at 35 ºC. Decreasing the RDF particle size from 2.2 mm to 1.1 had no beneficial effects on mesophilic temperatures, but methane yield was increased by 14% using thermophilic conditions. Regarding the effect of using a two-phase configuration, the same authors found that all these yields increased around 20%. They also tried to improve digester performance by using caustic soda as a pre-treatment. The aspects of pre-treatment are fully discussed in Chapter 8 of this book.

Raynal et al. (1998) studied the influence of pH, load and hydraulic retention time on the process performances at 35 ºC. The system involved several liquefaction laboratory-digesters, each of them treating one type of vegetable waste (potato peelings, green salad leaves, green beans mixed with carrots, apple pomace), linked to a central methane fixed-film reactor. As an average, except for apple pomace, hydrolysis yields were high (up to 80%) during the liquefaction step, whereas the mixture of the acidogenic effluent was degraded by up to 80% in the methanisation reactor. In a final run with average loading rates near 4 g COD l^{-1} day^{-1} and 17 days for hydraulic retention time, overall organic matter removal reached a value as high as 87%. Finally, another study supporting the advantages of two-phase anaerobic reactors was carried out at mesophilic conditions by Poirrier et al. (1999) using as substrate the solid waste from the brewery industry.

It is important to mention that, on an industrial scale, one-phase systems for OFMSW digestion are absolutely predominant. Technical advantages reported here seem not to be enough to justify a higher investment and higher maintenance costs (De Baere, 2000). This aspect is also considered in Chapter 5.

Another point that has raised interest in the literature is whether it is better to work in mesophilic conditions or in thermophilic ones. There are many industrial digesters working at both mesophilic and thermophilic temperatures (see Figure 1.4). Both ranges are possible for the successful operation of a digester and, in fact, there are many examples of industrial operation at meso- or thermophilic temperatures. However, there are substrates (that is, kinds of OFMSW) that seem more appropriate than others for digestion at one given range of temperature. With reference to the literature, thermophilic operation was found optimal for digesting the mechanically selected OFMSW by Cecchi et al. (1991) in a pilot-plant study. However, many problems were found when digesting source-sorted OFMSW at 55 ºC (Bernal et al., 1992). These authors concluded the inadequacy of this temperature to treat substrates with a high degree of biodegradability. Generally, a thermophilic operation produces more biogas than a mesophilic one. Against this surplus energy yield, thermophilic digestion obviously involves greater energy demand and thermal effort, which, in many cases, are comparable. Although yields and kinetics are more favourable at thermophilic temperature, optimal conditions depend on the type of substrate (biodegradability) and the type of system (one/two-phase) used. Two-phase operation is more advisable if thermophilic digestion is to be carried out with a highly biodegradable substrate (Pavan, 2000b). Biodegradability of solid wastes is a very important parameter when designing a digestion system. The next section is devoted to quantify its influence. Finally, Table 4.3 reports some recent studies of the performance of anaerobic digestion of solid wastes, which points out the level of activity in this research sector.

Table 4.3. Some studies concerning the performance of digestion of solid wastes.

Substrate	**Scale/Reactor type/Temperature**	**Reference**
Slaughterhouse and catering	Pilot / Mesophilic	Membrez et al. (1999)
Poultry mortalities	Lab/ Two phase (Leach bed + UASB) Mesophilic	Chen (1999)
OFMSW in Bamako (Mali)	Pilot (Leach bed + UASB) Psychrophilic	Ouedraogo (1999)
Sewage sludge	Lab/Two-phase/Mesophilic	García-Heras et al. (1999)
Mycelium waste (India)	Non-stirred digester Psychrophilic	Yeole and Ranade (1999)

Table 4.3. (*continued*).

OFMSW	Lab/ One and Two stages Psychro and Mesophilic	Wang and Banks (1999)
Coffee pulp	Lab/ Batch /Psychrophilic	Valdés et al. (1999)
Fish farming sludge	Lab/Batch/Mesophilic	Gebauer (1999)
OFMSW	Pilot / Two-phase / Thermophilic	Madokoro et al. (1999)
Food Wastes	Lab / Leach Bed / Mesophilic	Paik et al. (1999)
OFMSW	Lab / CSTR/ Mesophilic	Houbron et al. (1999)
Coffee pulp	Pilot / Plug flow / Mesophilic	Farinet and Pommares (1999)
OFMSW/Coffee pulp	Pilot / Two-phase/	Edelmann et al. (1999)
Activated sludge	Full-scale/ Digester/ Mesophilic to thermophilc	Zabranska et al. (2000)
Mixed sludge	Lab/Digester/Mesophilic coupled with thermophilic pre and post-treatments	Cheunbarn and Pagilla (2000)
Pineapple peel	Digester/ Mesophilic and thermophilic	Nakapadungrat et al. (1999)
Cattle manure	Lab/Batch completely mixed/Mesophilic and thermophilic	Sanchez et al. (2000)
Sludge	Lab/Batch/2-stage thermophilic/mesophilic	Gabb et al. (1999)
Mixed sludge	Digester/Thermophilic and mesophilic	Vandenburgh and Ellis (1999)
Biosolids	Digester/ Thermophilic and mesophilic	Ghosh et al. (1999)
Kraft-mill biosolids	Thermophilic and Mesophilic two-phase	Ghosh and Taylor (1999)
Wastewater derived from the production of protein isolates	Lab/fluidized-bed reactor/ Mesophilic	Borja et al. (2001)
Poultry mortalities	Leach bed-UASB/mesophilic	Chen and Wang (1998)
Distillery wastewater (wine vinasses)	Lab/anaerobic filter vs fluidised bed/thermophilic	Perez et al (1998)
Daily manure	Attached-film bioreactors/psychrophilic	Vartak et al (1997)
Instant-coffee wastewater	UASB/ thermophilic and mesophilic	Dinsdale et al. (1997)
Recalcitrant distillery wastewater	UASB (140L)/thermophilic	Harada et al. (1996)

Table 4.3. (*continued*).

Swine manure	Lab/Sequencing batch reactors/psychrophilic	Masse et al. (1997)
Human waste	Psychrophilic	Meher et al. (1994)
Swine manure	Thermophilic	Hansen et al. (1999)

4.6. BIODEGRADABILITY: AN IMPORTANT PARAMETER IN ANAEROBIC DIGESTION

When considering the yields reported in the literature of anaerobic digestion of the organic fraction of municipal solid waste, the different origin of the waste should be taken into account. At a given temperature, a given amount of volatile solids (VS) of a particular waste can be converted to a maximum amount of biogas, provided optimal conditions are prevalent. This conversion can be measured through what has been called 'ultimate methane yield' (B_0). Obviously, the higher the ultimate methane yield, the higher the biodegradability. However, the actual amount of biogas that is obtained in a digester at this given temperature is not this yield, but a function of it and the operating conditions of the digester. Thus, two factors should be taken into account when considering reported yields. The first factor is the biodegradability at a specified temperature, that is, its ultimate methane yield. This factor is intrinsic to the waste itself. Second, the operating conditions must be considered and these depend on kinetics, reactor configuration or, what amounts to the same thing, the flow pattern within the digester, digestion phases, etc. See section 3.5.1 for a more thorough discussion on the biodegradability concept.

Thus, if SMP is the specific methane production and B_0 the ultimate biogas yield, both expressed as m^3 methane/kg VS, a simple balance in a digester leads to the following equation (Mata-Alvarez et al., 1990):

$$SMP = B_0 - (1 - f_{VS})B'_0 \qquad (4.1)$$

where B'_0 is the ultimate biogas yield of the digester effluent and f_{VS} is the fraction of volatile solids biodegraded. It is clear that, if f_{VS} is 1, SMP = B_0 and if f_{VS} is 0, that is, no volatile solids are degraded, $B_0 = B'_0$ as, obviously, SMP = 0. However, although f_{VS} was not 1, which is the usual case, SMP can also reach the B_0 value, provided $B'_0 = 0$, that is, when the ultimate methane yield of the effluent becomes zero due to an extended digestion period (the substrate has biodegraded completely).

Assuming a first-order biodegradation rate, in a stirred digester, methane production rate (MPR, m^3/d) is proportional to the biogas potential of substrate in the digester, that is, its ultimate methane yield, which is equal to that of the effluent, B'_0:

$$MPR = k \cdot V \cdot VS_r \cdot B'_0 \tag{4.2}$$

where k is the first-order kinetic constant (d^{-1}), V, the digester volume (m^3) and VS_r the volatile solids content within the reactor ($kgVS/m^3$). If this expression is divided by $Q \cdot VS_f$, the product of the volumetric flow rate (m^3/d) and the volatile solids content of the feed (kg VS/m^3), then, taking into account that $MPR/(Q \cdot VS_f)$ is the specific methane production SMP, the following expression is obtained:

$$SMP = k \cdot HRT \cdot (1 - f_{VS}) \cdot B'_0 \tag{4.3}$$

Substituting B'_0 from equation 4.1 and rearranging:

$$SMP = B_0 / (1 + 1/(k \cdot HRT)) \tag{4.4}$$

This equation links specific methane production with both biodegradability and kinetic factors. This expression is of limited use, because a first-order model has been assumed and this will not always be true. However, using other common kinetic models, such as those described in Chapter 2, the qualitative trends that can be deduced from the equivalent equation 4.4 would be the same (see Figure 4.4). Hence, although quantitative information cannot be obtained, some general conclusions can be drawn from this analysis. It is clear that, the higher the HRT, the more biogas is produced and the closer SMP will approach B_0. This approach is more noticeable at high values of the kinetic constant. Thus, at low kinetics, it is necessary to operate the digester at longer HRT if high SMP is desired. When the kinetic constant is high, even at low HRT (obviously, over the washout value) the yields are high, and the profile SMP vs HRT is much flatter.

It is also important to notice that, when values of the specific methane production of a digester operation are reported, a lower value does not necessarily indicate a deficient performance of the bioreactor: it can simply be due to a lower biodegradability of the substrate (a low B0 value). As a consequence, attention should be paid when comparing reported yield values, especially regarding the units used and the source/type of waste.

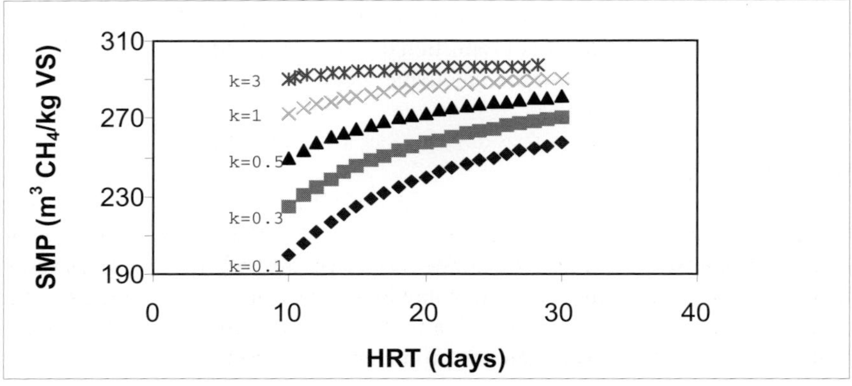

Figure 4.4. A simulated example for a substrate with B_0=300 m^3CH_4/kg VS. Kinetic constant k values are expressed in d^{-1}. Equation 4.4 has been used.

4.7. FINAL REMARKS

Anaerobic digestion of OFMSW has grown very much during the past five years, but the expectations for the near future are quite optimistic, as the potential is really very large. Today around 50% of municipal solid wastes are landfilled, with a content of around 30% of organic fraction (without considering paper and cardboard). Following in potential is the digestion of sewage sludge. Table 4.4, extracted from Tilche and Malaspina (1998) shows an estimate of the biogas potential of several wastes, among them solid ones. As can be seen, the growth potential for this technology is very important, especially because of the very important factor of the reduction in greenhouse gas emission agreed at the Kyoto Summit. This aspect takes on even greater significance considering that composting, the current biotechnology for OFMSW recycling, is more problematic, as discussed in this Chapter and especially in Chapter 10. For instance, considering data from Tilche and Malaspina (1998) and also those of Table 4.1, a daily reduction of 180,000 tons of CO_2 equivalents can be estimated, that is around 30% of the global emission reductions agreed in Kyoto. In addition, another factor that has to contribute to the consolidation of anaerobic digestion as a mainstream technology for the OFMSW is the fact that the digested residue can be considered quite stable organic matter with a very slow turnover of several decades (over 50 years), given adequate soil conditions. In this way the natural imbalance in CO_2 can be adjusted by restoring or creating organic rich soil (Verstraete et al., 2000). The removal of CO_2 constitutes an extra benefit that could help place AD among the most relevant technologies in this field (Verstraete et al. 2000).

From a more general point of view, the future of anaerobic technology should be sought in the integration of this process into an overall sustainable waste treatment perspective. As indicated in Chapter 10, Life Cycle Analysis applied to this technology offers several interesting features. The energy recovered (around 100–150 m^3/ton fresh waste) is a very important factor, particularly in developing countries.

Finally, it should also be remembered that the grey fraction, with an important content in organic matter, represents a very important factor of growth for anaerobic digestion of OFMSW, as this grey fraction will not be allowed to be landfilled. At the same time, but on a more long-term basis, all the old composting plants will be candidates for conversion to AD plants, given the ecological and, at that time economical, reasons discussed here and in Chapter 10.

Table 4.4. Biogas production from different sources in Europe.

Source	**Today estimate (10^6 m^3/d CH_4)**	**Estimated potential (10^6 m^3/d CH_4)**
OFMSW	4.5[a]	15[a]
SS	1.7	4
Industrial wastewater	0.8	3
Animal wastes	0.5[b]	10[b]
Total	7.5	32

[a] Estimates for today and potential (Tilche and Malaspina, 1998).
[a] Net methane emission reduction.

4.8. REFERENCES

Baldasano, J. M. and Soriano, C. (2000). Emission of greenhouse gases from anaerobic digestion processes. *Water Science and Technology* **41**(3), 275–282.

Bernal, O., Llabrés, P., Cecchi, F. and Mata-Alvarez, J. (1992). A comparative study of the thermophilic biomethanization of putrescible organic wastes. *Odpadní vody / Wastewaters* **1**(1), 197–206.

Borja, R., González, E., Raposo, F., Millan, F. and Martín, A. (2001). Performance evaluation of a mesophilic anaerobic fluidized-bed reactor treating wastewater derived from the production of proteins from extracted sunflower flour. *Bioresource Technol.* **76**(1), 45–52.

Cecchi, F., Traverso, P.G., Mata Alvarez, J. and Llabrés, P. (1987). pH and CO_2 as monitoring parameters of the anaerobic digestion process of municipal solid waste. *Ingegneria Sanitaria* **35**, 339–344.

Cecchi, F., Pavan, P., Mata-Alvarez, J., Bassetti, A. and Cozzolino, C. (1991). Anaerobic digestion of municipal solid waste: thermophilic versus mesophilic performance at high solids. *Waste Management & Research* **9**, 305–315.

Chen, T.-H. (1999). Anaerobic treatment of poultry mortalities. In: *Proceedings of the II Internat. Symposium on Anaerobic Digestion of Solid Wastes. Barcelona 15–18 June 1999*. Vol. 2 (ed. Mata-Alvarez, J., Tilche, A., Cecchi, F.), pp. 69–72. Printed by Gràfiques 92.

Chen, T.-H. and Wang, J.C. (1998). Performances of mesophilic anaerobic digestion systems treating poultry mortalities. *Journal of Environmental Sciences and Health* B **33**(4), 487–510.

Cheunbarn, T. and Pagilla, K.R. (2000). Anaerobic thermophilic/mesophilic dual-stage sludge treatment. *J. Envir. Eng.* **126**(9), 796–801.

De Baere, L. and Verstraete, W. (1984). Anaerobic fermentation of semisolid and solid substrate. EEC Conference on Anaerobic and Carbohydrate Hydrolisis of Waste (Luxembourg), pp. 95–208.

De Baere, L. (2000). Anaerobic digestion of solid waste: state-of-the-art. *Water Science and Technology* **41**(3), 283–290.

De Baere, L. and Boelens, J. (1999). The treatment of grey and mixed solid waste by means of anaerobic digestion: future developments. In: *Proceedings of the II Internat. Symposium on Anaerobic Digestion of Solid Wastes. Barcelona 15–18 June 1999*. Vol. 2 (ed. Mata-Alvarez, J., Tilche, A., Cecchi, F.), pp. 302–305. Printed by Gràfiques 92.

Dinsdale, R.M., Hawkes, F.R. and Hawkes, D.L. (1997). Mesophilic and thermophilic anaerobic digestion with thermophilic pre-acidification of instant-coffee production wastewater. *Water Research* **31**(8), 1931–1938.

Edelmann, W., Joss, A. and Engeli, H. (1999). Two step anaerobic digestion of organic solid wastes. In: *Proceedings of the II Internat. Symposium on Anaerobic Digestion of Solid Wastes. Barcelona 15–18 June 1999*. Vol. 2 (ed. Mata-Alvarez, J., Tilche, A., Cecchi, F.), pp. 50–153. Printed by Gràfiques 92.

Farinet, J.L. and Pommares, P. (1999). Anaerobic digestion of coffee pulp. A pilot study in Mexico. In: *Proceedings of the II Internat. Symposium on Anaerobic Digestion of Solid Wastes. Barcelona 15–18 June 1999*. Vol. 2 (ed. Mata-Alvarez, J., Tilche, A., Cecchi, F.), pp. 29–132. Printed by Gràfiques 92.

Gabb, D., Jenkins, D., Ghosh, S., Hake, J.M., De Leon, C. and Williams, D.R. (1999). Pathogen destruction efficiency in high temperature anaerobic digestion. WEFTEC'99 Annual Conf. Expo., 72nd, pp. 929–940.

García-Heras, J. L., Salaberría, A., Prevot, C. and Sancho, L. (1999). Increase of organic loading rate and specific methane production by splitting phases in anaerobic digestion of sewage sludge. In: *Proceedings of the II Internat. Symposium on Anaerobic Digestion of Solid Wastes. Barcelona 15–18 June 1999*. Vol. 2 (ed. Mata-Alvarez, J., Tilche, A., Cecchi, F.), pp. 77–80. Printed by Gràfiques 92.

Gebauer, R. (1999). Anaerobic digestion of fish farming sludge. In: *Proceedings of the II Internat. Symposium on Anaerobic Digestion of Solid Wastes. Barcelona 15–18 June 1999*. Vol. 2 (ed. Mata-Alvarez, J., Tilche, A., Cecchi, F.), pp. 101–104. Printed by Gràfiques 92.

Ghosh, S., Henry, M.P., Sajjad, A., Mensinger, M.C. and Arora, J.L. (2000). Pilot-Scale gasification of MSW by High-Rate & Two-Phase Anaerobic Digestion. *Water Science and Technology* **41**(3), 101–110.

Ghosh, S., Henry, M.P. and Sajjad, A. (1999). Effects of process-control parameters on hydrolysis and acidogenesis of biosolids. WEFTEC'99, Annu. Conf. Expo., 72nd, pp. 896–906.

Ghosh, S. and Taylor, D.C. (1999). Kraft-mill biosolids treatment by conventional and biphasic fermentation. *Water Science and Technology* **40**(11–12), 169–177.

Glauser, M., Aragno, M., Gandolla, M. (1987). Anaerobic digestion of urban wastes: sewage sludge and organic fraction of garbage. In: Bioenvironmental Systems, Vol. 3, D.L. Wise, Ed (CRC Press, Boca Raton, Florida), pp. 143–225.

Hansen, K.H., Angelidaki, I. and Ahring, B.K. (1999). Improving thermophilic anaerobic digestion of swine manure. *Water Research* **33**(8), 1805–1810.

Harada, H., Uemura, S., Chen, A.C. and Jayadevan, J. (1996). Anaerobic treatment of a recalcitrant wastewater by a thermophilic UASB reactor. *Bioresource Technology* **55**(3), 215–221.

Hofenk, G. , Lips, S.J.J., Rikjens, B.A. and Voetberg, J.W. (1984). Two-phase anaerobic digestion of solid organic wastes yielding biogas and compost. EEC Contract Final Report ESE-E-R-040-NL, pp. 57.

Houbron, E., Dumortier, R. and Delgenes, J.P. (1999). Food Solid Waste Solubilization using Methanogenic Sludge as Inoculo. In: *Proceedings of the II Internat. Symposium on Anaerobic Digestion of Solid Wastes. Barcelona 15–18 June 1999*. Vol. 2 (ed. Mata-Alvarez, J., Tilche, A., Cecchi, F.), pp. 117–120. Printed by Gràfiques 92.

Klein, M. and Rump, H. (1987). Anaerobic digestion of solids – example: organic fraction of municipal solid waste. In: *Biomass for Energy and Industry, 4th E.C. Conference* (ed. G. Grassi et al.), pp. 845–849. Elsevier Applied Science, London.

Le Roux, N.W. and Warkeley, D.S. (1978). The microbial production of methane from putrescible fractions of sorted household waste. *Conservation & Recycling* **2**, 163–169.

Le Roux, N.W., Warkeley, D.S. and Simpson, M.N. (1979). The microbial production of methane from household waste. *Conservation & Recycling* **3**, 165–174.

Madokoro, T., Ueno, M., Moro, M., Yamamoto, T. and Shibata, T. (1999). Anaerobic digestion system with micro-filtration membrane for kitchen refuse. In: *Proceedings of the II Internat. Symposium on Anaerobic Digestion of Solid Wastes. Barcelona 15–18 June 1999*. Vol. 2 (ed. Mata-Alvarez, J., Tilche, A., Cecchi, F.), pp. 105–108. Printed by Gràfiques 92.

Marty, B., Ragot, M., Ballester, J.M., Ballester, M. and Giallo, J. (1986). Semi-solid state thermophilic digestion of urban wastes. In: *EEC Contractor Meeting Anaerobic Digestion, Villeneuve D'Ascq (Lille), France, March 4–6 (1986).*

Masse, D.I., Droste, R.L., Kennedy, K.J., Patni, N.K. and Munroe, J.A. (1997). Potential for the psychrophilic anaerobic treatment of swine manure using a sequencing batch reactor. *Canadian Agricultural Engineering* **39**(1), 25–33.

Mata-Alvarez, J., Cecchi, F., Pavan, P. and Llabrés, P. (1990). The performances of digesters treating the organic fraction of municipal solid waste differently sorted. *Biological Wastes* **33**, 181–199.

Meher, K.K., Murthy, M.V.S., Gollakota, K.G. (1994). Psychrophilic anaerobic digestion of human waste. *Bioresource Technology* **50** (2), 103–106.

Membrez, Y., Schwitzguébel, J.P., Dubois, B., Wellinger, A., Descloux, D., Disetti, J.J., Heckly, C. (1999). Anaerobic treatment and valorization into animal feed of wastes from slaughterhouse and catering. In: Proceedings of the II Internat. Symposium on Anaerobic Digestion of Solid Wastes, Barcelona 15–18 June 1999, Vol. 2 (ed. Mata-Alvarez, J., Tilche, A. and Cecchi, F.), pp. 65–68. Printed by Gràfiques 92.

Nakapadungrat, Y., Chaiyo, S., Sumleekaew, N., Manatsirikiat, S., Lertaveesin, P. and Tanticharoen, M. (1999). The carbohydrate-utilizing bacteria in the mesophilic and thermophilic anaerobic digesters of pineapple peel. *Biol. Res. Trop.* **13**, 123–127.

Ouedraogo, A. (1999). Pilot-scale two-phase anaerobic digestion of the biodegradable organic fraction of Bamako district municipal solid waste. In: *Proceedings of the II Internat. Symposium on Anaerobic Digestion of Solid Wastes. Barcelona 15–18 June 1999*. Vol. 2 (ed. Mata-Alvarez, J., Tilche, A., Cecchi, F.), pp. 73–76. Printed by Gràfiques 92.

Paik, B.-C., Shin, H.-S., Han, S.-K., Song, Y.-Ch., Lee, Ch.-Y. and Bae, J.-H. (1999). Enhanced Acid Fermentation of Food Waste in the Leaching bed. In: *Proceedings of the II Internat. Symposium on Anaerobic Digestion of Solid Wastes. Barcelona 15–18 June 1999*. Vol. 2 (ed. Mata-Alvarez, J., Tilche, A., Cecchi, F.), pp. 109–112. Printed by Gràfiques 92.

Pauss, A., Nyns, E.J. and Naveau, H. (1984). Production of methane by anaerobic digestion of domestic refuse. In: *EEC Conference on Anaerobic and Carbohydrate Hydrolysis of Waste (Luxembourg)*.

Pavan, P., Battistoni, P., Traverso, P.G., Cecchi, F. and Mata-Alvarez-Alvarez, J. (2000a). Two-phase anaerobic digestion of source sorted of MSW: performance and kinetic study. *Water Science and Technology* **41**(3), 111–118.

Pavan, P., Battistoni, P., Mata-Alvarez-Alvarez, J. and Cecchi, F. (2000b). Performance of thermophilic semi-dry anaerobic digestion process changing the feed biodegradability. *Water Science and Technology* **41**(3), 75–82.

Perez, M., Romero, L.I., Sales, D. (1998). Comparative performance of high rate anaerobic thermophilic technologies treating industrial wastewater. *Water Research* **32**(3), 559–564.

Pipping, I. and Valo, R. (1988). Personal Communication.

Poirrier, P., Chamy, R. and Fernández, B. (1999). Optimization of the performance operation of a two-phase anaerobic reactor used in industrial solid waste treatment. In: *Proceedings of the II Internat. Symposium on Anaerobic Digestion of Solid Wastes. Barcelona 15–18 June 1999*. Vol. 2 (ed. Mata-Alvarez, J., Tilche, A., Cecchi, F.), pp. 99–106. Printed by Gràfiques 92.

Raynal, J., Delgenes, J.P. and Moletta, R. (1998). 2-Phase anaerobic-digestion of solid-wastes by a multiple liquefaction reactors process. *Bioresource Technology* **65**(1–2), 97–103.

Riggle, D. (1998). Acceptance improves for large-scale anaerobic digestion. *Biocycle* **39**(6), 51–55.

Saint-Joly, C., Desbois, S. and Lotti, J-P. (2000). Determinant impact of waste collection and composition on AD performance: industrial results. *Water Science and Technology* **41**(3), 291–298.

Sanchez, E., Borja, R., Weiland, P., Travieso, L. and Martin, A. (2000). Effect of temperature and pH on the kinetics of methane production, organic nitrogen and phosphorus removal in the batch anaerobic digestion process of cattle manure. *Bioprocess Engineering* **22**(3), 247–252.

Scherer, P.A.., Vollmer, G.-R.., Fakhouri, T. and Martensen, S. (2000). Development of a methanogenic process to degrade exhaustively municipal "residual refuse" (MRR) resp. ''grey waste'' under thermophilic and hyperthermophilic conditions. *Water Science and Technology* **41**(3), 83–92.

Tilche, A. and Malaspina, F. (1998). Biogas production in Europe. Paper presented at the 10th European Conference Biomass for Energy and Industry, Würzburg, Germany, 8–11 June, 1998.

Traverso, P.G. and Cecchi, F. (1988). Anaerobic digestion of the shredded organic fraction of municipal solid waste. *Biomass* **16**, 97–106.

Valdés, W., Díaz Portuondo, E.E., Duran Rodríguez, L. and Alvarez Hubert, I. (1999). Anaerobic digestion of coffee pulp. Influence of neutralization. In: *Proceedings of the II Internat. Symposium on Anaerobic Digestion of Solid Wastes. Barcelona 15–18 June 1999*. Vol. 2 (ed. Mata-Alvarez, J., Tilche, A., Cecchi, F.), pp. 93–96. Printed by Gràfiques 92.

Vandenburgh, S.R., Ellis, T.G. (1999). Effect of varying solids concentration and organic loading on the performance of TPAD to produce class a biosolids. WEFTEC'99, Annu. Conf. Expo., 72nd, pp. 918–928.

Vartak, D.R., Engler, C.R., McFarland, M.J. and Ricke, S.C. (1997). Attached-film media performance in psychrophilic anaerobic treatment of dairy cattle wastewater. *Bioresource Technology* **62**(3), 79–84.

Verstraete, W., Van Lier, J., Pohland, F., Tilche, A., Mata-Alvarez, J., Ahring, B., Hawkes, D., Cecchi, F., Moletta, R. and Noike, T. (2000). Developments at the Second International Symposium on Anaerobic Digestion of Solid Waste. *Bioresource Technology* **73**, 287–289.

Vieitez, E. and Ghosh, S. (1999). Biogasification of solid wastes by two-phase anaerobic fermentation. *Biomass Bioenergy* **16** (5), 299–309.

Wang, Z. and Banks, C.J. (1999). Effect of the temperature on the degradation of the organic fraction of municipal solid waste (OFMSW) using a conventional single stage anaerobic process and a two stage anaerobic–aerobic system. In: *Proceedings of the II Internat. Symposium on Anaerobic Digestion of Solid Wastes. Barcelona 15–18 June 1999*. Vol. 2 (ed. Mata-Alvarez, J., Tilche, A., Cecchi, F.), pp. 85–88. Printed by Gràfiques 92.

Yeole, T.Y. and Ranade, D.R. (1999). Anaerobic digestion of mycelium waste collected during antibiotic production. In: *Proceedings of the II International Symposium on Anaerobic Digestion of Solid Wastes. Barcelona 15–18 June 1999*. Vol. 2 (ed. Mata-Alvarez, J., Tilche, A., Cecchi, F.), pp. 81–84. Printed by Gràfiques 92.

Zabranska, J., Dohanyos, M., Jenicek, P. and Kutil, J., (2000). Thermophilic process and enhancement of excess activated sludge degradability–two ways of intensification of sludge treatment in the Prague central wastewater treatment plant. *Water Science and Technology* **41**(9), 265–272.

5

Types of anaerobic digester for solid wastes

P. Vandevivere, L. De Baere and W. Verstraete

ABBREVIATIONS AND DEFINITIONS

Biowaste	Mix of kitchen and garden waste, separated at the source (= VFG)
COD	Chemical oxygen demand
OFMSW	Mechanically sorted organic fraction of municipal solid wastes
OLR	Organic loading rate
OLR_{max}	Maximum sustainable OLR
TS	Total solids content
UASB	Upflow anaerobic sludge blanket reactor
VS	Volatile solids content
VFG	Vegetable-fruit-garden wastes (= biowaste)

Edited by J. Mata-Alvarez. ISBN: 1 900222 14 0

5.1. INTRODUCTION

If the literature on anaerobic digestion of solid wastes may at times appear confusing or difficult to summarize, one likely reason is that it is hard to find papers with similar experimental set-ups. In fact, it is precisely the appropriateness of a given reactor design for the treatment of particular organic wastes which forms the focus of most research papers. The comparison of research data and drawing of conclusions is difficult because the great diversity of reactor designs is matched by an as large variability of waste composition and choice of operational parameters (retention time, solids content, mixing, recirculation, inoculation, number of stages, temperature, ...). Empirical know-how is the rule and there certainly does not exist a consensus over the optimal reactor design to treat municipal solids. The reason most likely lies in the complexity of the biochemical pathways involved (Chapter 1) and the novelty of the technology.

The discussion and evaluation of reactor designs will greatly vary depending on whether one takes a biological, technical, economical, or environmental viewpoint. While the biologist is concerned with the rate, stability and completion of biochemical reactions, the engineer will rather focus on wear and maintenance of electro-mechanical devices. On the other hand, a seller will rate reactor designs based on fixed and operational costs while the environmentalist will consider emissions of pollutants and recovery of energy or materials. This chapter strives to address the technical and biological viewpoints in depth and highlight a few environmental and financial issues.

The scope of this chapter is limited to feedstocks consisting mainly in the organic fraction of municipal solid wastes sorted mechanically in central plants (OFMSW) or organics separated at the source, referred to here as biowaste (the vegetable-fruit-garden, or VFG, fraction). While this chapter specifically addresses the design of the biomethanization reactor, it should be kept in mind that the latter has many important implications on the need for specific pre- or post-treatment unit processes. Necessary pre-treatment steps may include magnetic separation, comminution in a rotating drum or shredder, screening, pulping, gravity separation or pasteurization (Figure 5.1). As post-treatment steps, the typical sequence involves mechanical dewatering, aerobic maturation, and water treatment but possible alternatives exist such as biological dewatering or wet mechanical separation schemes wherein various products may be recovered.

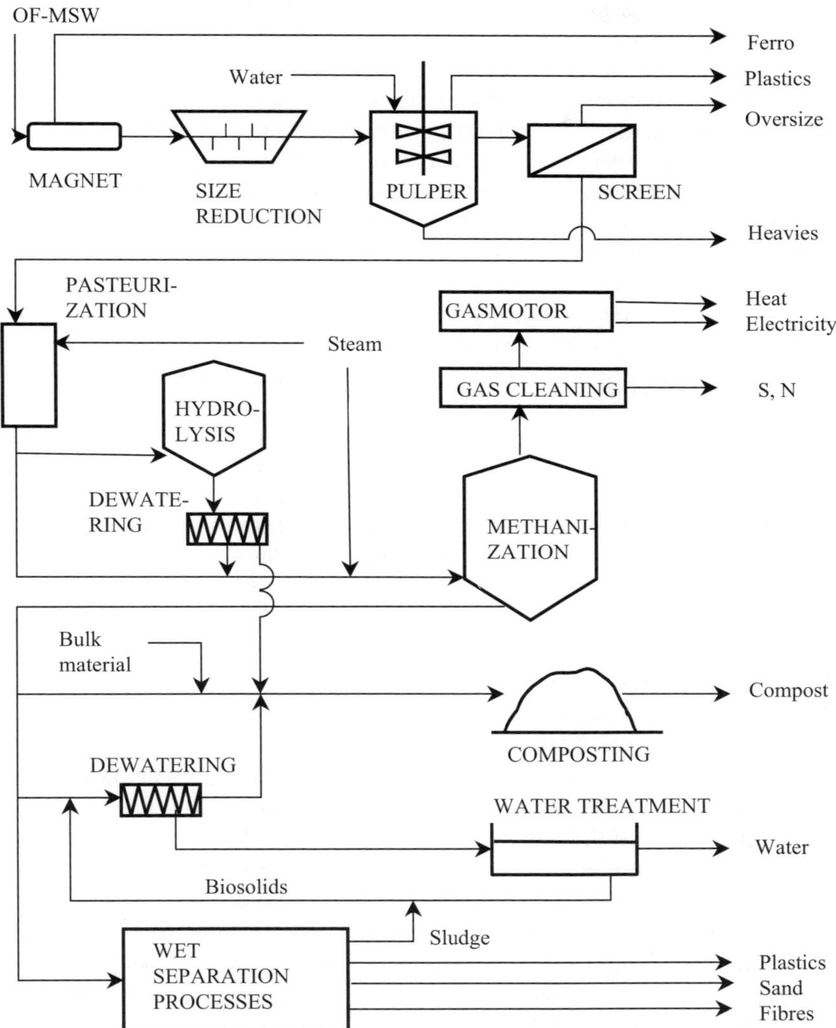

Figure 5.1. Examples of unit processes commonly used in conjunction with anaerobic digesters of solid wastes.

A plant treating municipal solids anaerobically is therefore best seen as a complex train of unit processes whereby wastes are transformed into a dozen products. Appropriate rating of given reactor designs should therefore also address the quantity and quality of these products (Table 5.1) as well as the need for additional pre- and post-treatments. These considerations are often decisive factors for the election of a technology for an actual project. The two main parameters chosen in this chapter to classify the realm of reactor designs are the number of stages and the concentration of total solids (% TS) in the fermenter because these parameters have a great impact on the cost, performance and reliability of the digestion process.

5.2. ONE-STAGE SYSTEMS

5.2.1. Introduction

The biomethanization of organic wastes is accomplished by a series of biochemical transformations, which can be roughly separated into a first step where hydrolysis, acidification and liquefaction take place and a second step where acetate, hydrogen and carbon dioxide are transformed into methane. In one-stage systems, all these reactions take place simultaneously in a single reactor, whereas in two- or multi-stage systems, the reactions take place sequentially in at least two reactors.

About 90% of the full-scale plants currently in use in Europe for anaerobic digestion of OFMSW and biowastes rely on one-stage systems and these are approximately evenly split between 'wet' and 'dry' operating conditions (De Baere, 1999). This industrial trend is not mirrored by the scientific literature, which reports as many investigations on two-, multi-stage or batch systems as on one-stage systems. A likely reason for this discrepancy is that two- and multi-stage systems afford more possibilities to the researcher to control and investigate the intermediate steps of the digestion process. Industrialists, on the other hand, prefer one-stage systems because simpler designs suffer less frequent technical failures and have smaller investment costs. Biological performance of one-stage systems is, for most organic wastes, as high as that of two-stage systems, provided the reactor is well designed and operating conditions carefully chosen (Weiland, 1992).

Table 5.1. Possible unit processes, products and quality standards involved in an anaerobic digestion plant for organics solids.

Unit processes	Reusable products	Standards or criteria
PRE-TREATMENT		
- Magnetic separation	- Ferrous metals	- Organic impurities
- Size reduction (drum or shredder)		- Comminution of paper, cardboard and bags
- Pulping with gravity separation	- Heavy inerts reused as construction material	- Organic impurities
- Drum screening	- Coarse fraction, plastics	- Calorific value
- Pasteurization		- Germs kill off
DIGESTION		
- Hydrolysis		
- Methanogenesis	- Biogas	- Norms nitrogen, sulphur
- Biogas valorization	- Electricity Heat (steam)	- 150–300 $kW.h_{elec}$/ton - 250–500 $kW.h_{heat}$/ton
POST-TREATMENT		
- Mechanical dewatering		- Load on water treatment
- Aerobic stabilization or Biological dewatering	- Compost	- Norms soil amendments
- Water treatment	- Water	- Disposal norms
- Biological dewatering	- Compost	- Norms soil amendments
- Wet separation	- Sand Fibres (peat) Sludge	- Organic impurities Norms potting media Calorific value

5.2.2. One-stage 'wet' complete mix systems

5.2.2.1. Technical evaluation

At first glance, the one-stage wet system appears attractive because of its similarity to the demonstrated technology in use for decades for the anaerobic stabilization of biosolids produced in wastewater treatment plants. The physical consistency of organic solid wastes is made to resemble that of biosolids, via pulping and slurrying to less than 15% TS with dilution water, so that a classical complete mix reactor may be used. One of the first full-scale plants for the treatment of biowastes, built in the city of Waasa, Finland, in 1989, is based on this principle (Figure 5.2). A pulper with three vertical auger mixers is used to shred, homogenize and dilute the wastes in sequential batches. To this end, both fresh and recycled process water are added to attain 10–15% TS. The obtained slurry is then digested in large complete mix reactors where the solids are kept in suspension by vertical impellers.

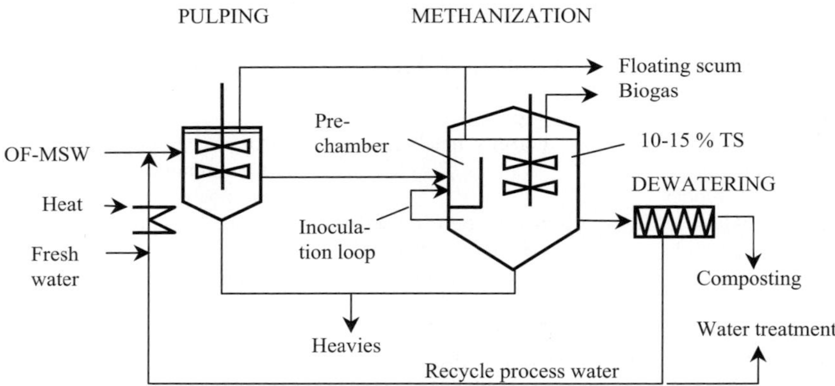

Figure 5.2. Typical design of a one-stage 'wet' system.

In contrast with the apparent simplicity of such one-stage wet process, many technical aspects need actually be taken into account and solved in order to guarantee a satisfactory process performance (Westergard and Teir, 1999; Farneti et al., 1999) (Table 5.2). The pre-treatment necessary to condition the wastes in a slurry of adequate consistency and devoid of coarse or heavy contaminants can be very complex, especially in the case of mechanically sorted OFMSW. To achieve the objective of removing these contaminants while at the same time keeping as much biodegradable wastes within the main stream, requires a complicated plant involving screens, pulpers, drums, presses, breakers, and flotation units (Farneti et al., 1999). These pre-treatment steps inevitably incur a 15–25% loss of volatile solids, with a proportional drop in biogas yield (Farneti et al., 1999).

Slurried wastes do not keep a homogenous consistency because heavier fractions and contaminants sink and a floating scum layer forms during the digestion process, resulting in the formation of three layers of distinct densities, or phases, in the reactor. The heavies accumulate at the bottom of the reactor and moreover may damage the propellers while the floating layer, several meters thick, accumulates at the top of the reactor and will hamper effective mixing. It is therefore necessary to foresee means to extract periodically the light and heavy fractions from the reactor. Since the heavies do also damage pumps, they must be removed as much as possible before they enter the reactor, either in specifically designed hydrocyclones or in the pulper which is designed with a settling zone.

Another technical drawback of the complete mix reactor is the occurrence of short-circuiting, i.e. the passage of a fraction of the feed through the reactor with a shorter retention time than the average retention time of the bulk stream. Not only does short-circuiting diminish the biogas yield, most importantly it impairs the proper hygienization of the wastes, i.e. the kill-off of microbial pathogens which requires a minimum retention time to complete. In the Waasa process, the advent of short-circuiting is somewhat alleviated by injecting the feed in a pre-chamber constructed within the main reactor (Figure 5.2). The piston flow occurring within the pre-chamber ensures at least a few days retention time. Since this compartmentalization hinders adequate inoculation of the feed, active biomass, drawn from the main compartment, is injected in the pre-chamber to speed up the digestion process. As the pre-chamber design seems however insufficient to guarantee satisfactory hygienization, it still remains necessary to pasteurize the wastes beforehand. To this end, steam is injected in the pulper to maintain the feed at 70 °C for one hour.

Table 5.2. Advantages and disadvantages of one-stage 'wet systems'.

Criteria	**Advantages**	**Disadvantages**
- Technical:	- Inspired from known process	- Short-circuiting - Sink and float phases - Abrasion with sand - Complicated pre-treatment
- Biological:	- Dilution of inhibitors with fresh water	- Particularly sensitive to shock loads as inhibitors spread immediately in reactor - VS lost with inerts and plastics
- Economic and Environmental:	- Equipment to handle slurries is cheaper (compensated by additional pre-treatment steps and large reactor volume)	- High consumption of water - Higher energy consumption for heating large volume

There exists a great variety of means to ensure adequate stirring of the digesting slurry within the reactor. For example, Weiland (1992) describes a pilot reactor with mechanical mixing ensured by downward movement in a centrally located draft tube enclosing a screw (loop reactor). An interesting advantage of this mixing mode is that it prevents the build-up of a floating scum layer. Since moving parts within a sealed reactor are technically challenging, several designs were developed that ensure adequate mixing without any mechanical moving parts within the reactor. For example the Linde process uses a loop reactor design where an ascending movement in a central compartment is created by injection of recirculated biogas at the bottom end of a central tube. Mixing modes using a combination of propellers and gas recirculation are also sometimes used (Cozzolino et al., 1992).

5.2.2.2. Biological performance

The three most important indicators of biological performance are

- the rate,
- the degree of completion,
- and the stability of the biochemical reactions.

Sensu strictu the degree of completion is quantified by comparing the biogas yield obtained in the reactor per unit mass substrate fed with the maximum biogas yield obtained in lab-scale batch reactors operated under optimal conditions. While this comparison is perhaps the most important test used in the industry, published reports almost invariably fail to mention what the maximum yield amounts to. Instead, publications refer simply to the biogas yield or alternatively to the % VS removal from the waste stream to assess the degree of completion of the methanization process. Biogas yield as such is however of very little use because it is much more dependent on waste composition than on process performance. For example, the methane yield in one full-scale plant varied between 170 and 320 Nm^3 CH_4/kg VS fed (40 and 75 % VS reduction) during the summer and winter months, respectively, as a result of the higher proportion of garden waste during summer months (Saint-Joly et al., 2000). Garden wastes are indeed known to yield much less biogas, relative to kitchen wastes, due to the higher proportion of poorly degradable lignocellulosic fibres. Similarly, Pavan et al. (1999b), using the same reactor configuration, observed a two-fold larger VS reduction with source-separated biowaste relative to mechanically sorted OFMSW. Such difference is not due to process performance but rather to the smaller biogas production potential of the mechanically sorted OFMSW which contains a greater proportion of poorly-degradable organic material such as plastic impurities.

A more useful criterion of biological performance is the maximum sustainable reaction rate, which can be expressed as a rate of substrate addition, i.e. the maximum organic loading rate OLR_{max} (kg VS/m^3 reactor.d), or as a rate of product formation, i.e. the volume of dry biogas or, better, of methane (under standard conditions of pressure and temperature) produced per unit time per unit reactor volume (Nm^3 CH_4/m^3 reactor.d). These indicators are more useful than the biogas yield or % VS reduction because they are less sensitive to the ill-defined composition of the waste and better reflect the level of biological activity that a given reactor design may sustain. Another parameter of use to quantify the rate is the retention time, which is roughly the inverse of the OLR when the OLR is expressed as mass wet substrate instead of mass substrate VS. This parameter does not give much biological information because it is too dependent on solid content and dilution with process water.

The only accurate way to compare the biological performance of different reactor designs requires the use of all three indicators simultaneously, however. The OLR_{max} indicates the degradative capacity of the system and the biogas yield its conversion efficiency, with 100% conversion efficiency being defined as the maximum biogas yield potential determined under optimal conditions in the laboratory. If the latter is unknown, the biogas yield remains a valid indicator only for comparisons between studies where wastes of similar origin and composition are used. Finally, and of foremost importance, only those data pertaining to reactors where stable performance is demonstrated should be considered.

Pavan et al. (1999b) examined the performance of the thermophilic one-stage wet system in a pilot reactor for the treatment of OFMSW and biowastes. The sustainable OLR_{max} for mechanically sorted OFMSW under thermophilic conditions was 9.7 kg VS/m^3.d. The same OLR was, however, unsustainable when the feed was switched to source-separated biowaste, for which the maximum OLR was 6 kg VS/m^3.d. Weiland (1992) found a similar OLR_{max} with various agro-industrial wastes under mesophilic conditions, provided these had C/N ratios greater than 20. Two plants were started in 1999 for the biomethanization of mechanically sorted OFMSW with wet processes. The one in Verona, Italy, was designed with an OLR of 8 kg VS/m^3.d (Farneti et al., 1999) while the one in Groningen, The Netherlands, has a design capacity of 5 kg VS/m^3.d (92,000 ton OFMSW per year in four reactors of 2,750 m^3 each).

It is not clear what the bottleneck is that determines these OLR_{max} values. Possible limiting factors are biomass concentration, mass transfer rate of substrates to bacteria, or accumulation of inhibitory substances. Since the feeding above the sustainable OLR_{max} typically leads to a decrease of biogas production, the bottleneck is most likely the concentration of inhibiting substances, such as fatty acids and ammonia. The high levels of Kjeldahl-N typical of biowastes (21 vs 14 g/kg TS for mechanically sorted OFMSW) leads to high levels of ammonia which decreases the methanogenic activity and affinity. This results in a rise of residual volatile fatty acids. Moreover, these fatty acids in turn inhibit the hydrolysis of polymers and acetogenesis of higher volatile fatty acids to acetate (Angelidaki, 1992). Inhibiting levels of fatty acids may also occur during overloads with substrates for which methanogenesis rather than hydrolysis is the limiting step, i.e. cellulose-poor substrates such as kitchen wastes.

Since inhibitors often limit the degradative capacity (OLR_{max}) of reactors treating OFMSW, the sensitivity of reactor designs toward inhibition is of particular concern. In this respect, the one-stage wet system suffers the disadvantage that the reactor contents are fully dispersed and homogenized which eliminates spatial niches wherein bacteria may be protected from transitory high concentrations of inhibitors. This disadvantage is, however, compensated by the fact that fresh water may be added to incoming wastes to lower the concentration of potential inhibitors. For example, in the above-mentioned study (Pavan et al., 1999b), the OFMSW was diluted two- to four-fold before feeding the reactor, apparently with tap water (no water recirculation was mentioned by the authors). The relevance of fresh water addition was demonstrated by Nordberg et al. (1992) in bench-scale reactors used to digest alfalfa silage. Process water produced in the dewatering stage was recycled to dilute the feed to a solid content of 6 % TS inside the reactor. However, the initially high biogas yield could be maintained only when a fraction of the recycled water was replaced by tap water in order to maintain the ammonium concentration below the threshold inhibitory level of 3 g/L. In the case of certain feed substrates, such as agro-industrial wastes with a C/N ratio below 20 and 60% biodegradable VS, the ammonium concentration cannot be brought under this threshold value, even when tap water is used to dilute the feed (Weiland, 1992). In this case, the one-stage wet process fails entirely and special two-stage processes need be applied.

5.2.2.3. Economic and environmental issues

The slurrying of the solid wastes brings the economic advantage that cheaper equipment may be used, e.g. pumps and piping, relative to solid materials. This advantage is, however, balanced by the higher investment costs resulting from larger reactors with internal mixing, larger dewatering equipment, and necessary pre-treatment steps. Overall, investment costs are comparable to those for one-stage 'dry' systems.

One drawback of ecological significance is the incomplete biogas recovery due to the fermentables removed with the floating scum layer and the heavy fraction. Another one is the relatively high water consumption necessary to dilute the wastes (about 1 m^3 tap water per ton solid waste). Water consumption is often a decisive factor in the selection process of a reactor design in full-scale projects because higher water consumption, aside from ecological considerations, also incurs higher financial costs for water purchase, treatment before disposal and discharge fees. The several-fold increase in waste volume due to dilution with water results in a parallel increase in steam consumption to heat up the reactor volume. This additional energy requirement does not, however, usually translate into larger internal use of produced biogas because the steam is usually recovered from the cooling water of the gas engines and exhaust fumes. In cases where the steam produced is exported to nearby factories, however, the yield will be lower.

5.2.3. One-stage 'dry' systems

5.2.3.1. Introduction

While the one-stage wet systems had initially been inspired from technology in use for the digestion of organic slurries, research during the 1980s demonstrated that biogas yield and production rate were at least as high in systems where the wastes were kept in their original solid state, i.e. not diluted with water (Spendlin and Stegmann, 1988; Baeten and Verstraete, 1993; Oleszkiewicz and Poggi-Varaldo, 1997). The challenge was not one of keeping biochemical reactions going at high TS values, but rather one of handling, pumping and mixing solid streams. While most industrial realizations built until the 1980s relied on 'wet' systems, the new plants erected during the last decade are evenly split between the wet and the dry systems (De Baere, 1999). No clear technology trend can be observed at this moment. Much will depend on the success of wet systems to deal with mechanically sorted OFMSW. 'dry' systems, on the other hand, have already proven reliable in France and Germany for the biomethanization of mechanically sorted OFMSW.

5.2.3.2. Technical evaluation

In dry systems, the fermenting mass within the reactor is kept at a solids content in the range 20–40% TS, so that only very dry substrates (>50% TS) need be diluted with process water (Oleszkiewicz and Poggi-Varaldo, 1997). The physical characteristics of the wastes at such high solids content impose technical approaches in terms of handling, mixing and pre-treatment which are fundamentally different from those of wet systems.

Transport and handling of the wastes is carried out with conveyor belts, screws, and powerful pumps especially designed for highly viscous streams. This type of equipment is more expensive than the centrifugal pumps used in wet systems and also much more robust and flexible inasmuch as wastes with solid content between 20 and 50% can be handled and impurities such as stones, glass or wood do not cause any hindrance. The only pre-treatment that is necessary before feeding the wastes into the reactor is the removal of the coarse impurities larger than ca. 40 mm. This is accomplished either via drum screens, as is typically the case with mechanically sorted OFMSW, or via shredders in the case of source-separated biowaste (Fruteau de Laclos et al., 1997; De Baere and Boelens, 1999; Levasseur, 1999). The heavy inert materials such as stones and glass which pass the screens or shredder need not be removed from the waste stream as is the case in wet systems. This makes the pre-treatment of dry systems somewhat simpler than that of their wet counterparts and very attractive for the biomethanization of OFMSW which typically contain 25% by weight of heavy inerts (Table 5.3).

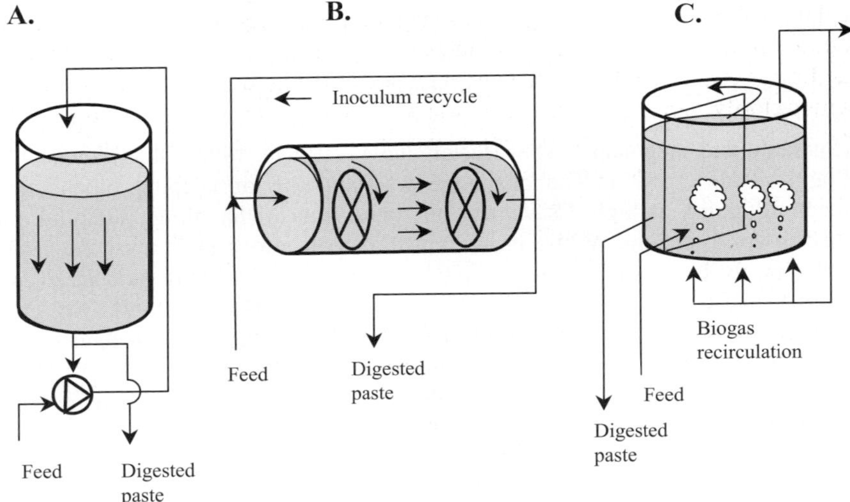

Figure 5.3. Different digester designs used in 'dry' systems (A illustrates the Dranco design, B the Kompogas and BRV designs, and C the Valorga design).

Because of their high viscosity, the fermenting wastes move via plug flow inside the reactors (Figure 5.3), contrary to wet systems where complete mix reactors are usually used. The use of plug flow within the reactor offers the advantage of technical simplicity as no mechanical devices need to be installed within the reactor. It leaves, however, the problem of mixing the incoming wastes with the fermenting mass, which is crucial to guarantee adequate inoculation and most of all to prevent local overloading and acidification. At least three designs have been demonstrated effective for the adequate mixing of solid wastes at the industrial scale (Figure 5.3). In the Dranco process, the mixing occurs via recirculation of the wastes extracted at the bottom end, mixing with fresh wastes (one part fresh wastes for six parts digested wastes), and pumping to the top of the reactor. This simple design has been shown effective for the treatment of wastes ranging from 20 to 50% TS. The Kompogas process works similarly, except that the plug flow takes place horizontally in cylindrical reactors. The horizontal plug flow is aided by slowly rotating impellers inside the reactors, which also serve for homogenization, degassing, and resuspending heavier particles. This system requires careful adjustment of the solid content around 23% TS inside the reactor. At lower values, heavy particles such as sand and glass tend to sink and accumulate inside the reactor while higher TS values cause excessive resistance to the flow. The Valorga system is quite different in that the horizontal plug flow is circular in a cylindrical reactor and mixing occurs via biogas injection at high pressure at the bottom of the reactor every 15 minutes through a network of injectors (Fruteau de Laclos et al., 1997). This elegant pneumatic mixing mode seems to work very satisfactorily since the digested wastes leaving the reactor need not be recirculated to dilute the incoming wastes. One technical drawback of this mixing design is that gas injection ports become clogged and maintenance of these is obviously cumbersome. As in the Kompogas process, process water is recirculated in order to achieve a solid content of 30% TS inside the reactor. The Valorga design is ill-suited for relatively wet wastes since sedimentation of heavy particles inside the reactor takes place at solid contents beneath 20% TS.

Owing to mechanical constraints, the volume of the Kompogas reactor is fixed and the capacity of the plant is adjusted by building several reactors in parallel, each one with a treatment capacity of either 15,000 or 25,000 ton/yr (Thurm and Schmid, 1999). On the other hand, the volume of the Dranco and Valorga reactors can be adjusted in function of the capacity required, though they are not made to exceed 3300 m^3 and a height of 25 m.

5.2.3.3. Biological performance

Given the relevance of inhibition of acetogenesis and methanogenesis in the one-stage ‘wet’ systems discussed in the previous section, even greater inhibition problems may be expected in the ‘dry’ designs since no fresh dilution water is added. The high OLR that are being achieved in both bench-scale and full-scale applications of one stage ‘dry’ systems indicate, however, that the ‘dry’ systems are not more sensitive to inhibition than the ‘wet’ systems. In fact, ‘dry’ systems can sustain at least as high OLR as ‘wet’ systems, without suffering inhibition (see below).

The sturdiness of the 'dry' systems toward inhibition was documented by Oleszkiewicz and Poggi-Varaldo (1997), but further research is needed in this area. Six and De Baere (1992) reported that no ammonium inhibition occurred in the thermophilic Dranco process for wastes having C/N ratios larger than 20. The same threshold value was noted by Weiland (1992) for mesophilic 'wet' systems, even though the latter system should yield much less of the toxic species NH_3 (assuming equal extent of ammonification). Threshold values for ammonium inhibition may also be expressed as ammonium concentration within the anaerobic reactor. The Valorga process running at 40 °C (Tilburg plant) sustains high OLR at ammonium concentration up to 3 g/l (Fruteau de Laclos et al., 1997) while the Dranco process running at 52 °C remains stable for ammonium concentrations up to 2.5 g/l. As these threshold values do not seem much higher than those commonly reported for 'wet' systems (though these are very disparate), one may speculate that the extent of ammonification is less in dry systems, leading to smaller production of inhibitory ammonium. Another possible explanation is that microorganisms within a dry fermenting medium are better shielded against toxicants since the absence of full mixing within the reactor limits the temporary shock loads to restricted zones in the digester, leaving other zones little exposed to transient high levels of inhibitors.

Table 5.3. Advantages and disadvantages of one-stage 'dry' systems.

Criteria	Advantages	Disadvantages
- Technical:	- No moving parts inside reactor - Robust (inerts and plastics need not be removed) - No short-circuiting	- Wet wastes (< 20 % TS) cannot be treated alone
- Biological:	- Less VS loss in pre- treatment - Larger OLR (high biomass) - Limited dispersion of transient peak concentrations of inhibitors	- Little possibility to dilute inhibitors with fresh water
- Economic and Environ-mental:	- Cheaper pre-treatment and smaller reactors - Complete hygienization - Very small water usage - Smaller heat requirement	- More robust and expensive waste handling equipment (compensated by smaller and simpler reactor)

In terms of extent of VS destruction, the three 'dry' reactor designs discussed above seem to perform very similarly, with biogas yields ranging from 90 Nm^3/ton fresh garden waste to 150 Nm^3/ton fresh food waste (Fruteau de Laclos et al., 1997; De Baere, 1999). These yields correspond to 210–300 Nm^3 CH_4 / ton VS, i.e. 50–70% VS destruction. Though as discussed above the biogas yield is not an accurate measure of a system performance, it can be noted that these values are comparable to those achieved with wet systems which fall in the range 40–70% VS destruction (Weiland, 1992; Pavan et al., 1999b; Westergard and Teir, 1999). A slightly greater biogas yield can however be expected with 'dry' systems compared to 'wet' systems since neither heavy inerts nor scum layer need be removed before or during the digestion.

Differences among the dry systems are more significant in terms of sustainable OLR. The Valorga plant at Tilburg, The Netherlands, treats peaks of 1,000 ton VFG wastes per week in two digesters of 3,000 m^3 each at 40 °C (Fruteau de Laclos et al., 1997). This corresponds to an OLR of 5 kg VS/m^3.d, a value comparable to the design values of plants relying on wet systems. Optimized 'dry' systems may, however, sustain much higher OLR such as the Dranco plant in Brecht, Belgium, where OLR values of 15 kg VS/m^3.d were maintained as an average during a one-year period (De Baere, 1999). This very high value is achieved without any dilution of the wastes, i.e. 35% TS inside the reactor, and corresponds to a retention time of 14 days during the summer months with 65% VS destruction. Typical design OLR values of the Dranco process are however more conservative (ca. 12 kg VS/m^3.d) but remain about twice as high as those for 'wet' systems. As a consequence, at equal capacity, the reactor volume of a Dranco plant is ca. two-fold smaller than that of a 'wet' system.

5.2.3.4. Economical and environmental issues

The economical differences between the 'wet' and 'dry' systems are small, both in terms of investment and operational costs. The higher costs for the sturdy waste handling devices such as pumps, screws and valves required for 'dry' systems are compensated by a cheaper pre-treatment and reactor, the latter being several times smaller than for 'wet' systems. The smaller heat requirement of 'dry' systems does not usually translate in financial gain since the excess heat from gas motors is rarely sold to nearby industries. As in the case of 'wet' systems, ca. 30 % of produced electricity is used within the plant.

Differences between the 'wet' and 'dry' systems are more substantial on environmental issues. While 'wet' systems typically consume one m^3 fresh water per ton OFMSW treated, the water consumption of their 'dry' counterparts is ca. ten-fold less. As a consequence, the volume of wastewater to be discharged is several-fold less for 'dry' systems. Another environmental advantage of 'dry' systems is that the plug flow within the reactor guarantees, at least under thermophilic conditions, the complete hygienization of the wastes and a pathogen-free compost as an end-product (Baeten and Verstraete, 1993).

5.3. TWO-STAGE SYSTEMS

5.3.1. Introduction

The rationale of two- and multi-stage systems is that the overall conversion process of OFMSW to biogas is mediated by a sequence of biochemical reactions which do not necessarily share the same optimal environmental conditions (Chapter 1). Optimizing these reactions separately in different stages or reactors may lead to a larger overall reaction rate and biogas yield (Ghosh et al., 1999). Typically, two stages are used where the first one harbors the liquefaction-acidification reactions, with a rate limited by the hydrolysis of cellulose, and the second one harbours the acetogenesis and methanogenesis, with a rate limited by the slow microbial growth rate (Liu and Ghosh, 1997; Palmowski and Müller, 1999). With these two steps occurring in distinct reactors, it becomes possible to increase the rate of methanogenesis by designing the second reactor with a biomass retention scheme or other means (Weiland, 1992; Kübler and Wild, 1992). In parallel, it is possible to increase the rate of hydrolysis in the first stage by using microaerophilic conditions or other means (Capela et al., 1999; Wellinger et al., 1999). The application of these principles has led to a great variety of two-stage designs.

The increased technical complexity of two-stage relative to single-stage systems has not, however, always been translated in the expected higher rates and yields (Weiland, 1992). In fact, the main advantage of two-stage systems is not a putative higher reaction rate, but rather a greater biological reliability for wastes which cause unstable performance in one-stage systems (Table 5.4). It should be noted however that, in the context of industrial applications, even for the challenging treatment of highly degradable biowastes, preference is given to technically-simpler one-stage plants. Biological reliability is then achieved by adequate buffering and mixing of incoming wastes, by precisely controlled feeding rate and, if possible, by resorting to co-digestion with other types of wastes (Weiland, 2000). Industrial applications have up to now displayed little acceptance for two-stage systems as these represent only ca. 10 % of the current treatment capacity (De Baere, 1999).

A distinction is made in this chapter between two-stage systems with and without a biomass retention scheme in the second stage. The reason for using this criterion is that the retention of biomass within a reactor is an important variable in determining the biological stability of the digester. Unstable performance can be caused either by fluctuations of OLR, due to wastes heterogeneity or discontinuous feeding, or by wastes excessively charged with inhibiting substances such as nitrogen. All types of two-stage system, regardless of whether biomass is accumulated or not, provide some protection against the fluctuations of OLR. However, only those two-stage systems with biomass retention schemes display stable performance with wastes excessively charged with nitrogen or other inhibitors (Weiland, 1992). Most commercial two-stage designs propose a biomass retention scheme in the second stage.

Table 5.4. Advantages and disadvantages of two-stage systems.

Criteria	Advantages	Disadvantages
- Technical:	- Design flexibility	- Complex
- Biological:	- More reliable for cellulose-poor kitchen waste - Only reliable design (with biomass retention) for C/N <20	- Smaller biogas yield (when solids not methanogenized)
- Economic and Environmental:	- Less heavy metal in compost (when solids not methanogenized)	- Larger investment

5.3.2. Without biomass retention

5.3.2.1. Technical evaluation

The most simple design of two-stage systems, used primarily in laboratory investigations, are two complete mix reactors in series (Pavan et al., 1999a; Scherer et al., 1999). The technical features of each reactor are comparable to those presented above for the one-stage 'wet' system. The wastes are shredded and diluted with process water to ca. 10% TS before entering the first digester.

Another possible design is the combination in series of two plug-flow reactors, either in the 'wet–wet' or 'dry–dry' mode, as illustrated by the Schwarting-Uhde and BRV processes, respectively. The source-sorted biowaste, finely chopped and diluted to 12% TS, rises upward through a series of perforated plates placed within the reactors (Figure 5.4). Uniform upward movement is imparted by pulsating pumps which also ensure localized short-term mixing via time-controlled impulses creating rapid rising of the liquid column (Trösch and Niemann, 1999). The impulses also push the biogas through the plate apertures. This elegant design, applied under 'wet' thermophilic conditions, is able to ensure, without any internal moving parts, adequate mixing and a plug flow mode which guarantees complete hygienization since short-circuiting is avoided. Moreover this design is not conducive to the formation of the thick floating scum layer commonly plaguing wet reactors. However, its sensitivity to clogging of the perforated plates limits the Schwarting-Uhde process to relatively clean, highly biodegradable biowastes.

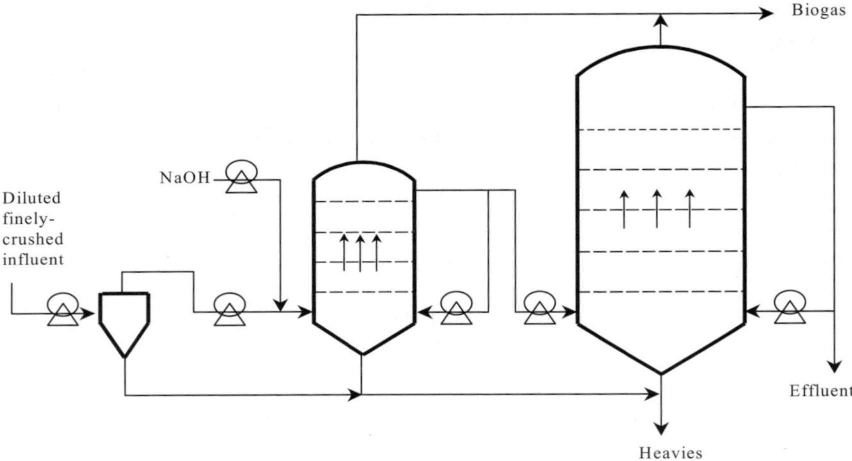

Figure 5.4. The Schwarting-Uhde process, a two-stage 'wet–wet' plug-flow system applicable to source-sorted biowastes, finely chopped (ca. 1 mm) and diluted to 12% TS.

In the BRV process, the source-separated biowastes, adjusted to 34% TS, pass through an aerobic upstream stage where organics are partly hydrolysed and ca. 2% lost through respiration. The reason for conducting the hydrolysis stage under microaerophilic conditions is that the loss of COD due to respiration is more than compensated by a higher extent of liquefaction, which, moreover, proceeds faster than under anaerobic conditions (Wellinger et al., 1999; Capela et al., 1999). After a two-day retention time, the pre-digested wastes are pumped through methanogenic reactors in a horizontal plug flow mode. The digestion lasts 25 days at 55 °C and 22% TS. The primary advantages of this system are the use of 'dry' conditions which reduces the size of the digesters and the use of piston flow which affords complete hygienization without a pasteurization step. The horizontal flow requires, however, the use of floor scrapers to eliminate the heavy material from the reactor and mixing equipment inside the reactor to prevent the formation of a crust layer.

5.3.2.2. Biological performance

The main advantage of the two-stage system is the greater biological stability it affords for very rapidly degradable wastes like fruits and vegetables (Pavan et al., 1999a). The reason commonly invoked is that the slower metabolism of methanogens relative to acidogens would lead to inhibiting accumulation of acids. Theoretically, however, this reasoning seems illogical as it would suffice to adjust the OLR of a one-stage system to the rate which can be handled by the methanogens to avoid any risk of acid accumulation. The OLR chosen in this manner for a one-stage system would not be inferior to that of a two-stage system.

In the practice, however, the greater reliability of two-stage systems has indeed at times been observed, at least in discontinuously-fed laboratory set-ups. For example, Pavan et al. (1999a) compared the performances of the one- and two-stage systems, using pilot complete mix reactors fed with very rapidly hydrolysable biowastes from fruit and vegetable markets. Although the one-stage system failed at 3.3 kg VS/m^3.d, the performance of the two-stage plant remained stable at an overall system OLR of 7 kg VS/m^3.d. This departure from theoretical predictions can be explained by the fact that actually applied OLR vary a great deal with time and space due to the heterogeneity of wastes and due to the discontinuous working of the feeding pump (feeding occurred only four times daily in the Pavan study). In cases where special care is taken to mix the feed thoroughly and dose it at constant OLR, one-stage ‘wet’ systems are as reliable and performable as two-stage systems even for highly degradable agro-industrial wastes, provided these have a C/N above 20 (Weiland, 1992).

The short-lived fluctuations of the actually applied OLR may lead to short-lived overloading in the one-stage system. In a two-stage system, however, these OLR fluctuations are somewhat buffered by the first stage, so that the OLR applied to the second stage is more uniform in time and space. In fact, this buffering of OLR in the first stage is somewhat similar to the effect of the plug flow pattern often used in the one-stage ‘dry’ systems because a plug flow with external mixing leaves large zones in the digester unexposed to transient high concentrations of inhibitors. Highly biodegradable kitchen wastes can indeed be digested in single-stage reactors provided these are thoroughly mixed before feeding and provided feeding occurs continuously, or at least five days per week as in the one-stage ‘dry’ Dranco plant in Salzburg, Austria. This plant, which treats kitchen wastes, achieves a mean OLR of 5.0 kg VS/m^3.d with 80% VS destruction.

As pointed out by Edelman et al. (1999), the OLR buffering taking place in a pre-digester is beneficial and useful only for the treatment of cellulose-poor wastes for which methanogenesis rather than hydrolysis-acidification is the rate-limiting step. For most wastes, however, hydrolysis of cellulose is the rate-limiting step (Noike et al., 1985), and shock loads are not conducive to inhibition.

The second type of inhibition, resulting from unbalanced average composition of feed rather than from transient shock load, is, however, as deleterious to two-stage systems as it is to one-stage systems, except in cases where two-stage systems are equipped with a biomass retention scheme in the second stage, e.g. via attached growth on a fixed bed (see below).

In terms of biogas yields and OLR_{max}, little difference can be noted between one- and two-stage systems, at least for these two-stage systems without biomass retention discussed in this section. For example, the BRV plant in Heppenheim is designed with an OLR of 8.0 kg $VS/m^3.d$ while the Schwarting-Uhde process seems to sustain an OLR_{max} up to 6 kg $VS/m^3.d$ (Trösch and Niemann, 1999).

5.3.3. With a biomass retention scheme

5.3.3.1. Technical evaluation

In order to increase rates and resistance to shock loads or inhibiting substances, it is desirable to achieve high cell densities of the slowly growing methanogenic consortium in the second stage. There are two basic ways to achieve this.

The first method to increase the concentration of methanogens in the second stage is to uncouple the hydraulic and solids retention time, thereby raising the solid content in the methanogenic reactor. These accumulated solids represent active biomass only in the case of wastes leaving no more than 5–15% of their original solid content as residual suspended solids inside the reactor. This design will therefore be effective only for highly hydrolysable kitchen or market wastes (Weiland, 1992; Madokoro et al., 1999). One way to uncouple the solid and hydraulic retention times is to use a contact reactor with internal clarifier (Weiland, 1992). Another way is to filter the effluent of the second stage on a membrane and return the concentrate in the reactor in order to retain the bacteria (Madokoro et al., 1999). Plugging of the microfiltration membranes can be avoided using a high cross-flow velocity achieved via reinjection of biogas. Excessive biomass was purged in a separate outlet line. Further upscaling of these two interesting designs, which up to now could only be tested in small pilot plants, may face technical challenges such as the crushing of the feed down to 0.7 mm.

Another method to increase the concentration of slowly growing methanogens in the second stage is to design the latter with support material allowing attached growth, high cell densities and long sludge age. The prerequisite of this design avenue is however that the feed to the attached growth reactor be very little charged with suspended particles, which means that the suspended solids remaining after the hydrolysis (first) stage should be removed. Two industrial processes, the BTA and Biopercolat designs, are based on these principles.

In the BTA 'wet–wet' process, illustrated in Figure 5.5, the 10% TS pulp exiting the pasteurization step is dewatered and the liquor directly sent to the methanogenic reactor (Kübler and Wild, 1992). The solid cake is resuspended in process water and hydrolysed in a complete mix reactor under mesophilic conditions (HRT 2–3 d). The pH within the hydrolysis reactor is maintained in the range 6-7 by recirculating process water from the methanogenic reactor. The output stream of the hydrolysis reactor is once more dewatered and the liquor fed to the methanogenic reactor. The latter, receiving only liquid effluents, is designed as a fixed film loop reactor in order to increase biomass concentration and age. From a technical point of view, this design shares the same limitations as the one-stage 'wet' system, i.e. short-circuiting, foaming, sinking of heavies, fouling of the impeller blades with plastic foils, obstruction of pipes with long objects such as sticks, and loss of 10–30% of the incoming VS caused by the removal of the rake fraction in the hydropulper (Kübler and Wild, 1992). The major drawback of the 'wet–wet system remains, however, its technical complexity as four reactors are necessary to achieve what other systems achieve in a single reactor.

The Biopercolat follows the same principles as the BTA process, with the difference that the first stage is carried out under 'dry' and microaerophilic conditions and is continuously percolated with process water to accelerate the liquefaction reaction (Edelmann et al., 1999; Wellinger et al., 1999). The flush water, containing up to 100 g COD/l, is fed to an anaerobic plug-flow filter filled with a support material. The separate optimization of the first stage, via aeration, and of the second stage, via biofilm growth, allows the system to run at the exceedingly low overall retention time of 7 days.

The Biopercolat system is quite innovative from a technical point of view. In order to prevent the channelling and clogging typically occurring in 'dry' percolated systems (see section 'batch design'), percolation occurs in large slowly-rotating (1 rpm) sieve drums with 1 mm mesh openings. In the methanogenic filter, a pulsating motion is imparted to the horizontal plug flow in order to prevent plugging of the support material, improve mass transfer of substrates to biofilm, and improve degasification. Moreover, the 'dry' design of the percolation hydrolysis stage avoids the troublesome pulping stage required in 'wet' or 'wet–wet' systems. This system, however, awaits validation in the first full-scale plant currently planned in Germany (Garcia and Schalk, 1999).

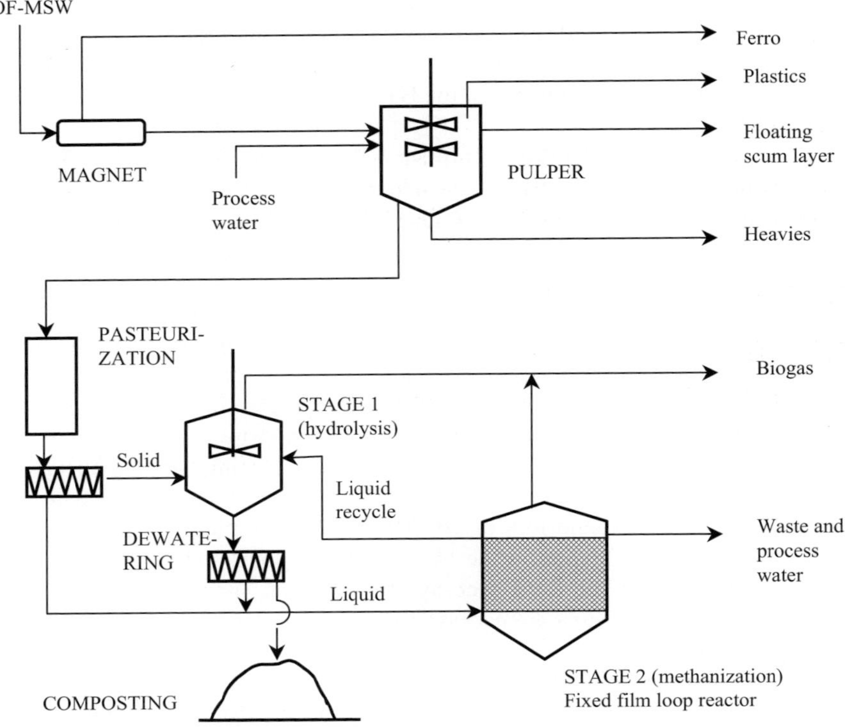

Figure 5.5. Two-stage 'wet–wet' design with a biomass retention scheme in the second stage (BTA process). The non-hydrolysed solids are not sent to the second stage.

5.3.3.2. Biological performance

As a consequence of the higher biomass concentration in two-stage designs with attached growth, greater resistance toward inhibiting chemicals is achieved. Weiland (1992) compared one- and two-stage 'wet' pilot plants for the treatment of highly biodegradable agro-industrial wastes. Although the one-stage system failed at OLR of 4 kg VS/m^3.d for those wastes which yielded ca. 5 g NH_4^+/l due to ammonium inhibition, the same wastes could be processed in the two-stage system at OLR of 8 kg VS/m^3.d without impairment of methanogenesis. The stability of the methanogenesis at such elevated ammonium concentration was attributed to the higher bacterial concentration and age which could be obtained in the contact reactor with internal clarifier used in the second stage.

Another consequence of two-stage systems with biomass retention is the possibility of applying higher OLR in the methanogenic reactor, with values up to 10 and 15 kg $VS/m^3.d$ reported for the BTA and Biopercolat processes, respectively (Kübler and Wild, 1992; Wellinger et al., 1999). These relatively high rates were, however, only achieved at a cost of 20–30% lower biogas yields, because the coarse solid particles remaining after the short hydrolysis stage, which still contain residual biodegradable polymers, are not fed to the methanogenic digester (Kübler and Wild, 1992; Garcia and Schalk, 1999).

5.4. BATCH SYSTEMS

5.4.1. Introduction

In batch systems, digesters are filled once with fresh wastes, with or without addition of seed material, and allowed to go through all degradation steps sequentially in the 'dry' mode, i.e. at 30–40% TS. Though batch systems may appear as nothing more than a landfill-in-a-box, they in fact achieve 50- to 100-fold higher biogas production rates than those observed in landfills because of two basic features. The first is that the leachate is continuously recirculated, which allows the dispersion of inoculant, nutrients, and acids, and in fact is the equivalent of partial mixing. The second is that batch systems are run at higher temperatures than that normally observed in landfills.

Batch systems have up to now not succeeded in taking a substantial market share. However, the specific features of batch processes (Table 5.5), such as a simple design and process control, robustness towards coarse and heavy contaminants, and lower investment cost make them particularly attractive for developing countries (Ouedraogo, 1999).

5.4.2. Technical evaluation

The hallmark of batch systems is the clear separation between a first phase where acidification proceeds much faster than methanogenesis and a second phase where acids are transformed into biogas. Three basic batch designs may be recognized, which differ in the respective locations of the acidification and methanogenesis phases (Figure 5.6).

In the single-stage batch design, the leachate is recirculated to the top of the same reactor where it is produced. This is the principle of the Biocel process, which is implemented in a full-scale plant in Lelystad, The Netherlands, treating 35,000 ton/yr source-sorted biowaste (ten Brummeler, 1999). The waste is loaded with a shovel in fourteen concrete reactors, each of 480 m^3 effective capacity and run in parallel. The leachates, collected in chambers under the reactors, are sprayed on the top surface of the fermenting wastes. One technical shortcoming of this and other batch systems, is the plugging of the perforated floor, resulting in the blockage of the leaching process. This problem is alleviated by limiting the thickness of the fermenting wastes to four meters in order to limit compaction and by mixing the fresh wastes with bulking material (one ton dewatered digested wastes and 0.1 ton wood chips added per ton fresh waste) (ten Brummeler, 1992). The addition of dewatered digested wastes, aside from acting as bulking material, also serves the purpose of inoculation and dilution of the fresh wastes. Safety measures need to be closely observed during the opening and emptying of the batches, as explosive conditions can occur.

Table 5.5. Advantages and disadvantages of batch systems.

Criteria	Advantages	Disadvantages
- Technical:	- Simple - 'Low-tech' - Robust (no hindrance from bulky items)	- Clogging - Need for bulking agent - Risk explosion during emptying of reactors
- Biological:	- Reliable process due to niches and use of several reactors	- Poor biogas yield due to channelling of percolate - Small OLR
- Economic and Environmental:	- Cheap, applicable to developing countries - Small water consumption	- Very large land acreage required (comparable to aerobic composting)

In the sequential batch design, the leachate of a freshly filled reactor, containing high levels of organic acids, is recirculated to another more mature reactor where methanogenesis takes place (Figure 5.6). The leachate of the latter reactor, freed of acids and loaded with pH buffering bicarbonates, is pumped back to the new reactor. This configuration also ensures cross-inoculation between new and mature reactors which eliminates the need to mix the fresh wastes with seed material. The technical features of the sequential batch design are similar to those of the single-stage design.

Finally, in the hybrid batch-UASB design, the mature reactor where the bulk of the methanogenesis takes place is replaced by a UASB reactor. The UASB reactor, wherein anaerobic microflora accumulates as granules, is well suited to treat liquid effluents with high levels of organic acids at high loading rates (Figure 5.6) (Anderson and Saw, 1992; Chen, 1999). This design is in fact very similar to the two-stage systems with biomass retention such as the Biopercolat system discussed above, with the difference that the first stage is a simple fill-and-draw (batch) instead of fully mixed design.

5.4.3. Biological performance

The Biocel plant in Lelystad achieves an average yield of 70 kg biogas/ton source-sorted biowaste. This is ca. 40% smaller biogas yield than that obtained in continuously fed one-stage systems treating the same type of waste (Saint-Joly et al., 2000; De Baere, 1999). This low yield is the result of leachate channelling, i.e. the lack of uniform spreading of the leachate which invariably tends to flow along preferential paths. The OLR of the Biocel process is however not exceedingly less than continuously fed systems, as might have been expected from the simple design. The design OLR of the Lelystad plant was 3.6 kg VS/m^3.d at 37 °C and peak values of 5.1 kg VS/m^3.d during summer months seem sustainable (ten Brummeler, 1999).

In the sequential batch design, the conversion of the acids in a separate mature reactor ensures the rapid depletion of the produced acids, thus a more reliable process performance and less variable biogas composition (O'Keefe et al., 1992; Silvey et al. 1999). At OLR of 3.2 kg VS/m^3.d, biogas yields equivalent to 80–90% of the maximal yield could be obtained in pilot reactors at 55 °C (O'Keefe et al., 1992; Silvey et al., 1999), which is considerably more than the yield reported in the Biocel plant. While the Biocel data were obtained from a full-scale plant treating compacted poorly-structured source-sorted biowaste at 40% TS, the impressive biogas yields reported for the sequential batch design were obtained in pilot plants treating either unsorted MSW or mechanically sorted OFMSW at 60% TS with high levels of paper and cardboard and low bulk density (280 kg/m^3). The coarser structure and lesser degree of compaction of these wastes render these less conducive to the channelling and plugging phenomena responsible for poor biogas yields.

5.4.4. Economic and environmental issues

Because batch systems are technically simple, the investment costs are significantly (ca. 40%) less than those of continuously fed systems (ten Brummeler, 1992). The land area required by batch processes is, however, considerably larger than for continuously-fed 'dry' systems, since the height of batch reactors is about one fifth and their OLR half, resulting in a ten-fold larger required footprint per ton of treated wastes. Operational costs, on the other hand, seem comparable to those of other systems (ten Brummeler, 1992).

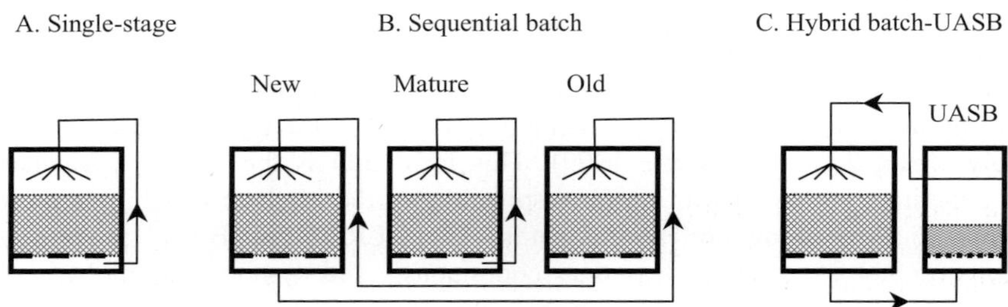

Figure 5.6. Configuration of leachate recycle patterns in different batch systems.

5.5. PERSPECTIVES AND CONCLUSIONS

In the past 25 years, a remarkable evolution has occurred in the attitude towards in-reactor digestion of solid wastes. The scepticism with respect to the feasibility has changed towards a general acceptance that various digester types are functioning at the full scale in a reliable way.

Most existing full-scale plants were designed with a single-stage reactor and reflect the relative newness of the technology. It can be expected that one-stage systems will continue to dominate the market, but that the reactor designs will be improved and matched to more specific substrates. This should provide far more reliable plants. Many companies also offer several versions of one technology, or propose both 'wet' and 'dry' systems. Two-stage systems may start playing a more and more important role, especially if treatment of industrial wastes is to be combined with that of biowaste and hygienization may require a separate treatment step at higher temperatures. Batch systems also still need to make a breakthrough, but chances are that hygienization as well as safety requirements will make these systems more difficult to introduce. Batch systems may be more successful in developing countries owing to the low investment costs.

At present, it is not possible to single out specific processes as all-round and optimally suited under all circumstances. Indeed, as discussed, many variables have to be taken into consideration and a final evaluation for a specific site will need to be made. There is and will continue to be room for technical diversity in this domain of waste treatment. Yet, practice shows that initial investment costs are of crucial importance. This factor, rather than the overall operating costs or performance characteristics often determines he outcome of a public tender. In view of the fact that such investments should serve at least several decades, it is to be hoped that decision makers will learn to have long-term foresight in these matters.

It must be recognized that anaerobic digestion of solid wastes still has to compete vigorously with aerobic composting. This is in part related to the fact that composting is a long-established technology which generally requires less initial investment. However, current energy prices and targeted reduction of fossil fuel combustion in the coming decades will draw increasingly more attention towards anaerobic digestion. Indeed, the amount of gas potentially recovered from the solid wastes is substantial at the level of a country. In the framework of the Kyoto agreements, many countries in Europe have agreed to stimulate the production of methane from wastes, e.g. by subsidizing the electricity from biogas by as much as 0.1 euro/kWh. The latter certainly will be a major support for anaerobic digestion of complex wastes. The European Union has set the goal to increase the fraction of electricity produced with renewable resources (excluding large hydroplant) from 3.2% in 1997 to 12.5% in 2010. Electricity generated from municipal solid waste by means of anaerobic digestion can make a significant contribution towards this goal.

5.6. REFERENCES

Anderson, G.K. and Saw, C.B. (1992). Leach-bed two-phase anaerobic digestion of municipal solid waste. In *Proc. Int. Symp. on Anaerobic Digestion of Solid Waste, Venice, 14–17 April 1992* (ed. F. Cecchi, J. Mata-Alvarez and F.G. Pohland), pp. 171–179. Int. Assoc. on Wat. Poll. Res. and Control.

Angelidaki, I. (1992). Anaerobic thermophilic process: the effect of lipids and ammonia. Ph.D. thesis, Dept of Biotechnology, Technical Univ. of Denmark, Lyngby, DK.

Baeten, D. and Verstraete, W. (1993). In-reactor anaerobic digestion of MSW-organics. In *Science and engineering of composting: design, environmental, microbiological and utilization aspects* (ed. Hoitink, H.A.J. and Keener, H.M.), pp. 111–129. Renaissance Publications, Worthington, OH.

Capela, I.F., Azeiteiro, C., Arroja, L. and Duarte, A.C. (1999). Effects of pre-treatment (composting) on the anaerobic digestion of primary sludges from a bleached kraft pulp mill. In *II Int. Symp. Anaerobic Dig. Solid Waste*, held in Barcelona, June 15–17, 1999 (eds. J. Mata-Alvarez, A. Tilche and F. Cecchi), vol. 1, pp. 113–120, Int. Assoc. on Wat. Qual.

Chen, T.-H. (1999). Anaerobic treatment of poultry mortalities. In *II Int. Symp. Anaerobic Dig. Solid Waste, Barcelona, June 15–17, 1999* (ed. J. Mata-Alvarez, A. Tilche and F. Cecchi), vol. 2, pp. 69–72. Int. Assoc. on Wat. Qual.

Cozzolino, C., Bassetti, A. and Rondelli, P. (1992). Industrial application of semi-dry anaerobic digestion process of organic solid waste. In *Proc. Int. Symp. on Anaerobic Digestion of Solid Waste, Venice, 14–17 April, 1992* (ed. F. Cecchi, J. Mata-Alvarez and F.G. Pohland), pp. 551–555, Int. Assoc. on Wat. Poll. Res. and Control.

De Baere, L. (1999). Anaerobic digestion of solid waste: state-of-the art. In *II Int. Symp. Anaerobic Dig. Solid Waste, Barcelona, June 15–17, 1999* (ed. J. Mata-Alvarez, A. Tilche and F. Cecchi), vol. 1, pp. 290–299, Int. Assoc. on Wat. Qual.

De Baere, L. and Boelens, J. (1999). The treatment of greyand mixed solid waste by means of anaerobic digestion: future developments. In *II Int. Symp. Anaerobic Dig. Solid Waste, Barcelona, June 15–17, 1999* (ed. J. Mata-Alvarez, A. Tilche and F. Cecchi), vol. 2, pp. 302–305, Int. Assoc. on Wat. Qual.

Edelmann, W., Joss, A. and Engeli, H. (1999). Two-step anaerobic digestion of organic solid wastes. In *II Int. Symp. Anaerobic Dig. Solid Waste, Barcelona, June 15–17, 1999* (ed. J. Mata-Alvarez, A. Tilche and F. Cecchi), vol. 2, pp. 150–153, Int. Assoc. on Wat. Qual.

Farneti, A., Cozzolino, C., Bolzonella, D., Innocenti, L. and Cecchi, C. (1999). Semi-dry anaerobic digestion of OFMSW: the new full-scale plant of Verona (Italy). In *II Int. Symp. Anaerobic Dig. Solid Waste*, held in Barcelona, June 15–17, 1999 (eds. J. Mata-Alvarez, A. Tilche and F. Cecchi), vol. 2, pp. 330–333, Int. Assoc. Wat. Qual.

Fruteau de Laclos, H., Desbois, S. and Saint-Joly, C. (1997). Anaerobic digestion of municipal solid organic waste: Valorga full-scale plant in Tilburg, The Netherlands. In *Proc. 8th Int. Conf. on Anaerobic Dig., Sendai, May 25–29 1997*, vol. 2, pp. 232–238. Int. Assoc. on Wat. Qual.

Garcia, J.L. and Schalk, P. (1999). Biopercolat-procedure. In *II Int. Symp. Anaerobic Dig. Solid Waste*, held in Barcelona, June 15–17, 1999 (eds. J. Mata-Alvarez, A. Tilche and F. Cecchi), vol. 2, pp. 298–301. Int. Assoc. on Wat. Qual.

Ghosh, S., Henry, M.P., Sajjad, A., Mensinger, M.C. and Arora, J.L. (1999). Pilot-cale gasification of MSW by high-rate and two-phase anaerobic digestion. In *II Int. Symp. Anaerobic Dig. Solid Waste, Barcelona, June 15–17, 1999* (ed. J. Mata-Alvarez, A. Tilche and F. Cecchi), vol. 1, pp. 83–90, Int. Assoc. on Wat. Qual.

Kübler, H. and Wild, M. (1992). The BTA-process high rate biomethanisation of biogenous solid wastes. In *Proc. Int. Symp. on Anaerobic Digestion of Solid Waste, Venice, 14–17 April, 1992* (ed. F. Cecchi, J. Mata-Alvarez and F.G. Pohland), pp. 535–538, Int. Assoc. on Wat. Poll. Res. and Control.

Levasseur, J.-P. (1999). Anaerobic digestion of organic waste:from theory to industrial practice. In *II Int. Symp. Anaerobic Dig. Solid Waste*, held in Barcelona, June 15–17, 1999 (ed. J. Mata-Alvarez, A. Tilche and F. Cecchi), vol. 2, pp. 334–337, Int. Assoc. on Wat. Qual.

Liu, T. and Ghosh, S. (1997). Phase separation during anaerobic fermentation of solid substrates in an innovative plug-flow reactor. In *Proc. 8th Int. Conf. on Anaerobic Dig., Sendai, May 25–29, 1997*, vol. 2, pp. 17–24. Int. Assoc. on Wat. Qual.

Madokoro, T., Ueno, M., Moro, M., Yamamoto, T. and Shibata, T. (1999). Anaerobic digestion system with micro-filtration membrane for kitchen refuse. In *II Int. Symp. Anaerobic Dig. Solid Waste, Barcelona, June 15–17, 1999* (ed. J. Mata-Alvarez, A. Tilche and F. Cecchi), vol. 2, pp. 105–108, Int. Assoc. on Wat. Qual.

Noike, T., Endo, G., Chang, J., Yaguchi, J. and Matsumoto, J. (1985). Characteristics of carbohydrate degradation and the rate-limiting step in anaerobic digestion. *Biotechnol. Bioeng.* **27**, 1482–1489.

Nordberg, A., Nilsson, A. and Blomgren, A. (1992). Salt accumulation in a biogas process with liquid recirculation. Effect on gas production, VFA concentration and acetate metabolism. In *Proc. Int. Symp. on Anaerobic Digestion of Solid Waste, Venice, 14–17 April, 1992* (ed. F. Cecchi, J. Mata-Alvarez and F.G. Pohland), pp. 431–435. Int. Assoc. on Wat. Poll. Res. and Control.

O'Keefe, D.M., Chynoweth, D.P., Barkdoll, A.W., Nordstedt, R.A., Owens, J.M. and Sifontes, J. (1992). Sequential batch anaerobic composting. In *Proc. Int. Symp. on Anaerobic Digestion of Solid Waste, Venice, 14–17 April, 1992* (ed. F. Cecchi, J. Mata-Alvarez and F.G. Pohland), pp. 117–125. Int. Assoc. on Wat. Poll. Res. Control.

Oleszkiewicz, J.A. and Poggi-Varaldo (1997). High-solids anaerobic digestion of mixed municipal and industrial wastes. *J. Environ. Eng.* **123**, 1087–1092.

Ouedraogo, A. (1999). Pilot scale two-phase anaerobic digestion of the biodegradable organic fraction of Bamako district municipal solid waste. In *II Int. Symp. Anaerobic Dig. Solid Waste, Barcelona, June 15–17, 1999* (ed. J. Mata-Alvarez, A. Tilche and F. Cecchi), vol. 2, pp. 73–76. Int. Assoc. on Wat. Qual.

Palmowski, L. and Müller, J. (1999). Influence of the size reduction of organic waste on their anaerobic digestion. In *II Int. Symp. Anaerobic Dig. Solid Waste, Barcelona, June 15–17, 1999* (ed. J. Mata-Alvarez, A. Tilche and F. Cecchi), vol. 1, pp. 137–144. Int. Assoc. on Wat. Qual.

Pavan, P., Battistoni, P., Cecchi, F., Mata-Alvarez, J. (1999a). Two-phase anaerobic digestion of source-sorted OFMSW: performance and kinetic study. In *II Int. Symp. Anaerobic Dig. Solid Waste, Barcelona, June 15–17, 1999* (ed. J. Mata-Alvarez, A. Tilche and F. Cecchi), vol. 1, pp. 91–98. Int. Assoc. on Wat. Qual.

Pavan, P., Battistoni, P., Mata-Alvarez, J. (1999b). Performance of thermophilic semi-dry anaerobic digestion process changing the feed biodegradability. In *II Int. Symp. Anaerobic Dig. Solid Waste, Barcelona, June 15–17, 1999* (ed. J. Mata-Alvarez, A. Tilche and F. Cecchi), vol. 1, pp. 57–64. Int. Assoc. on Wat. Qual.

Saint-Joly, C., Desbois, S. and Lotti, J.P. (2000). Determinant impact of waste collection and composition on anaerobic digestion performance: industrial results. *Wat. Sci. Technol.* **41**(3), 291–297.

Scherer, P.A., Vollmer, G.-R., Fakhouri, T. and Martensen, S. (1999). Development of a methanogenic process to degrade exhaustively municipal 'residual refuse' (MRR) and 'grey waste' under thermophilic and hyperthermophilic conditions. In *II Int. Symp. Anaerobic Dig. Solid Waste, Barcelona, June 15–17, 1999* (ed. J. Mata-Alvarez, A. Tilche and F. Cecchi), vol. 1, pp. 65–74, Int. Assoc. on Wat. Qual.

Six, W. and De Baere, L. (1992). Dry anaerobic conversion of municipal solid waste by means of the Dranco process at Brecht, Belgium. In *Proc. Int. Symp. on Anaerobic Digestion of Solid Waste, Venice, 14–17 April, 1992* (ed. F. Cecchi, J. Mata-Alvarez and F.G. Pohland), pp. 525–528. Int. Assoc. on Wat. Poll. Res. Control.

Silvey, P., Blackall, L., Nichols, P. and Pullammanappallil, P. (1999). Microbial ecology of the leach bed anaerobic digestion of unsorted municipal solid waste. In *II Int. Symp. Anaerobic Dig. Solid Waste, Barcelona, June 15–17, 1999* (ed. J. Mata-Alvarez, A. Tilche and F. Cecchi), vol. 1, pp. 17–24. Int. Assoc. on Wat. Qual.

Spendlin, H.-H. and Stegmann, R. (1988). Anaerobic fermentation of the vegeable, fruit, and yard waste. In *Proc. 5th Int. Solid Wastes Conf., Copenhagen, September 11–16, 1988* (ed. L. Andersen and J. Moller), vol. 2, pp. 25–31. Academic Press, London.

ten Brummeler, E. (1999). Full scale experience with the Biocel-process. In *II Int. Symp. Anaerobic Dig. Solid Waste, Barcelona, June 15–17, 1999* (ed. J. Mata-Alvarez, A. Tilche and F. Cecchi), vol. 1, pp. 308–314. Int. Assoc. on Wat. Qual.

ten Brummeler, E. (1992). Dry anaerobic digestion of solid waste in the Biocel process with a full scale unit. In *Proc. Int. Symp. on Anaerobic Digestion of Solid Waste, Venice, 14–17 April, 1992* (ed. F. Cecchi, J. Mata-Alvarez and F.G. Pohland), pp. 557–560. Int. Assoc. on Wat. Poll. Res. Control.

Thurm, F. and Schmid, W. (1999). Renewable energy by fermentation of organic waste with the Kompogas process. In *II Int. Symp. Anaerobic Dig. Solid Waste, Barcelona, June 15–17, 1999* (ed. J. Mata-Alvarez, A. Tilche and F. Cecchi), vol. 2, pp. 342–3345. Int. Assoc. on Wat. Qual.

Trösch, W. and Niemann, V. (1999). Biological waste treatment using the thermophilic schwarting-Uhde process. In *II Int. Symp. Anaerobic Dig. Solid Waste, Barcelona, June 15–17, 1999* (ed. J. Mata-Alvarez, A. Tilche and F. Cecchi), vol. 2, pp. 338–341. Int. Assoc. on Wat. Qual.

Weiland, P. (1992). One- and two-step anaerobic digestion of solid agroindustrial residues. In *Proc. Int. Symp. on Anaerobic Digestion of Solid Waste, Venice, 14–17 April, 1992* (ed. F. Cecchi, J. Mata-Alvarez and F.G. Pohland), pp. 193–199. Int. Assoc. on Wat. Poll. Res. and Control.

Weiland, P. (2000). Cofermentation of organic wastes – concepts, processes and new developments. In *Abstr. of the lecture group Environ. Technol., Int. Meeting Chem. Eng., Environ. Protec., Biotechnol., Frankfurt am Main, May 2000*, pp. 192–194. Dechema, Frankfurt am Main, D.

Wellinger, A., Widmer, C. and Schalk, P. (1999). Percolation – a new process to treat MSW. In *II Int. Symp. Anaerobic Dig. Solid Waste, Barcelona, June 15–17, 1999* (ed. J. Mata-Alvarez, A. Tilche and F. Cecchi), vol. 1, pp. 315–322. Int. Assoc. on Wat. Qual.

Westergard, R. and Teir, J. (1999). The Waasa process integrated in the eco-cycling society. In *II Int. Symp. Anaerobic Dig. Solid Waste, Barcelona, June 15–17, 1999* (ed. J. Mata-Alvarez, A. Tilche and F. Cecchi), vol. 2, pp. 310–313. Int. Assoc. on Wat. Qual.

6

Characteristics of the OFMSW and behaviour of the anaerobic digestion process

F. Cecchi, P. Traverso, P. Pavan, D. Bolzonella and L. Innocenti

6.1. INTRODUCTION

The collection of municipal solid wastes represents their first 'treatment'. In fact, the different choices made by municipalities about the way of separating the municipal solid waste (MSW) fractions before any technological treatment are strictly linked to the material characteristics. This is particularly evident when speaking of the organic fraction of municipal solid wastes (OFMSW). The mechanical selection from the unsorted waste, the separate collection and the

Edited by J. Mata-Alvarez. ISBN: 1 900222 14 0

source sorting actually represent the three main pathways to obtain the OFMSW. Each of them provides different kinds of OFMSW that, obviously, lead to different yields and behaviours of the subsequent treatment process. These concepts are described in this chapter, beginning from the analytical description of the substrates characteristics (both physical–chemical and biological), followed by a presentation of the process monitoring parameters and their behaviour according to the typical substrate and by the different results obtainable by the anaerobic digestion process. A short example of anaerobic digester design is also given.

6.2. THE OFMSW FROM MECHANICAL SORTING (MS-OFMSW)

The MS-OFMSW coming from the unsorted collected waste was probably the earliest category of organic fraction recovered and used for biological processes. This recovery approach was widely applied in the past 20 years, with the dual purpose of obtaining a good quality organic fraction and a highly calorific material to be used as RDF ('Refuse Derived Fuel').

Sorting plants for the production of these materials can be classified into three main groups, as described in Table 6.1 (CITEC, 2000).

Table 6.1. Main groups of plants for the OFMSW mechanical sorting (CITEC, 2000).

Plant type	Features
'Simplified'	Simple plants consisting of a primary shredder, a trommel screen (hole size 50–100 mm) and a magnetic belt from which three streams are obtained: the over-sieve (paper, plastics and small amounts of putrescible matter) to send to incineration, the organic fraction which can be sent for biological treatment, and the iron material fraction for recycling.
'Medium complex'	Plants designed with a more complex operational sequence: at least one size reduction step, iron separation, and more than one screen operation. These plants allow material of over-sieve size to be captured (to send to incineration or landfills), the more 'pure' OFMSW coming from the screen operations (to send for biological treatment), and the iron material fraction (for recycling).
'Complex'	Plants designed with a more complete sorting line: size reduction, iron separation, screenings, dry matter shredder and pellet formation for RDF production. The products are more 'pure', and the OFMSW obtained is a more suitable substrate for biological processes.

The MS-OFMSW characteristics are connected to the type of sorting plant and, obviously, to the incoming materials quality. Only few data of systematic characterisation are reported in literature for these substrates; thus the results obtained by the authors will be reported as typical for a sorting plant which can be considered as a 'complex plant'. The flow sheet is shown in Figure 6.1.

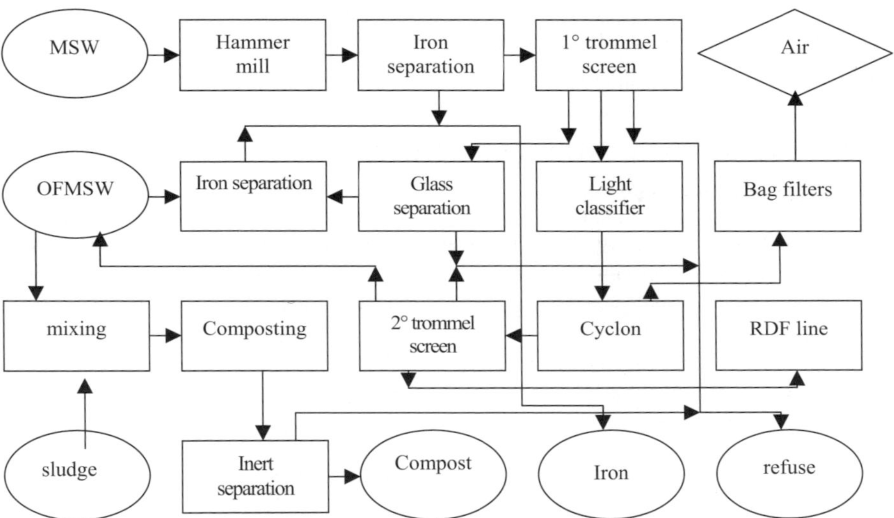

Figure 6.1. Flow sheet of a complex sorting plant for the production of a substrate for composting process (Cecchi et al., 1991).

This first kind of substrate considered here has been studied for several years in terms of both composition and chemical–physical parameters. Table 6.2 summarises these results.

As can be seen from these results, the MS-OFMSW is characterised by a high content of dry solids. This is due to the inert fraction of the unsorted waste, which is incompletely separable with this sorting approach. This aspect is much more evident if the total volatile solids (TVS) content is considered; in fact an average value less than 50% is reached. As for the specific composition, Table 6.3 and Figure 6.2 show the percentage of each fraction in terms of both total and volatile solids.

Table 6.2. Chemical–physical characteristics of MS-OFMSW sorted in 'complex' plants.

	average	max	min	Nr of samples	Std. Dev.
TS, g/Kg	763.0	952.0	513.1	210	81.3
TVS, %TS	43.9	57.4	29.1	210	5.4
TCOD, TS	59.6	90.4	23.3	41	17.4
TOC, %TS	19.3	34.4	7.5	187	5.3
IC, %TS	1.3	2.7	0.3	187	0.5
TKN, %TS	2.2	3.4	1.2	59	0.5
P, %TS	0.11	0.22	0.05	59	0.03

TS = total solids; TVS = total volatile solids; TCOD = total chemical oxygen demand; TOC = total organic carbon; IC = inorganic carbon; TKN = total Kjieldahl nitrogen; P = total phosphorus.

Table 6.3. Composition and characteristics of MS-OFMSW sorted in 'complex' plants in terms of percentage of TS and TVS of each typical fraction.

	%TS	%TVS
Putrescible	59.0	78.0
Paper	4.6	7.1
Wood	1.1	2.2
Plastic	1.8	3.4
Inerts	33.5	9.3
Total	**100.0**	**100.0**

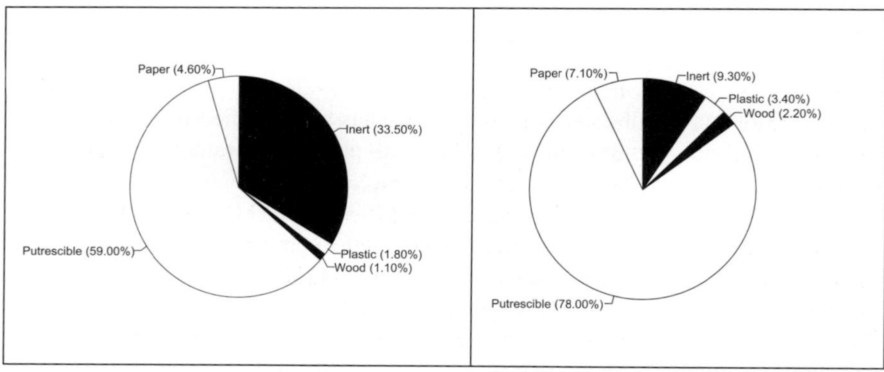

Figure 6.2. TS and TVS distribution in MS-OFMSW sorted in 'complex' plants.

Five fractions were considered: putrescible matter, paper, wood, plastic and inert materials. The high content of inert materials is widely confirmed by these data: more than 40% of the substrate is actually unusable as feed for the anaerobic process. Furthermore, these inert materials will be present in the digester effluent sludge, making any agronomic recovery more difficult. Since about 80% of the substrate TVS content comes from the putrescible fraction it is underlined as the other fractions are only partly involved in the biological process.

Another more recent set of data (2001) concerns the results of an author's research on municipal wastes that were treated to recover a dry fraction to use as primary product for incineration plants. The OFMSW obtained in this case is only a secondary product. The plant can be classified according to Table 6.1 as 'medium complex' or 'simplified' plant.

This sorting line worked in two different operational conditions with two types of trommel screen (with wide and narrow holes). Table 6.4 reports a summary of the results obtained.

Table 6.4. Chemical–physical characteristics of the substrate obtained with the 'medium complex plant'.

	TS, kg/kg	**TVS, kg/kg**	**TVS/TS, %**	**TCOD, kgO_2/kg**	**TKN, %TS**	**P, %TS**
Narrow holes screen	0.54	0.27	47.0	0.6	1.1	0.1
Wide holes screen	0.50	0.33	68.0	1.0	0.7	0.4

As for the results obtained with the narrow holes screen, it can be seen that the substrate characteristics are very similar to those obtained with the 'complex' plant (see Table 6.1). Furthermore no important differences exist between the two substrates, except for the higher TCOD/TS and TCOD/TVS ratios (1.25 and 2.5 respectively vs. 0.7 and 1.5 in the MS-OFMSW coming from the 'complex' plant). Considering both the increase of this ratio in the samples obtained with the wide holes screen (TCOD/TS from 1.25 to 2.0, TCOD/TVS from 2.5 to 2.96 respectively) and the increase of the organic fraction, mainly due to the paper higher content, this kind of substrate can probably be pointed at as more productive in terms of biogas conversion. Looking at the data related to the wide holes screen, the TVS content has increased significantly (from 47 to 68%, see Table 6.4). This can be explained taking paper content into account, which is higher, as also outlined by the composition analysis and confirmed by the lower nitrogen content.

Table 6.5 reports the results obtained by the analysis of the composition of the fraction obtained with the narrow holes screen, and Figures 6.3 shows the distribution of TS and TVS. The first three columns in Table 6.5 concern the TS and TVS concentration of each fraction, while the last three report the related distribution.

Table 6.5. Composition of the waste obtained with the narrow holes screen in the 'medium complex' line.

Fraction	TS Kg/Kg	TVS Kg/Kg	TVS/TS %	Distribution Wet weight %	Distribution TS %	Distribution TVS %
Cellulose	0.5	0.3	67.0	19	17	27
Organic	0.4	0.3	62.0	57	43	58
Plastic	0.9	0.5	56.0	9	13	11
Inerts	0.9	0.7	7.0	15	26.6	4

The distributions confirm the chemical–physical data that is to say large amounts of plastic and inert materials (about 40% on a TS basis, see Figure 6.5). Table 6.6 reports the results of the samples obtained with the wide holes screen and Figure 6.4 show the distributions.

Table 6.6. Composition analysis of the substrate obtained with the wide holes screen.

Fraction	TS Kg/Kg	TVS Kg/Kg	TVS/TS %	Distribution Wet weight %	Distribution TS %	Distribution TVS %
Cellulose	0.6	0.5	82.0	47	48	55
Organic	0.4	0.3	72.0	35	24	25
Plastic	0.8	0.7	88.0	10	15	19
Inerts	0.9	0.03	3.0	7.4	13	1

The results clearly show the typical effect deriving from the use of a mechanical selection strategy starting from a completely unsorted waste. In both the cases studied, independently from the screen used (wide or narrow holes), the inert fractions are present in large amount. This situation is completely different in wastes that are source or separately collected, as described in the next paragraph.

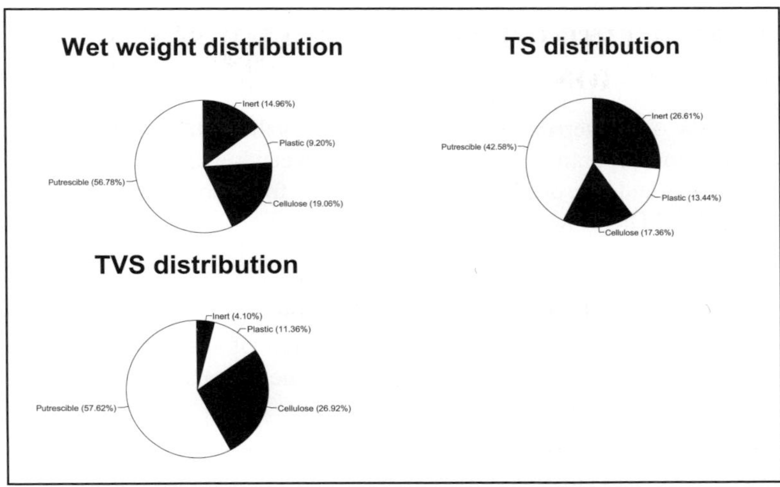

Figure 6.3. Distribution of composition of the product obtained with the narrow holes screen in the 'medium complex' plant.

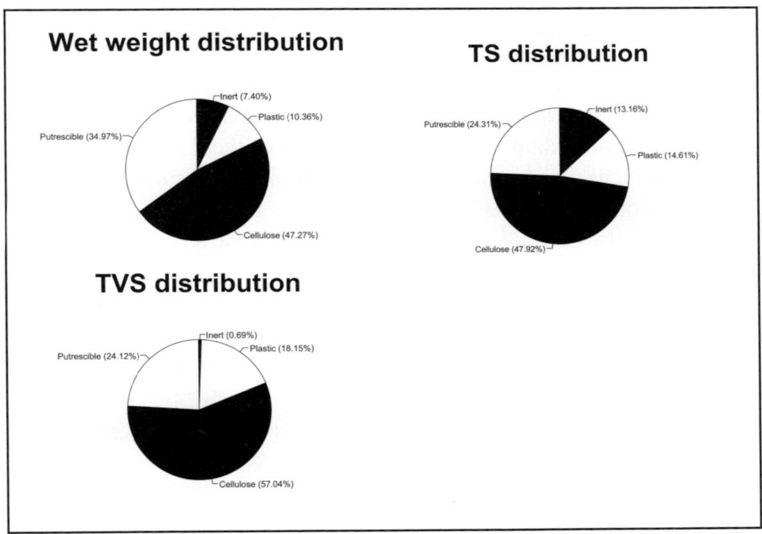

Figure 6.4. Distribution of composition of the product obtained with the wide holes screen in the 'medium complex' plant.

6.3. OFMSW FROM SEPARATE AND SOURCE COLLECTION (SC-OFMSW AND SS-OFMSW)

The OFMSW from separate collection can be split into two categories: the organic fraction separately collected from markets, canteens, restaurants, etc. (SC-OFMSW) and the organic fraction coming from domestic source sorting (SS-OFMSW). These two kind of substrate can be assembled and described together although it is possible, on an analytical basis, to underline some different characteristics.

The SC-OFMSW is normally characterised by a high grade of separation, probably thanks to the information and education of the population done in recent years. As an example, Table 6.7 reports the characteristics of the waste collected in the area around Milan (Italy).

Table 6.7. Degree of separation of SC-OFMSW in an area around Milan (Italy) (Cecchi et al., 1997). The data were collected in 24 municipalities.

Size of municipalities (nr. inhabitants)			Non-compostable materials (%)			Compostable materials (%)		
Average	Min.	Max.	Average	Min.	Max.	Average	Min.	Max.
24304	3808	119.187	2.2	0.3	5.0	97.8	93.0	99.7

It has to be noted that the percentage of non-compostable material seldom reaches 5% of the total weight collected. The data reported in Table 6.8 concern a summary of literature data on the characteristics of these materials: as it can be observed, the typical range of total solids content in these substrates is 15–25%, while the TVS content reaches 70–90% TS. The nutrients content is in the range 2.5–3.5%TS for nitrogen and 0.5–1.0%TS for phosphorus, quite similar to those already encountered for MS-OFMSW.

Table 6.8. Characteristics of SC-OFMSW reported in literature.

Reference	TS, %	TVS, %TS	N, %TS	P, %TS
De Baere, 2000	31	70	—	—
Kubler et al., 1999	29	63	2.2–3.4	0.4–0.6
CITEC, 2000(*)	17–25	70–90	—	—
CITEC, 2000(**)	7–15	80–90	1.5–3	1–3

* SS-FORSU. ** SC-FORSU.

Similar values were recently found by the authors when analysing wastes from the separate collection in Treviso city (Northern Italy; data not published at the moment). The waste, mainly constituted by kitchen residues (bread, pasta, vegetables and fruit), had a TS content of 25%, with a percentage of volatile matter of 80%TS. Furthermore, Table 6.9 reports data regarding the SC-OFMSW collected from canteens. This substrate was particularly rich in dry food residues, as bread, which increase the dry matter content in the waste.

Table 6.9. Characteristics of SC-OFMSW collected in canteens (Cecchi et al., 1997).

Parameter	Range	Typical value
TS, %	21.4–27.4	25.6
TVS, % TS	91.3–99.7	96.5
TCOD, gO_2/gTS	1.2–1.3	1.2
TKN, % TS	2.6–3.7	3.2
Total P, % TS	0.13–0.28	0.2

On the contrary, the organic fractions from fruit and vegetables markets are very rich in water content: this is the reason why the average dry matter content often does not reach 10%, as outlined by the data reported in Table 6.10, regarding a substrate coming from a fruit and vegetables markets.

Table 6.10. Average characteristics of SC-OFMSW from fruit and vegetables market (Pavan et al., 2000).

	Average	Max.	Min.	Nr. of samples	Std. Dev.
TS, g/Kg	81.8	132.7	54.4	96	15.7
TVS, %TS	81.9	92.0	78.2	96	11.3
TCOD, gO_2/g TS	1.0	1.5	0.7	32	18.1
TKN, %TS	2.1	3.3	1.4	23	0.5
P_{tot}, %TS	2.8	3.3	1.3	23	0.5

Thus, it is evident that dry matter content in the OFMSW coming from different separate collection approaches can sensibly change, in relation to the presence of kitchen and fruit and vegetables residues: in particular, the TS content can be significantly influenced by the presence of garden wastes. Some examples coming from experimental research are reported in Table 6.11.

Table 6.11. Characteristics of SS-OFMSW obtained in different experiments.

	Cecchi et al., 1989	Sans et al., 1995
TS, g/Kg	200	163.9
TVS, %TS	88	90
TCOD, gO_2/gTS	1.1	1.1
TKN, %TS	3.2	2.1
P_{tot}, %TS	0.4	2.6

Trying to make an overall scenario about the data presented, in terms of chemical–physical parameters, it can be said that the OFMSW from the separate sorting approach shows a dry matter content from 10% (as typical for fruit and vegetable wastes) to 20–25% for kitchen waste mixed with garden wastes. The volatile solids percentage can be considered in the range of 85–90% for both residues, while the nitrogen content is about 2–3%TS and P is negligible (0.2–0.5%TS).

An additional consideration has to be made about the influence of seasonal variations on the substrate characteristics. Table 6.12 shows a summary of the data obtained during one year of experimentation (Zorzi, 1997; Cecchi et al., 1994; Pavan et al., 1998).

Table 6.12. Chemical–physical characteristics of SC-OFMSW monitored during one year (Zorzi, 1997).

	Oct–Jan	**Feb–May**	**May–June**	**June–July**	**July–Aug**	**Aug–Sept**	**Sept–Nov**	**Nov–Dec**	**Dec–Feb**
T°C	12.6	7.5	21.4	22.2	24.5	23.1	15.1	8.8	7.1
PH	4.7	4.9	5.0	4.7	4.0	4.32	4.3	4.3	5.0
TS, g/Kg	95.4	93.6	96.3	94.7	88.6	105.8	103.4	102.6	97.3
TVS,%TS	91.5	94.5	89.2	88.9	90.0	92.2	90.6	90.9	91.2
TCOD, g/Kg	99.2	101.3	100.4	88.7	95.5	108.6	106.7	108.5	101.7
SCOD, g/Kg	39.3	53.3	45.1	44.4	42.7	49.7	51.7	51.2	60.1
TVFA, mgCOD/l	4092	4256	3948	3410	4062	7563	3023	3931	4563
TKN, g/Kg	—	—	—	—	23.2	23.4	23.1	23.2	21.5
Ptot, g/Kg	—	—	—	—	3.9	3.7	3.7	3.8	3.5

The content of VFAs produced by fermentation processes in the feedstock tanks (1–2 hours retention time) was monitored to evaluate the putrescibility of these materials in relation to the seasonal variations of temperature. The most important observation that can be drawn from this concerns the fact that seasonal variations seem not to affect the SC-OFMSW characteristics.

A final remark has to be made on the biological characteristics of these substrates in terms of their response to the anaerobic digestion process. The most appropriate parameter to consider seems to be the so-called 'ultimate biodegradation potential': B_0. Typical values for B_0 are given in Table 6.13, in which the biodegradability of different OFMSW is compared (Mata-Alvarez et al., 1992).

Table 6.13. B_0 values for different substrates (Mata-Alvarez et al., 1992).

Reference	B_0, m^3 CH_4/KgTVS
MS-FORSU (Valorga, 1985)	0.301
MS-FORSU (De Baere and Verstraete, 1984)	0.321
MS-FORSU (Pauss et al., 1984)	0.397
MS-FORSU (Roux and Wakerley, 1978)	0.381
MS-FORSU (Cecchi et al., 1989b)	0.158
SC-FORSU (Mata-Alvarez and Cecchi, 1989)	0.445
SS-FORSU (Cecchi et al., 1986)	0.401
SS-FORSU (Mata-Alvarez et al., 1992)	0.489

From literature data the range of production can be evaluated, both in terms of methane and biogas production. Considering a methane percentage of 55%, these values are the ones reported in Table 6.14: the lower biogas potentiality of the MS-OFMSW reveals.

Table 6.14. Ultimate methane and biogas production for the three kinds of OFMSW.

Substrate:	MS-FMSW	SC-OFMSW	SS-OFMSW
B_0, m^3CH_4/kgTVS	016–0.37	0.45–0.49	0.37–0.40
G_0*, m^3/KgTVS	0.29–0.66	0.81–0.89	0.67–0.72

*: G_0= ultimate biogas production.

6.4. OPERATIONAL PARAMETERS FOR THE MANAGEMENT OF AN ANAEROBIC DIGESTER

The main parameters for managing the anaerobic digestion process can be split into two groups: the operational parameters of the reactor and the stability parameters of the biological process.

6.4.1. Managing parameters of the anaerobic digestion reactor

These managing parameters define the operational mode of the reactor, in terms of solid retention time, biomass concentration and yields of biogas production both related to the reactor volume (GPR) and to the substrate characteristics (SGP). The substrate is generally defined according to the content of total solids (TS), of total volatile solids (TVS), of chemical oxygen demand (COD) and of biochemical oxygen demand (BOD).

These terms are defined as follows (APHA, 1995):

TS: total solids, or residue upon water evaporation, is the content of dry matter in the substrate sample, after 48 hours drying at 105 °C (constant weight). This is a raw estimation of all the organic and inorganic matter content in the original sample.

TVS: total volatile solids, is the fraction of solid matter that can be oxidised and driven off as gas at 550 °C for 24 hours (constant weight). This is, as an approximation, the organic fraction of the dry matter determined at 105 °C, the total solids. The difference between the TS and TVS values is the inert (mineral) fraction, mainly due to inorganic matter.

COD: chemical oxygen demand. This is the oxygen equivalent of the organic matter that can be oxidised. It is measured by using a strong chemical oxidising agent in an acidic medium. This is a measure of all the organic matter in the substrate and also gives an idea of the oxidation level of the substrate and, thus, how a substrate could be further on oxidised by a biological process, anaerobic digestion like.

BOD_5: biochemical oxygen demand. This measure involves the measurement of the dissolved oxygen used by microorganisms in the biochemical oxidation of organic matter for 5 days. Thus, this is an indirect determination of the biologically removable organic matter content of the substrate.

BOD_L: biochemical oxygen demand for 28 days. Theoretically speaking, this is a measure of the total content of the biodegradable organic matter in the sample. It is assumed, here, that all the organic matter will be biologically oxidised within the 28 day span.

The operational parameters of the anaerobic digestion reactor are the following:

a) Hydraulic retention time (HRT), [days];
b) Solids retention time (SRT), [days];
c) Organic loading rate (OLR), [kg substrate / m^3 reactor day];
d) Organic loading on biomass in the reactor (biomass) (F/M), [kg substrate / kg TVS day];
e) Specific gas production (SGP), [m^3 biogas / kg feed substrate];
f) Gas production rate (GPR), [m^3 biogas / m^3 reactor day];
g) Substrate removal yield, [%].

In particular:

a) Hydraulic retention time (HRT)

The HRT is the ratio of the reactor volume to the flow rate of the influent substrate:

$$HRT = \frac{V}{Q}$$

where:

HRT,	hydraulic retention time, [days];
V,	reactor volume, [m^3];
Q,	flow rate, [m^3/day].

Therefore, it is the time that a fluid element spends in the reactor. This is strictly true for ideal reactors.

b) Solids retention time (SRT)

The average residence time of solids into the reactor is the ratio between the content of total solids in the reactor and the solids flow rate extracted from the reactor. If the quantity of biomass extracted from the reactor is equal to the biomass produced in the reactor the solids concentration in the reactor, as biomass, will be constant in a given time and it can be said that the reactor is operating in steady-state conditions.

Analytically:

$$SRT = \frac{V * X}{W}$$

where:

SRT,	solid retention time, [days];
V,	reactor volume, [m^3];
X,	volatile solids concentration in the reactor, [kg TVS / m^3];
W,	flow rate of the extracted volatile matter from the reactor, [kg TVS /day].

c) Organic loading rate (OLR)

The organic loading rate is the substrate quantity introduced into the reactor volume in a given time:

$$OLR = \frac{Q * S}{V}$$

where:

OLR, organic loading rate, [kg substrate / m^3 reactor day];
Q, substrate flow rate, [m^3/day];
S, substrate concentration in the inflow, [kg/m^3];
V, reactor volume, [m^3].

Here, the substrate can be defined by means of different measured parameters, as: TS, TVS, COD or BOD.

d) Organic loading to the biomass in the reactor (F/M)

This is the fed substrate in a unit time to the biomass in the reactor ratio. Here, the total volatile solids concentration in the reactor is assumed as a measure of the biomass content:

$$F / M = \frac{Q * S}{V * X}$$

where:

F/M, substrate loading to biomass, as volatile matter, in the reactor, [kg substrate / kg TVS day];
Q, inlet flow rate, [m^3/day];
S, substrate concentration in the feed, [kg TVS / m^3];
V, reactor volume, [m^3];
X, concentration of volatile solids in the reactor, [kg TVS / m^3].

This parameter is difficult to estimate because of the impossibility to distinguish the active biomass and the substrate in the reactor by means of the volatile solids measurement.

e) Specific gas production (SGP)

This parameter indicates the biogas produced by a unit of mass of substrate, in terms of the total volatile solids in the feed, as $m^3_{biogas}/kg_{substrate\ fed}$. This index is strictly linked both to the biodegradability of the fed substrate and to the process attitude. The SGP value is often used to compare the performances of different anaerobic processes.

Analytically:

$$SGP = \frac{Q_{biogas}}{Q * S}$$

where,

SGP,	specific gas production, [m^3 biogas / kg substrate fed];
Q_{biogas},	biogas flow rate, [m^3/day];
Q,	inlet flow rate, [m^3/day];
S,	substrate concentration in the influent, [kg substrate / m^3].

f) Gas production rate (GPR)

It is the produced biogas/reactor volume ratio, in a given time:

$$GPR = \frac{Q_{biogas}}{V}$$

where:

GPR,	gas production rate, [m^3 biogas / m^3 reactor day];
Q_{biogas},	biogas flow rate, [m^3/day];
V,	reactor volume, [m^3].

g) Substrate removal effectiveness

The removal efficiency or substrate conversion, although it represents an unquestionable parameter, can be expressed in several ways.

In fact, different methods to measure substrates or different equations can be found. This is because in anaerobic digestion processes mass balances are difficult to close and researchers developed different equations to fit the data.

Generally, the simplest and most used equation is the following:

$$\eta\% = \frac{Q*S - Q*Se}{Q*S} \times 100$$

where:

$\eta\%$,	total volatile solids removed, as percentage, [%];
Q,	inlet and outlet flow rate, [m^3/day];
S,	total volatile solids concentration in the inlet flow rate, [kg/m^3];
Se,	concentration of total volatile solids in the effluent flow rate, [kg/m^3].

It is important to observe that some researchers prefer the measurement of the substrate in terms of total solids or total volatile solids (Bhattacharya et al., 1996), whereas others prefer the use of the COD (Brunetti et al., 1988).

According to Ross et al. (1992) the efficiency removal in terms of volatile percentage of the total solids can be expressed as:

$$\text{Removal}_{VS\%} = \frac{VS_{in} - VS_{out}}{VS_{in} - (VS_{in} * VS_{out})} \times 100$$

where:

VS_{in}	percentage of volatile matter in the inflow, [%];
VS_{out}	percentage of volatile matter in the outlet, [%];

With this approach only the biodegradable fraction of the fed volatile solids is considered at the denominator of the equation.

6.4.2. Monitoring of parameters to evaluate the stability of the anaerobic digestion process

Control strategies should address the maintenance of the optimal operational conditions and the stability of the process.

In the anaerobic digestion process this is particularly true, since the methanogenesis process is sensitive to the environmental variations in the bulk.

Some parameters, such as the pH, the VFA concentration, the alkalinity, the VFA:alkalinity ratio, the production and composition of the biogas and the temperature, are of particular importance in the process control.

Some details and explanations about the meaning of these parameters to the process are here below given. Anyway, it is important to underline that all these parameters have to be simultaneously considered, since only a global approach helps to perfectly manage the anaerobic digestion process. In fact, the variation of a single parameter is not meaningful to understand the behaviour of the anaerobic digestion process.

pH

The pH gives some information about the stability of the medium since its variation depends on the buffer capacity of the medium itself. Variations in pH are related to variations of the species involved in the trophic chain of the anaerobic digestion process (VFAs, CO_2, …).

In the pH range 6.5–7.5 the anaerobic digestion process is generally stable.

The pH value is basically determined by the CO_2 concentration in the medium and thus by the partial pressure of its gaseous phase. Also the ammonia and VFAs concentrations influence the pH values. Therefore, this parameter is an indicator of a complex equilibrium system, where several chemical species are involved. Even though its determination is really useful, it is absolutely poor by itself. It is absolutely important to relate its value to other process parameters, e.g. alkalinity, VFAs concentration, biogas production and composition. Moreover, it has to be clear that this parameter gives information about an unstable process with a delay time: when pH variations are monitored the reaction medium has already lost its stability. In fact, when a pH variation is measured it means that the following equilibrium has already changed to the right side:

$$H{-}R \Leftrightarrow R^- + H^+ + NaHCO_3 \Leftrightarrow NaR + H_2O + CO_2$$

where, H−R stands for a generic organic acid.

This dynamic system is explained in Figure 6.5: here the pH and alkalinity profiles versus the acidity, as VFAs equivalent, in the medium are shown.

As can be seen, the alkalinity variation is faster than the pH variation: so, when pH variations are observed, the alkalinity of the system, and thus buffer capacity, was already lost. Therefore, it is of particular importance to monitor also other parameters such as alkalinity, VFA concentration, and biogas production and composition.

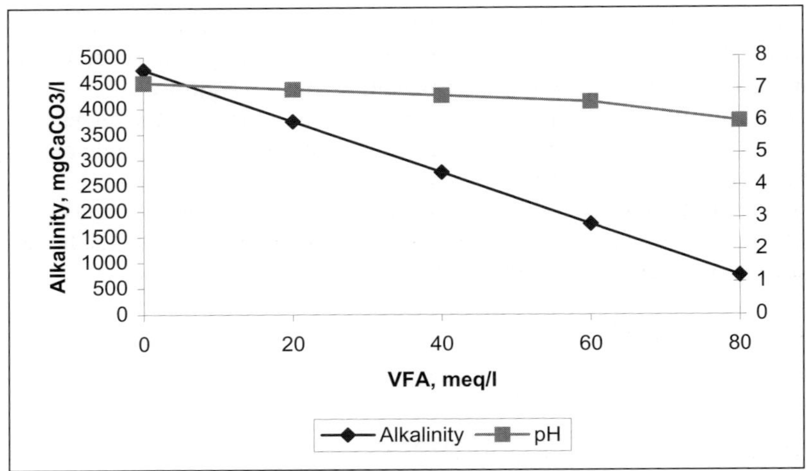

Figure 6.5. pH and alkalinity profiles versus VFA concentration in the reaction medium.

Alkalinity (buffer) (TA)

Alkalinity is the acid-neutralising capacity of a medium. That is the capacity to resist changes in pH caused by the addition (or increase) of acids in the medium. It results from the presence of hydroxides, carbonates and bicarbonates of elements such as calcium, magnesium, sodium potassium or ammonia. In the case of the anaerobic digestion medium the VFA presence, beside borates, silicates and phosphates, also contributes to the alkalinity.

It is determined by titrimetric method by adding hydrochloric acid in a sample of the liquid phase of the sludge from an anaerobic digester (APHA, 1998).

Typical values of alkalinity in anaerobic digesters are in the range 2000–4000 mg $CaCO_3$ per litre.

The titration is firstly taken to pH 6: at this point the bicarbonate has been converted into CO_2 which gases off from the system. The second step is the addition of acidic solution to pH 4, to determine all the other bases such as volatile fatty acids and inorganic anions in the sample.

The difference between the values of alkalinity at pH 4 and 6 gives a raw idea of the content of VFAs in the medium, expressed in terms of acidic equivalents (IRSA-CNR, 1982).

This parameter is of particular importance in the control of the stability of anaerobic processes. In fact, methanogenic microorganisms show a slow growing capacity, therefore an increase in the organic loading rate, and, thus, in VFAs concentration, could determine an unbalanced development of the trophic chain. Finally, the concentration of fatty acids will increase and the pH will drop down. So, the alkalinity of the system becomes particularly important because it represents the buffer capacity of the system, that is the capability to resist variations in pH. The buffer capacity, in an anaerobic digester, is due to the presence of ammonia, from the degradation of proteins, and bicarbonate, from the carbon dioxide solubilisation in the liquid phase.

The contemporary presence of ammonia and bicarbonate gives rise to the formation of a buffer system, the so-called carbonate-acetic buffer. That is the formation of NH_4HCO_3 through the following dynamic equilibriums:

$$CO_2 + H_2O \Leftrightarrow HCO_3^- + H^+$$

$$HCO_3^- + NH_4^+ \Leftrightarrow NH_4HCO_3$$

The presence of this salt determines a high level of alkalinity in the medium, and, thus, the stability of the process, even though a high concentration of VFAs is present.

Volatile fatty acids (VFA)

The fatty acids are generally represented as:

$$R\text{–}COOH$$

where, R is an alkyl moiety with general formula

$$CH_3(CH_2)_n$$

Generally, hydrolytic and acetogenic microorganisms produce fatty acids where R contains from 0 to 3 atoms of carbon (short-chain volatile fatty acids, SC-VFA). The VFA concentration in anaerobic digesters depends on the substrate fed to the digester, and generally is in the range 200–2000 mgAc/l. This value is generally expressed in terms of acetic acid or COD concentration. Anyway, it is important to underline that it is not the absolute concentration of the VFAs that determines the process upset but its fast increase. This means that the trophic chain is going toward an acidogenic cells population rather than the needed methanogenic one. The increase in fatty acids is generally due to the increase in the organic loading rate (OLR). This increase is generally buffered by the alkalinity of the system and no pH variations are measured. So, these three parameters, namely, the VFA concentration, the alkalinity and the pH, have to be considered as a single descriptor of the stability of the anaerobic process. Moreover, the values of these parameters have to be considered besides the biogas production and its composition, in terms of methane and carbon dioxide percentages, in order to gain as much information as possible about the process stability.

Volatile fatty acids : alkalinity ratio

The fatty acids and the alkalinity concentrations show fast changes when the stability of the anaerobic digestion process is upset. Because, when the process is not stable, the volatile fatty acids concentration increases, whereas the alkalinity decreases, the ratio of these two parameters can be a good tool for the observation of the stability of the anaerobic process. In this ratio, the VFAs are expressed in terms of acetic acid concentration and the alkalinity in terms of calcium carbonate concentration. Values of the ratio around 0.3 are an index of good stability of the process, whereas larger values suggest an upset of the stability process.

Biogas production and composition

The measurement of the quantity and composition of the biogas produced, in terms of methane and carbon dioxide content, is of fundamental importance to evaluate the stability of the process (Stafford et al., 1980). When the process is stable the amount and composition of the biogas are stable too. A decrease in biogas production contemporary to an increase in CO_2 content can indicate an inhibition of the methanogenesis of the system, i.e. because of the presence of high level of volatile fatty acids or ammonia. Therefore, the analysis of the biogas production and composition should be performed besides the determination of VFAs and alkalinity concentrations in the medium. In fact these parameters are strictly linked one to each other, but only a global evaluation can give an explanation of the stability of the process.

In particular three different situations can be observed:

1. when the VFA concentration is low, e.g. < 1000 mgCOD/l, and the biogas production high, with a carbon dioxide content around the 25–33%, the process is stable and the trophic chain is correctly balanced;
2. when VFA concentration in the medium and the carbon dioxide content in biogas simultaneously increase, the process is going to be upset and the acidifying microorganisms are prevailing on the methanogenic ones; therefore, VFAs are accumulating in the medium;
3. when the VFAs concentration increases and the biogas production decreases inhibition or toxicity problems can be the cause.

Hydrogen content in biogas

According to some authors, hydrogen in biogas should be constantly monitored to evaluate the stability level of the process (Stafford et al., 1980). Unfortunately, hydrogen concentration in biogas is low and difficult to determine, so this measurement is generally absent in industrial application of the anaerobic digestion processes.

Temperature

Since the anaerobic digestion process is due to the interaction and equilibrium of different microorganism populations along the trophic chain, temperature stability is of particular importance. A variation of some 2–3 °C in temperature can give rise to a change of the system: in fact, different temperature ranges determine totally different bacterial populations rather than a shift of the original bacterial population. Several different microorganisms can survive only in restricted temperature ranges. Thus temperature should be carefully measured so as to perfectly adjust the heat supply to the reactor. Generally speaking mesophilic processes operate well in the range 30–40 °C, whereas thermophilic processes operate in the range 45–60 °C (optimal around 55 and 37 °C, respectively).

6.5. ANAEROBIC DIGESTION PROCESS OF SUBSTRATES WITH DIFFERENT BIODEGRADABILITY

The discussion of the digester behaviour using different substrates as feed can be divided into two main aspects: the yields and typical values of stability parameters in steady-state conditions, and the different behaviour of the process in transient conditions.

6.5.1. Process behaviour in steady state conditions

As for the first point, to get a more complete evaluation of the description of the different substrates previously considered and their influence on the anaerobic process behaviour the results of pilot-scale studies using different OFMSW as digester feed are reported (Pavan et al., 2000; Mata-Alvarez et al., 1990).

6.5.1.1. Thermophilic steady-state conditions

The studies concerned the application of the semi-dry single-phase thermophilic anaerobic digestion process to a mixture of OFMSW (MS- and SS-), considering a progressive increase in the feed biodegradability, so as to evaluate the process behaviour changing from an undifferentiated collection of waste to a separate collection (Pavan et al., 2000). The experiments were done using 3 m^3 and 1 m^3 CSTR pilot-scale reactors. The main characteristics of the substrates used during the experiments are shown in Table 6.15.

Table 6.15. Comparison between MS-and SS-OFMSW used in the experiments.

	MS-OFMSW(*)	SS-OFMSW(**)
TS, g/Kg	647.2	163.9
TVS, %TS	46.5	90.6
TCOD, KgO_2/Kg	0.5	1.1
TKN, %TS	1.4	2.1
P, %TS	1.9	2.1

(*) Average value of 45 samples. (**) Average values of 115 samples.

The SS-OFMSW had a solid content significantly lower than MS-OFMSW (TS = 160 g/Kg and 650 g/Kg respectively). The organic content of the two substrates averaged TVS/TS ratios considerably different: about 45%, for the former and 90% for the latter. The results were discussed referring to an experimental period when the only MS-OFMSW was fed to the digester (Run 0) and to periods when the only SS-OFMSW was fed to the digester (Run 4); these situations corresponded to cases of lowest and highest biodegradability.

The feed characteristics (Table 6.16) showed that the TVS content increased with the increase of the SS-OFMSW presence. The increase in biodegradability was also emphasized by the values of the parameters related to the soluble fraction: the percentage of STS (Soluble Total Solids) in the MS-OFMSW alone was about 7%TS, while in the mixtures rich in SS-OFMSW (Runs 1, 2, 3, 4) was up to 40%TS. A similar trend was observed for SCOD/TCOD (Soluble COD/Total COD) as indicative parameter of the hydrolysed organic fraction. The trend of the feed biodegradability influenced the process yields: increasing the percentage of SS-OFMSW fed to the digester, the specific gas production increased from 0.32 m^3/KgTVSf (Run 1) up to 0.78 m^3/KgTVSf (Run 5). The direct proportionality between the TVS/TS ratio in the feed and the yields obtained in the Runs 0, 1, 2, 3 clearly showed the range of applicability of this kind of process: in fact, the stability parameters in the digester (Table 6.16) during these periods assured the process fine management.

Table 6.16. Operational conditions adopted and results obtained during the study

Run	0	1	2	3	4
SS-OFMSW %TS	0	15	30	50	100
Operational conditions					
T, °C	56.2	55.5	55.9	55.7	55.1
HRT, d	11.7	12.5	11.6	11.2	11.8
OLR, KgTVS/m^3 d	9.7	7.5	12.1	12.6	6.0
Yields					
GPR, m^3/m^3 d	3.1	3.1	5.9	6.2	4.9
SGP, m^3/KgTVS d	0.32	0.42	0.49	0.50	0.78
TVS rem, %	37.3	42.8	59.3	57.9	82.2
COD rem, %	34.8	44.2	47.4	53.2	79.8
Feed characteristics					
TS, g/Kg	252.5	140.3	202.2	185.9	98.2
TVS, %TS	45.6	65.3	69.3	75.8	81.9
STS, %TS	7.4	17.5	26.5	37.0	41.7
TCOD, g/Kg	140.0	117.8	192.5	177.8	95.1
SCOD, g/Kg	21.6	16.8	43.9	51.3	33.4
TVFA, gHAc/l	13.0	15.5	7.8	8.9	7.6
PH	6.4	5.7	5.2	4.4	4.4
Reactor characteristics					
TS, g/Kg	142.4	82.1	86.6	96.7	52.1
TVS, %TS	54.2	53.7	54.3	55.2	55.4
STS, %TS	5.5	14.9	14.2	17.2	24.6
TCOD, g/Kg	99.1	55.9	58.3	66.4	31.9
SCOD, g/Kg	5.3	4.8	5.7	14.1	5.0
TVFA, gHAc/l	0.4	1.3	1.8	6.3	1.4
PH	7.6	7.6	7.8	7.8	7.9
TA(4), g$CaCO_3$/l	10.0	7.9	10.8	15.8	10.7
NH4-N, mg/l	806	826	1150	2750	1200

* Each value is the average of at least two complete HRT after reaching the steady-state condition.

However, in Run 4 the limits of system seemed to be reached in terms of OLR (12.6 KgTVSf/m^3d) because of the substrate rich in biodegradable substances (SS-OFMSW/MS-OFMSW = 50%), although the process performed with very high yields (GPR=6.2 m^3/m^3d, SGP=0.5 m^3/KgTVSf). These were comparable with the yields obtained in Run 2, when the OLR applied was 12.1 KgTVSf/m^3d and the percentage of SS-OFMSW in the feed was 30%TS. Observing the characteristic parameters of the sludge inside the digester, the total concentration of VFAs in Run 3 was about 6.3 gHAc/l: in particular the concentration of propionic acid (C3) was about 3.5 gHAc/l. These data were indicative of the possible approaching to the overloading of the digester.

Also, the difference between the alkalinity measured at pH = 6 and at pH = 4, which is related to the VFA concentration, was higher than in Run 2 (8 $gCaCO_3/l$ vs. 4.8 $gCaCO_3$). During Run 4, although only SS-OFMSW was treated, the VFA concentration in reactor was about 1.4 gHAc/l: this may be interpreted considering both the lower OLR (6.0 $KgTVS/m^3d$) and the importance of the HRT in feedstock tanks and external temperature (Cecchi et al., 1992). In fact, this period was carried out during summer time, when the high external temperature could have caused, in the case of long residence time, a pre-hydrolysis of the substrate in the feed stock tank. This means that the process scheme turned into a two-phase-like system, although the fermentation step was uncontrolled. Yields in Run 4 were particularly high (GPR = 4.9 m^3/m^3d, SGP = 0.78 $m^3/KgTVSf$), because of the substrate and the related bacterial speciation.

Conclusions that were drawn from these studies were the following:

- Operating on HRT = 12 days, the thermophilic semi-dry single-phase anaerobic digestion process was stable using medium biodegradable substrates (TVS/TS $\leq$ 0.70) and organic loading rates up to 12 $KgTVS_{feed}/m^3d$, obtaining high yields in terms of GPR and SGP (respectively 6 m^3/m^3d and 0.5 $m^3/KgTVS_{feed}$)
- With easily biodegradable substrates (TVS/TS > 0.70) it seemed convenient to use organic loading rates not higher than 6 $KgTVS_{feed}/m^3d$, because stability parameters showed the limit of the process (especially VFA in Run 3: 6.3 gHAc/l). In fact, although the OLR increased, the GPR and SGP did not increase, while the digester overloading was clear.
- In the case of the only SS-OFMSW, it seemed more profitable to use a two-phase anaerobic digestion process. In fact, the separation of the hydrolytic - fermentative phase from the methanogenic one enabled the bacterial speciation. It permitted to use high values of OLR without running into instability problems.

6.5.1.2. Mesophilic steady-state conditions

The mesophilic anaerobic digestion of MS-OFMSW and SS-OFMSW has been studied and reported on by Cecchi et al. (1986, 1989b) and by Mata-Alvarez et al. (1990). As expected, the yields were quite different. Table 6.17 compares the performances obtained at two similar HRTs.

Although the OLR was higher and thus the digester more stressed, the yield were much better when SS-OFMSW was used; percentage TVS removal was nearly two times greater and SGP was nearly three times.

Table 6.17. Operational conditions, yields, feed, reactor and gas characteristics for SS-OFMSW and MS-OFMSW mesophilic anaerobic digestion.

Type of OFMSW	MS-OFMSW	SS-OFMSW
Operational conditions		
HRT, d	15.6	13.6
OLR, KgTVS/m^3 d	6.8	4.2
Yields		
TVS removal, %	36.5	67.1
SVS removal, %	57.6	89.7
TVFA removal, %	91	81
SGP, m^3/KgTVS d	0.225	0.637
SMP, m^3CH_4/kgTVS	0.114	0.398
Feed characteristics		
TVS, kg/m^3	107.0	57.1
SVS, kg/m^3	11.8	16.3
TVFA, kg/m^3	3.18	4.30
Reactor characteristics		
TVS, kg/m^3	67.9	18.8
SVS, kg/m^3	5.0	1.5
TVFA, kg/m^3	0.29	0.80
Gas characteristics		
CH_4 content, %	50.6	62.5

Another remarkable difference was the amount of soluble VS: the percentage in SS-OFMSW was about 30 (with respect to the TVS) whereas the percentage was as low as 13 in MS-OFMSW. Additionally, the quality of biogas produced was upgraded when the refuse was selected at origin.

In general it was observed that digester performances, referred to the TVS removal percentages, were highly sensitive to the quality of the feed: more than half of the TVS of the MS-OFMSW were non-biodegradable, which accounted for the lower TVS removal and specific gas production yields as compared with those of SS-OFMSW (Table 6.17). To obtain high biogas yields the non-biodegradable portion needs to be removed at source so as to improve the process economics.

6.5.2. Process behaviour in transition conditions

The behaviour of the process in transient conditions is a very difficult topic to discuss since literature is very poor about it. However, some information can be given by the study done by the authors. The wider set of data concerns MS-OFMSW, whereas only a little information is given about the other substrates. The main parameters considered for digester stability monitoring are pH, alkalinity, VFA, %CO_2 and biogas production.

A brief description of parameters change during four transient condition is given in the following discussion, concerning MS-OFMSW digestion in semi-dry process according to Pavan et al. (1994) that is:

1) Transient A: transition of the temperature range from mesophilic to thermophilic conditions;
2) Transient B: transition with light loading rates (from very stable conditions to low OLR in thermophilic conditions);
3) Transient C: transition with medium loading rates (from low to medium OLR in thermophilic conditions);
4) Transient D: transition with high loading rates (from medium to high OLR in thermophilic conditions).

Table 6.18 summarizes the variations of the operational conditions applied to study the semi-dry process in transient conditions according to a quantitative approach.

Table 6.18. Variations of the operational conditions applied to study the semi-dry process in transient conditions (Pavan et al., 1994).

	Transient A			**Transient B**	**Transient C**	**Transient D**
	from	**to**	**% variation°**	**% variation°**	**% variation°**	**% variation°**
OLR	0	6–7	—	+30 : 40	+45 : 60	+60 : 80
HRT	—	14	—	−3 : 4	−18 : 24	−33 : 44
T	37	55	—	—	—	—
TS feed	—	—	—	+38 : 42	+33 : 50	+29 : 55
GPR	0.2	1.8	+800	—	—	—
CH_4	50	55	+10	—	—	—
TA(4)	6.9	7.5	+9	—	—	—
PH	7.6	7.3	−3.3	—	—	—
TVFA	1.1	0.3	−73	—	—	—

(°) Calculated on the basis of the first steady state condition values.

The whole period that was analysed along one year in Pavan et al. (1994) consisted of six steady-state conditions (SSC), one in the mesophilic range and five in the thermophilic range according to the timeline progression shown in Figure 6.6.

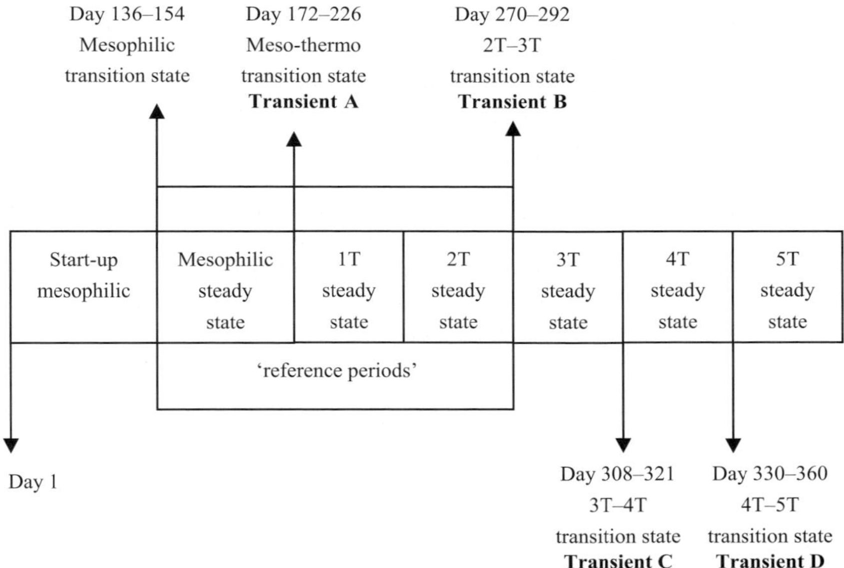

Figure 6.6. Time-line progression of the complete study on transient conditions.

The first recognizable transient is the mesophilic transition state that connected the start-up of the mesophilic phase and the mesophilic steady state (Pavan, 1988). After the acclimation of the digester to the mesophilic range, the OLR was gradually increased by increasing the amount of substrate daily fed and maintaining the TS content constant (around 20%). Steady-state conditions were reached without any problem as for the dynamic equilibrium inside the digester and in a very fast way (from day 136 to day 154) as it can be also seen from Figure 6.7.

The initial relatively high concentration of TVFA in the reactor, caused by the sudden increase of OLR (Figure 6.7) stabilized after only 25–30 days to the steady-state values. The pH never gave rise to particular problems of instability and remained in the typical range for mesophilic anaerobic digestion.

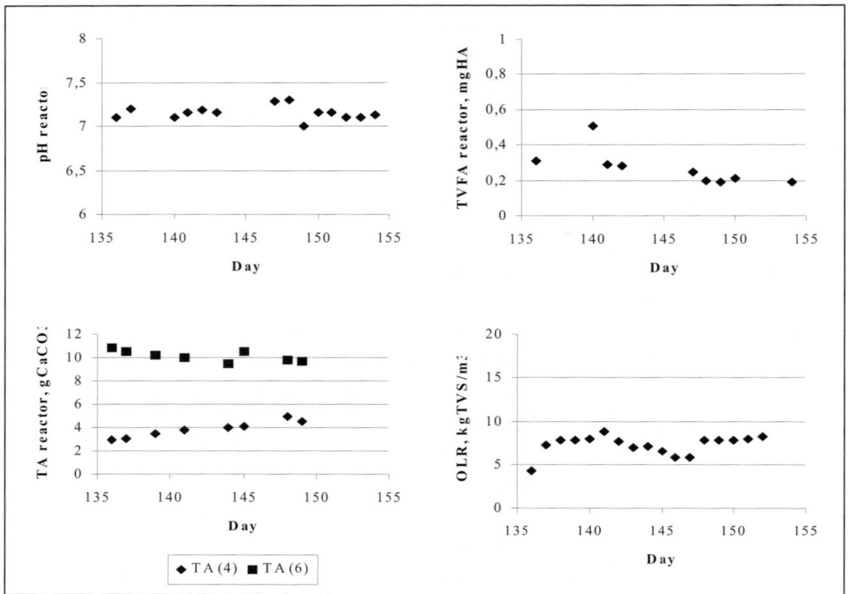

Figure 6.7. pH, TVFA and TA profiles during the mesophilic transient (Pavan et al., 1994

The reference periods were the first mesophilic (M) and the two thermophilic steady states and started from Transient A (from mesophilic to the first thermophilic operational thermal conditions) and continued in the following second thermophilic steady state (2T) that were obtained by operating, as described in Cecchi et al. (1993), in a very fast way if compared with literature (Rimkus et al., 1982) and precisely:

- feed break for 10 days;
- temperature rising from 35 to 55 °C after the fifth day, reaching the operational temperature in 48 hours;
- feed restart with OLR = 4–5 kgTVS/m^3d;
- full loading re-application after 5 days.

Two subsequent states were achieved: the first (pseudo-steady-state) with lower yields, after 30 days from the beginning of the thermal transient and the second (real steady state) after 60 days.

The achievement of a first pseudo steady state was quite easy but it was not the ultimate one even though it was characterised by good stability. Only after 4 HRTs was it possible to define the period as stationary; in fact the operational conditions after 60 days are the same as observed after longer times and this let suppose the real formation of a stable culture. Furthermore it confirms that a relation does exist between the concentration of volatile matter in the reactor ($TVS_{reactor}$) and the viable biomass.

In transient A the digester was immersed in an extreme unbalanced situation, due to the initial low presence of thermophilic consortia. The imbalance is clearly shown by GPR trend (Figure 6.8) in the days immediately after the change of temperature. Even if the situation could be surely considered of extreme stress, the recovery of the digester, from the stability point of view, occurred in only few days after the change of temperature, maintaining constant the new operational conditions applied without using any chemicals. These results had shown that very long periods reported in literature for the acclimatization of the process can be avoided. The other monitoring parameters (e.g. SGP, pH, TA, CH_4 %, etc.) do not bring any reliable information.

Transient B was the connection between two low stress steady state conditions (Figure 6.6). The OLR was increased by 30–40% according to the TS content in the feed that increased by 38–42% (Figure 6.9). These alterations, together with a contemporary decrease of ca. 1°C in temperature (Figure 6.9), only caused a small decrease in the process yields that remained at a comparable level with the following steady state (Thermophilic 3). The only evidence of the system overloading was the accumulation of TVFA (up to 1200 mgHAc/l, Figure 6.9) that decreased to normal levels in few days. This fact was explained by speculating on the higher thermophilic biomass content in the digester in respect to the 'reference period'.

In Transient C (medium load conditions, Figure 6.6) the system was moved from low load conditions and, before the beginning of the new operational conditions, the OLR was reduced to ca. 3 kgTVS/m^3d to simulate a short drop in substrate availability (i.e. a weekend) (Figure 6.10). As a consequence the gas production reduced from 3 to 1 m^3/m^3d but recovered after only two days (Figure 6.10). Therefore the OLR decline of two days did not affect the digester behaviour, producing only small transitory and quickly reversible effects on the process performances. Afterwards, the OLR was gradually increased of 15–20% by the increase of TS content of the feed (5–8%) maintaining HRT constant. The main observed effect was the faster rising of GPR than in the 'reference period'.

Transient D (Figure 6.6), from medium to high load conditions, is the most interesting period because it lead to many anomalous situations. The transient was split in two sub-periods: the first, with a technical one-week stop of feeding and the second when the system was fed again and the heating system had a shutdown causing a temperature decrease of about 10 °C.

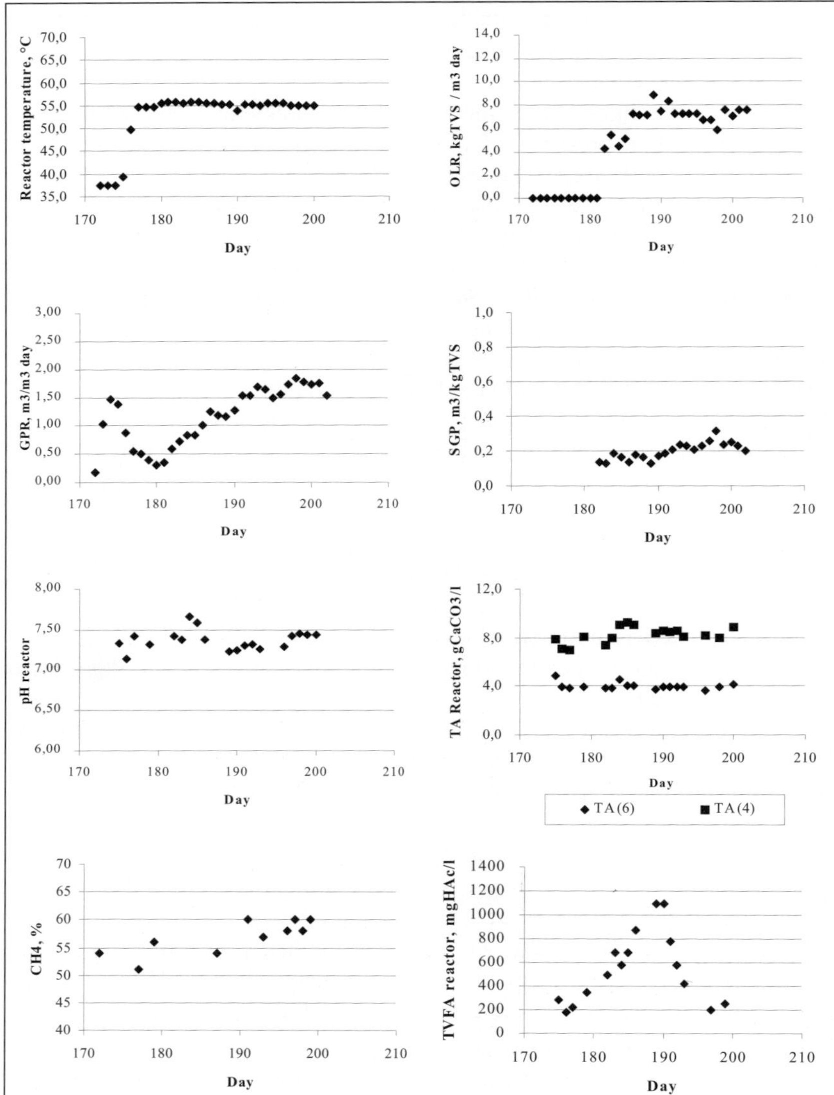

Figure 6.8. Temperature, OLR, GPR, SGP, pH, Total alkalinity, methane percentage and TVFA profiles during Transient A (from mesophilic to thermophilic (1) operational conditions) (Cecchi et al., 1993).

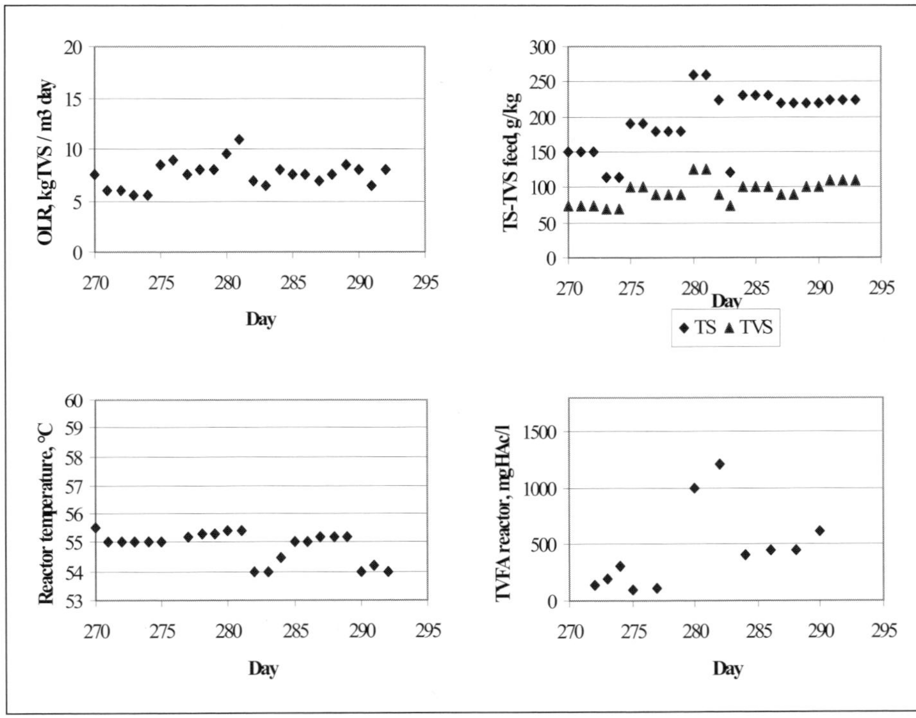

Figure 6.9. OLR, TS/TVS in the feed, reactor temperature and TVFA profiles during Transient B (from very stable conditions to low OLR in thermophilic conditions) (Cecchi et al., 1993).

In this transient (D) the recovery of process performances was more gradual than other transients but in agreement with the 'reference period'. In fact, in both cases, the time required for the achievement of GPR values comparable to the stationary ones was longer than two weeks. This meant that the temperature decreased played a major role. Furthermore, it was observed that, when considerable stress conditions are applied, the digester behaved as in fast start-up conditions but reached steady-state conditions in a faster way.

As a conclusion for the MS-OFMSW treatment in semi-dry and dry processes (the extension to semi-dry processes seems reasonable), significant variations of the organic loading (from 15 to 40% in 48 hours) don't involve any permanent effect on the process performances. Wider variations can generate instability, that can be, however, recovered and recognised by monitoring parameters as VFA, pH, alkalinity and production/composition of the biogas.

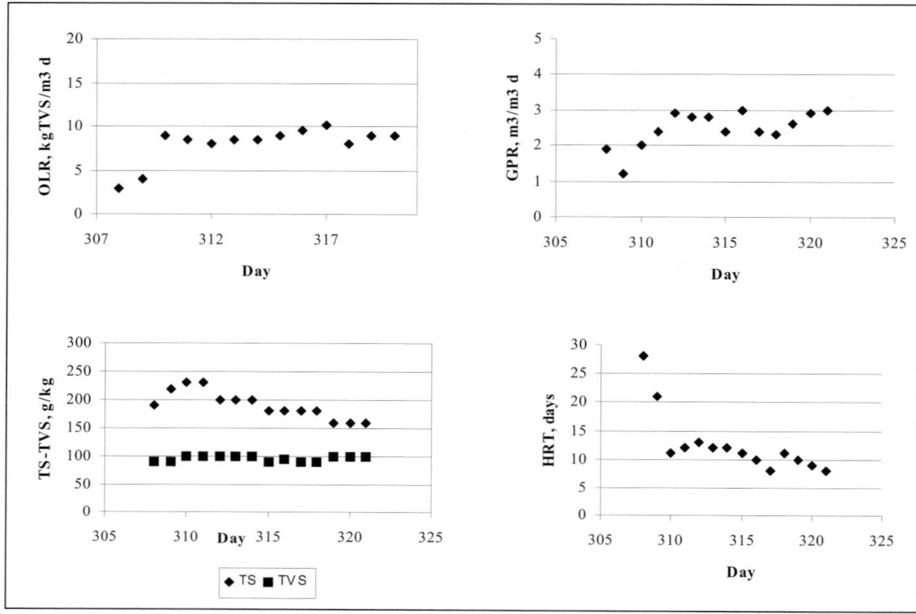

Figure 6.10. OLR, GPR, TS/TVS in the feed and HRT profiles during Transient C (transition with medium loading rates from low to medium OLR in thermophilic conditions) (Cecchi et al., 1993).

Since the analysis of available data in the literature for more easily degradable materials (SS- and SC-OFMSW) would represent a considerable and quite hard job, the authors will consider it for a future study; however here a general statement has been made as tentative on the basis of biogas production trend between two feeding operations, comparing the profiles obtained with different substrates (Figure 6.12).

What is clear from this figure is that more biodegradable substrates are characterised by higher slopes because of the larger values of the kinetic constants of biodegradation (Cecchi et al., 1991). In other words, more biodegradable substrates promote faster transformation processes and, as a consequence, the effects of either loading or operational conditions variations will be faster too and more evident if compared with 'poorer' substrates like MS-OFMSW. However, this is just a hypothesis that should be further studied and demonstrated.

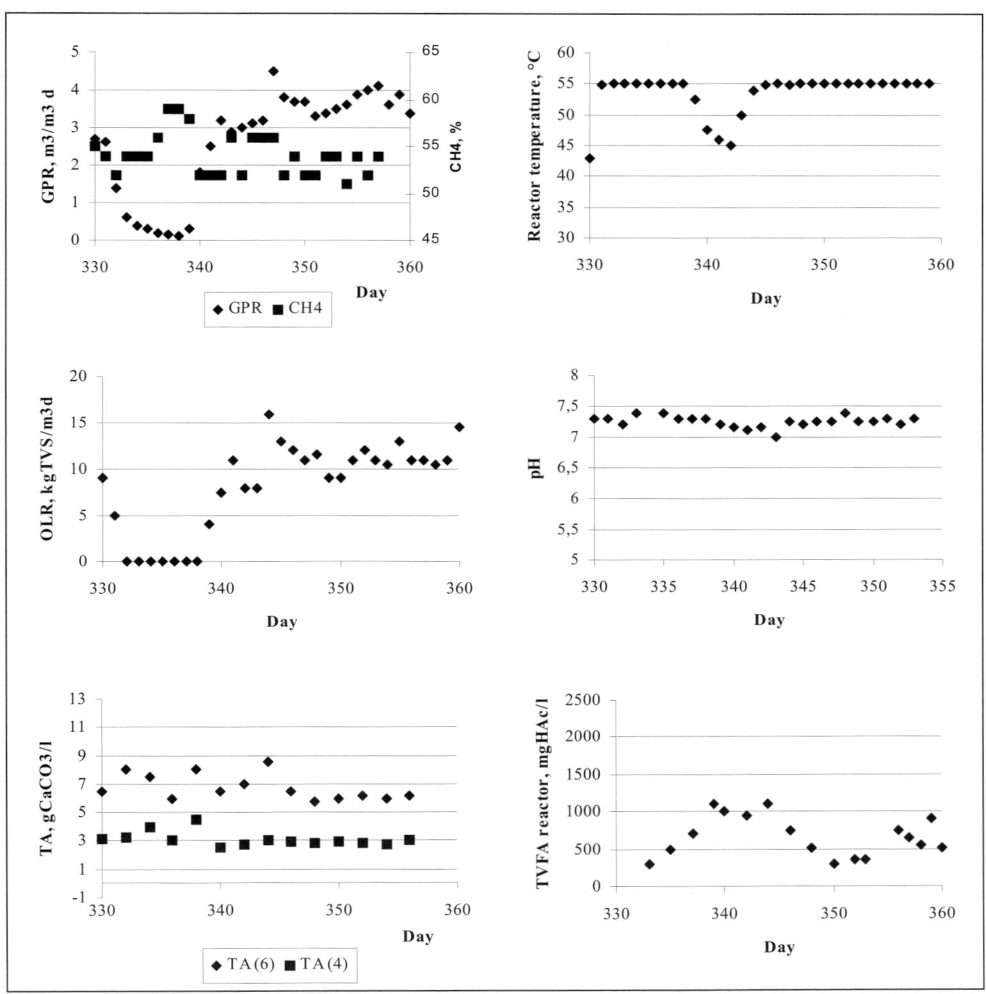

Figure 6.11. GPR, CH_4 %, reactor temperature, OLR, pH, total alkalinity and reactor TVFA profiles during Transient D (transition with high loading rates from medium to high OLR in thermophilic conditions) (Cecchi et al., 1993).

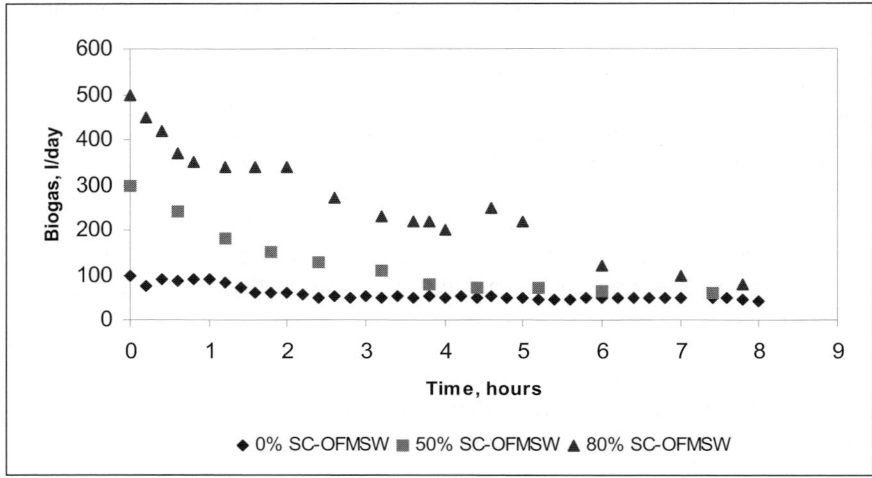

Figure 6.12. Biogas production profiles observed during the time elapsed (8 h) between one feed to the digester and the next one, with different mixtures of sewage sludge (SS) and separately collected organic fraction of municipal solid waste (SC-OFMSW) (Mata-Alvarez et al., 1990).

6.6 SIZING OF AN ANAEROBIC DIGESTER FOR THE TREATMENT OF THE OFMSW ON THE BASIS OF THE OPERATIONAL PARAMETERS AND LOADING FACTORS

According to the previous paragraphs, important information can be derived as for the comparison among the different classes of substrates and the different applicable processes. In the future, the probable evolution of these aspects will lead to better tools for sizing but, at present, a relatively simple approach is widely used.

The use of substrates like the OFMSW offers substantial benefits such as the energetic high yields deriving from biogas production and is not subject to the limitations typical for the anaerobic treatment of waste sludge coming, especially, from biological nutrients removal plants (Bolzonella et al., 2001). Furthermore, the common approach of applying post-composting to the digester effluent involves sizing parameters that enhance biogas production rather than sludge high grade of stabilisation. These two aspects lead the way for the application of high organic loading rates and short retention times using, as substrates, those compounds characterised by high utilisation rate (compound types A and B in the step-diffusional model: Cecchi et al., 1991), maximising the specific yield / m^3 reactor (GPR) rather than the one specific to the biomass (SGP).

On this basis, the sizing of an anaerobic digester for the OFMSW treatment can be substantially carried out starting from two approaches:

1. considering only the daily loading rate: in this case the only parameter that should be taken into account is the hydraulic retention time;
2. considering the organic loading rate that fits the process: the reactor should be sized on the basis of the amount of substrate that is enough for the biomass.

Obviously, the best and most reliable approach for the future management of the process is the latter but, to keep a safety range, both criteria should be taken into account. The estimate is made of consecutive approximations and can be summarised as follows.

1. Determination of the influent mass flow

Usually, as a basic hypothesis, the served population (PE) is already known and the specific production, expressed as OFMSW produced per PE per day, is estimated on the basis of the actual waste production in the area. Then, the influent mass flow, in terms of raw wastes, will be:

(Per capita daily production) × (population) = (OFMSW daily production)

Since the content of total and volatile solids will differ depending on the type of waste collection, the inflow in terms of TS and TVS will be:

(OFMSW daily inflow) × (%TS) = (TS daily inflow)

(TS daily inflow) × (TVS/TS) = (TVS daily inflow)

which is the total daily organic loading rate for the digester.

2. Determination of the organic loading rate per m^3 of reactor and computation of the effective volume

On the basis of the available data, an optimal organic loading rate is chosen: this can be done considering some literature results both for loading ranges and

HRTs. An indication can be given by Table 6.20, where these data are given for single-phase anaerobic digestion processes.

Thus the digester volume required will be:

(TVS daily inflow) / (OLR [kgTVS/m^3 d]) = (m^3 reactor)

3. Check of the operational conditions

The volume previously determined, even if theoretically correct, could be unsatisfactory to maintain the certain HRTs. To make sure the estimate is correct, it is necessary to know the volume of the substrate to be fed which should be estimated taking into account potential dilutions (i.e. water addition to achieve a concentration of 20%TS in the semi-dry process). Thus, knowing the density:

(Diluted OFMSW) / (density [ton/m^3]) = (m^3 OFMSW per day)

HRT = (digester volume, [m^3]) / (m^3 OFMSW per day)

Following this approach, an HRT that is too short could come out: it is reasonable, then, to repeat the calculations decreasing the loading rate until an optimal compromise is reached.

The decrease in organic loading rate is always an additional safety factor since the system can work in less stressing conditions.

4. Energy assessment

A daily biogas production can be calculated by the following equation:

SGP, m^3/KgTVS × KgTVS/day = m^3 biogas produced per day

Estimating a calorific power less than 5500 Kcal/m^3, the daily energy power will be:

5500 Kcal/m^3 × m^3 biogas/day = Kcal/day

The energy requirement for heating is due to two losses:
the influent waste heating;
heat losses from the digester.

Table 6.19. Range of operational conditions for anaerobic digestion processes.

Process	Operational conditions	Substrate
Mesophilic Wet	HRT: 14–30 d OLR: 2.6–4.0 kgTVS/m^{3}d	MS-OFMSW
	HRT: 14–30 d OLR: 1–4 kgTVS/m^{3}d	SS-OFMSW
Mesophilic Semi-dry	HRT: 15–20 d OLR: 6–8 kgTVS/m^{3}d	MS-OFMSW
	HRT: 12–14 d OLR: 3–4 kgTVS/m^{3}d	SC-OFMSW
Thermophilic Semi-dry	HRT: 6–15 d OLR: 6–20 kgTVS/m^{3}d	MS-OFMSW
	HRT: 12–14 d OLR: 8–12 kgTVS/m^{3}d	SS-OFMSW
Mesophilic Dry	HRT: 17–30 d OLR: 6–9 kgTVS/m^{3}d	MS-OFMSW
	HRT: 17–25 d OLR: 4–6 kgTVS/m^{3}d	SC-OFMSW
	HRT: 17–25 d OLR: 3–4 kgTVS/m^{3}d	SS-OFMSW
Thermophilic Dry	HRT: 12–20 d OLR: 9–15 kgTVS/m^{3}d	MS-OFMSW
	HRT: 12–16 d OLR: 6–9 kgTVS/m^{3}d	SC-OFMSW
	HRT: 12–16 d OLR: 4–6 kgTVS/m^{3}d	SS-OFMSW
Co-digestion	HRT:14–16 d OLR: 1.9–3.9 KgTVS/m^{3}d (from 25–80% SC-OFMSW)	SC-OFMSW CO-SUBSTRATE

6.7. REFERENCES

APHA (1995). Standard Methods for Water and Wastewater Examination (1995). 19th Ed. American Public Health Association/Water Works Association/Water Environment Federation, Washington D.C., USA.

APHA–AWWA–WPCF (1998). Standard methods for the examination of water and wastewater.

Bhattacharya, S., Madura, R., Walling, D. and Farrell, J. (1996). Volatile solids reduction in two-phase and conventional anaerobic sludge digestion. *Wat. Res.* **30**(5), 1041–1048.

Bolzonella, D., Innocenti, L., Cecchi, F. (2001). BNR wastewater treatments and sewage sludge anaerobic mesophilic digestion performances. In: *Proc. Specialised Conf. on Sludge Management: regulation, treatment, utilisation and disposal, Acapulco, Mexico, October 25–27, 2001* (ed. B. Jimenez and L. Spinosa), pp. 220–227.

Brunetti, A., Lore, F. and Lotito, V. (1988). Methanogenic potential of substrate in anaerobic digestion of sewage sludge. *Env. Tech. Letters* **9**, 753–762.

Cecchi F., Traverso, P.G. and Cescon, P. (1986). Anaerobic digestion of the organic fraction of municipal solid waste. Digester performance. *The Science of Total Environment* **56**, 183–197.

Cecchi, F., Pavan, P., Mata-Alvarez, J. and Vallini, G. (1989a). Anaerobic mesophilic digestion – co-composting research in Italy. *Bio-Cycle* **30**(7), 68–71.

Cecchi, F., Traverso, P.G., Perin, G. and Vallini, G. (1989b). Comparison of codigestion performance of two differently collected organic fractions of municipal solid waste with sewage sludges. *Env. Tech. Lett.* **9**, 391–400.

Cecchi, F., Mata-Alvarez, J., Marcomini, A. and Pavan P. (1991). First order and step diffusional kinetic models in simulating the mesophilic anaerobic digestion of complex substrate. *Bioresource Technology* **36**, 261–269.

Cecchi F., Pavan P., Musacco A., Mata-Alvarez J., Vallini G. (1993). Digesting the organic fraction of municipal solid waste: moving from mesophilic to thermophilic conditions. *Waste Management and Research* **11**, 403–414.

Cecchi, F., Battistoni, P., Pavan, P., Fava, G. and Mata-Alvarez, J. (1994). Anaerobic digestion of OFMSW and BNR processes: a possible integration. *Wat. Sci. Tech.* **30**(8), 65–71.

Cecchi, F., Pavan, P. and Mata-Alvarez, J. (1997). Kinetic study of the thermophilic anaerobic digestion of the fresh and pre-composted mechanically selected organic fraction of MSW. *Journal of Environmental Sciences and Health* **A32**(1), 195–213.

CITEC (2000). Le linee guida per la progettazione, la realizzazione e la gestione degli impianti a tecnologia complessa per lo smaltimento dei rifiuti urbani. A cura di A. Magagni. Ed. Hyper.

De Baere, L. and Vestraete W. (1984). High rate anaerobic composting with biogas recovery. *Biocycle* **25**, 30.

De Baere, L. (2000). Anaerobic digestion of solid waste: state of the art. *Wat. Sci. Tech.* **41**(3), 283–290.

IRSA–CNR (1982). Metodi Analitici per le Acque. Quaderno 55. Istituto Poligrafico e Zecca dello Stato, 1982, pp. 342.

Kubler, H., Hoppendeidt, K., Horsch, P., Kottmair, A., Nimmrichter, R., Nordsieck, H., Mucke, W. and Swerev, M. (1999). Full scale co-digestion of organic waste. In: *Proc. II ISAD-SW, Barcelona, Spain, 15–17 June 1999* (ed. J. Mata-Alvarez) pp. 175–182

Mata-Alvarez, J. and Cecchi, F. (1989). Joint anaerobic digestion of sewage sludge and sorted organic fraction of municipal solid waste to attain the energetic autonomy in wastewater treatment plants. Workshop of the FAO-CNRE: Biogas Production Technologies. Zaragoza, 10–13 April.

Mata-Alvarez, J., Cecchi, F., Pavan, P. and Llabres, P. (1990). The performances of digesters treating the organic fraction of municipal solid wastes differently sorted. *Biological Wastes* **33**, 181–199.

Mata-Alvarez, J., Cecchi, F. and Pavan, P. (1992). Substrate utilization kinetic models in the semi-dry thermophilic anaerobic digestion of municipal solid waste. *J. Environ. Sci. & Health.* A **27**(7), 1967–1986.

Pauss, A., Nyns, E.J. and Naveau, H. (1984). Production of methane by anaerobic digestion of domestic refuse. In: *Proc. E.E.C. Conference on Anaerobic and Carbohydrate Hydrolysis of Waste, 8–10 May 1984, Luxembourg.*

Pavan, P. (1988). Studio e sperimentazione su impianto pilota del processo di digestione anaerobica applicato alla frazione organica dei rifiuti solidi urbani. Tesi di laurea.

Pavan, P., Battistoni, P., Traverso, P., Musacco, A. and Cecchi, F. (1998). Effect of addition of anaerobic fermented OFMSW on biological nutrient removal process: preliminary results. *Wat. Sci. Tech.* **38**(1), 327–334.

Pavan, P., Musacco, A., Cecchi, F., Bassetti, A. and Mata-Alvarez, J. (1994). Thermophilic semi-dry anaerobic digestion process of the organic fraction of MSW during transient conditions. *Environmental Technology* **15**, 123–133.

Pavan, P., Battistoni, P., Mata-Alvarez, J. and Cecchi, F. (2000). Performance of thermophilic semi-dry anaerobic digestion process changing the feed biodegradability. *Wat. Sci. Tech.* **41**(3), 64–72.

Rimkus, R.R., Ryan, J.M. and Cook E.J. (1982). Full scale thermophilic digetion at the west south-west sewage treatment works, Chicago, Illinois. *J. Wat. Control Poll. Fed.* **54**, 1447–1457.

Ross, W.R., Novella, P.H., Pitt, A.J., Lund, P., Thomson, B.A., King, P.B. and Fawcett K.S. (1992). Anaerobic digestion of wastewater sludge. WRC Project no. 390, TT 55/92, Pretoria, South Africa.

Roux, N.W. and Wakerley, D.S. (1978). The microbial production of methane from putrescible fractions of sorted household waste. *Conservation & Recycling* **2**, 163–169.

Sans, C., Mata-Alvarez, J., Cecchi, F., Pavan, P. and Bassetti, A. (1995). Volatile fatty acids production by mesophilic fermentation of mechanically sorted urban organic wastes in a plug flow reactor. *Bioresource Technology* **51**, 89–96.

Stafford D., Hawkes D., Horton R. (1980). *Methane production from waste organic matter*. CRC Press, Boca Raton, Florida (USA), pp. 285.

Valorga (1985). Waste recovery as a source of methane and fertilizer. The Valorga process. In: *Proc. 2nd Annual Int. Symposium. on Industrial Resource Management, February 17–20 1985, Philadelphia, USA.*

Zorzi, G. (1997). Influenza delle condizioni operative sul processo di fermentazione anaerobica della FORSU, tesi di laurea.

7

Co-digestion of the organic fraction of municipal waste with other waste types

H. Hartmann, I. Angelidaki and B.K. Ahring

7.1. INTRODUCTION

Several characteristics make anaerobic digestion of the organic fraction of municipal solid waste (OFMSW) difficult. By co-digestion of OFMSW with several other waste types it will be possible to optimize the anaerobic process by waste management. The co-digestion concept involves the treatment of several waste types in a single treatment facility. By combining many types of waste it will be possible to treat a wider range of organic waste types by the anaerobic digestion process (Figure 7.1).

Edited by J. Mata-Alvarez. ISBN: 1 900222 14 0

Furthermore, co-digestion enables the treatment of organic waste with a high biogas potential that makes the operation of biogas plants more economically feasible (Ahring et al., 1992a). Thus co-digestion gives a new attitude to the evaluation of waste: since anaerobic digestion of organic waste is both a waste stabilization method and an energy gaining process with production of a fertilizer, organic waste becomes a valuable resource.

Co-digestion treatment has been successfully applied to several agricultural and industrial organic waste types in recent years. In Denmark, for example, the co-digestion concept has been successfully used since the mid-1980s for the treatment of livestock waste and industrial organic waste in Joint Biogas Plants (Danish Energy Agency, 1995). However, at present only 7% of the overall OFMSW treated by anaerobic digestion in Europe was done so by means of co-digestion (De Baere, 2000). In this chapter we will show that co-digestion of OFMSW has several benefits which can be used for establishing a wider application of the anaerobic treatment of OFMSW.

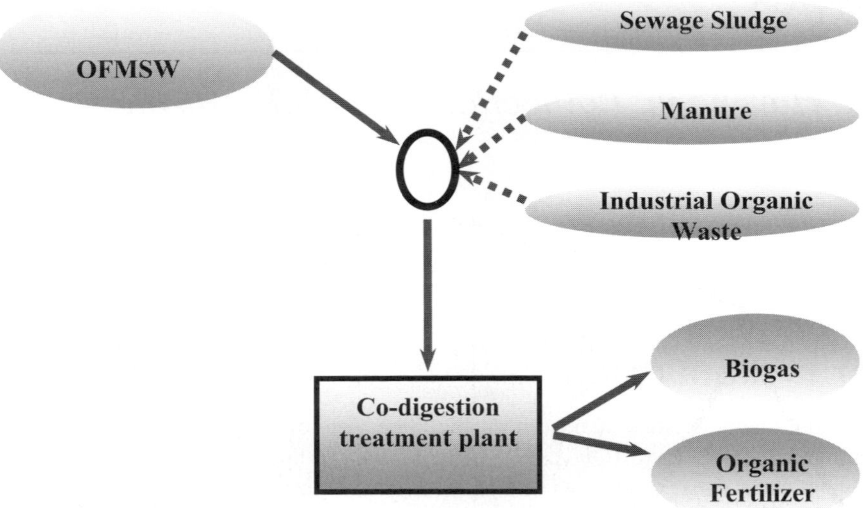

Figure 7.1. Principle of co-digestion of OFMSW.

7.2. GENERAL ASPECTS OF CO-DIGESTION

Co-digestion is defined as anaerobic treatment of a mixture of at least two different waste types. The mixing of several waste types has positive effects both on the anaerobic digestion process itself and on the treatment economy.

The profit of co-digestion in the anaerobic degradation process is mainly within the following areas:

- Increasing the methane yield.
- Improving the process stability.
- Achieving a better handling of the waste.

Waste treatment by co-digestion is economically more favourable due to:

- Combination of different waste streams in one common treatment facility.
- Treatment of larger waste amounts in centralized large-scale facilities.

Co-digestion does further ensure a stable treatment of organic waste that vary significantly during the year, both in quantity and characteristics. (Angelidaki and Ahring, 1997; Ahring, 1995; Gavala et al., 1999; Bozinis et al., 1996; Hamzawi et al., 1998).

Generally, the key for co-digestion lies in balancing several parameters in the co-substrate mixture (Figure 7.2). Some qualities of each co-substrate can be advantageous for use in the biogas process whereas other qualities can hinder the degradation solely of this waste type.

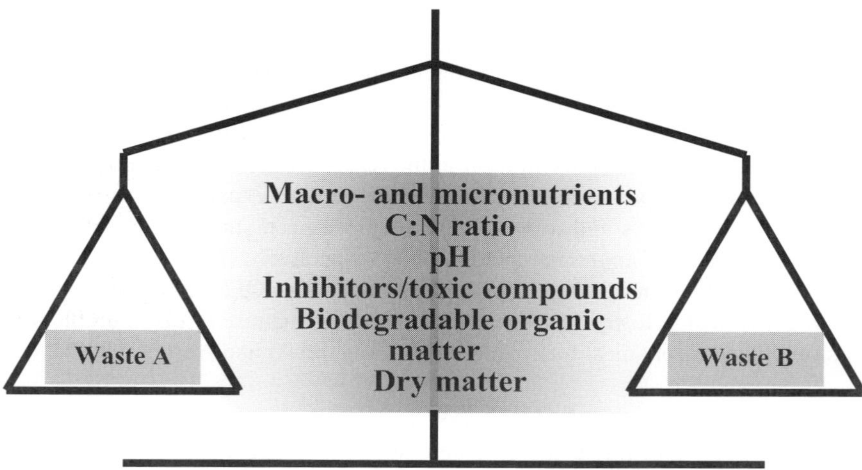

Figure 7.2. The balance of co-digestion.

The balance of nutrients, an appropriate C:N ratio and a stable pH are prerequisites for a stable process performance. High C:N ratio will lead to nitrogen deficiency, whereas ammonia toxicity is the principal problem with low C:N ratio. Nutrient deficiency of a given waste can be adjusted by co-digestion together with a nutrient-rich waste type. The problem associated with ammonia toxicity can be corrected by dilution of the ammonia concentration in the liquid phase, or by adjusting the C:N ratio of the feedstock (Kayhanian and Tchobanoglous, 1992). The pH can be balanced by addition of waste with a high buffer capacity, which protects the process against failure due to pH drop when the VFA concentration increases. Referring to the effect on the degradation of toxic substances by co-digestion, it is not only the dilution by addition of other waste that serves as a benefit (Hamzawi et al., 1998). Furthermore, detoxification of toxic compounds can be achieved in the co-substrate mixture by, for example, co-metabolic mechanisms, where a compound is transformed along with the general metabolism of microbes using a primary substrate. For example, it has been shown that waste containing tetrachloroethene (PCE) in concentrations up to 100 ppm can be degraded in co-digestion with manure (Ahring et al., 1996).

In the treatment of organic waste with a high content of recalcitrant organic matter (i.e. lignocellulose), the co-digestion with waste rich in easily biodegradable organic matter will be advantageous for obtaining a higher biogas yield. Organic industrial waste is usually characterized by high concentrations of easily degradable substrates such as carbohydrates, lipids and proteins, having a high biogas potential (Ahring et al., 1992a). Besides achieving a better economical feasibility of the treatment, the addition of easily degradable material has been shown to stabilize the anaerobic digestion process if added in a controlled fashion (Mathrani et al., 1994). This effect could partly be due to a higher active biomass concentration in the reactor, which will be more resistant to inhibitory compounds. Furthermore, the inorganic parts of some organic wastes, such as clays and iron compounds, have been shown to counteract the inhibitory effect of ammonia and sulphide, respectively (Ahring et al., 1992b). Finally, the dilution of waste with a high TS such as OFMSW by co-digestion with waste with a lower TS concentration such as manure resolves problems of pumping and mechanical treatment of SW (Angelidaki and Ahring, 1997).

7.2.1. Co-digestion of OFMSW

Generally, the application of the anaerobic digestion process to OFMSW faces problems due to the following substrate characteristics (Kayhanian and Rich, 1996; Demirekler and Anderson, 1998; Kayhanian and Rich, 1995; Rintala and Järvinen, 1996; Mathrani et al., 1994; Ahring and Johansen, 1992):

- High total solids (TS) content of typically 30–50%
- High C:N ratio
- Deficiency in macro- and micronutrients
- Content of toxic compounds (heavy metals, phthalates)

The benefit on the other hand of using OFMSW in an anaerobic co-digestion process is mainly the high content of easily biodegradable organic matter. This leads to a methane yield of up to 330 l/kgVS (Rintala and Järvinen, 1996; Diaz et al., 1981; Six and De Baere, 1992; Rivard et al., 1990).

Owing to the high total solids content of OFMSW both dry (TS content higher than 20%) and wet (TS below 20%) anaerobic processes have been developed (Cecchi et al., 1988; Poggi-Varaldo et al., 1997; Six and De Baere, 1992; Rintala and Järvinen, 1996). Co-digestion is used for both types of treatment, but mainly for the wet process. In high-solids systems the addition of co-substrates improves process performance due to a better nutrient balance. In wet digestion systems, co-digestion of OFMSW with more diluted waste streams enables a reduction of the TS concentration of OFMSW below 20%. The lower TS content of the wet treatment system offers an easier mechanical handling of the waste and allows for the use of a continuous stirred tank reactor for the anaerobic digestion.

The C:N ratio and the nutrient content of OFMSW vary significantly due to the composition of the single fractions of OFMSW. The different organic fractions (food waste, yard waste, paper, newspaper, etc.) have different C:N ratios, which makes the treatment of OFMSW in itself a co-digestion process. C:N ratio based on biodegradable organic carbon is of food and yard waste below 20 and of mixed paper it is more than 100 (Kayhanian and Tchobanoglous, 1992). In high solids anaerobic digesters, the optimum C:N ratio for methane production with no adverse effect on the performance was found to be in the range of 25 to 30, based on biodegradable carbon. (Kayhanian and Tchobanoglous, 1992; Kayhanian and Hardy, 1994; Kayhanian and Rich, 1995). For mixtures with less than 50% OFMSW, an optimal C:N ratio between 25 and 80 has been reported (Hamzawi et al., 1998; Kayhanian and Tchobanoglous, 1992). As can be seen from the different C:N ratios in food waste and yard waste compared to mixed paper, the optimum C:N ratio can thus be also adjusted by the right mixture of mixed paper to food and yard waste. It also shows that anaerobic digestion solely of food waste can lead to ammonia toxicity. Because of the different specific biodegradability of the single fraction the specific combination of the different fractions is one way to achieve optimal reactor operation (Hamzawi et al., 1998). Most MSW processing technologies using modern separation techniques, however, result in the separation and removal of the food and yard waste fraction (Rivard et al., 1990). This does not only lead to a higher TS content, but also reduces the nutrient value of the produced OFMSW when used as a feedstock for anaerobic digestion, and results in a high cellulose content for the processed MSW caused by increasing the overall percentage of wood, paper and cardboard. Therefore, using this kind of processed MSW necessitates nutrient supplementation for effective biological conversion. OFMSW of such quality is recommended to be treated by co-digestion.

Recently, the level of toxic compounds found in OFMSW has been discussed. Investigations in Denmark have shown that cadmium contamination of OFMSW is derived from garden waste and not so much household waste, while the latter contributes to a higher degree with phthalates such as bis-(2-ethylhexyl)-phthalate (DEHP) contamination (Kjølholt et al., 1998).

Impurities (plastics, metal, glass) found in OFMSW do often hinder the application of co-digestion if the effluent is intended to be used as fertilizer (De Baere, 2000). This is both because fertilizer should meet requirements for the content of toxic compounds (e.g. phthalates, heavy metals) and because farmers will not accept it if it has no clean appearance. The content of impurities and of toxic compounds found in OFMSW is much dependent on the collection system. In terms of the heavy metal content, for example, source-sorted household waste has generally a much better quality than OFMSW separated from municipal solid waste (MSW) by processing technology (Richard and Woodbury, 1992; Ahring and Johansen, 1992). The content of DEHP was reported to be influenced by the collecting material (plastic or paper bags), the acceptance of other material than the organic fraction in the collecting system, the collecting frequency, the temperature and the moisture of the waste (Kjølholt et al., 1998). It is therefore a prerequisite to introduce an adequate collection system in order to supply the treatment plant with clean OFMSW that can be recycled to soil.

The potential of co-digestion, on the other hand, to improve the detoxification abilities in the anaerobic degradation process is an important quality to enable the processing of an environmentally safe fertilizer product from OFMSW in the anaerobic digestion process.

Most investigations in the field of co-digestion of OFMSW have been made together with sewage sludge or livestock waste. Several large-scale plants are nowadays in operation using co-digestion of these substrates.

7.2.2. Co-digestion of OFMSW with sewage sludge

With the large amount of sewage sludge (SS) produced in wastewater treatment plants and the large number of existing anaerobic digesters to stabilize it, the anaerobic co-digestion of OFMSW with sewage sludge is especially attractive (Hamzawi et al., 1998). The co-digestion can be applied at existing treatment facilities without great investments and it combines the treatment of the two largest municipal waste streams. First attempts to dispose of garbage through a wastewater treatment plant took place in 1923. No or little attention was paid to this approach until the 1930s and 1940s, when the introduction of the home garbage grinder made the co-digestion of garbage and sewage more feasible (Diaz et al., 1981).

Co-digestion of OFMSW together with sewage sludge is beneficial due to a number of substrate characteristics of both waste types that are complementary in their combination (Figure 7.3).

The addition of high solids concentration of OFMSW to a sewage sludge digester operated with sludge having a low TS concentration will be possible even in rather high concentrations. The higher concentration of macro- and micronutrients in the sludge solids will compensate the lack of nutrients in OFMSW. The stabilizing effect of sludge on the digestion of OFMSW has been confirmed with sludge doses between 8 and 20% of feedstock volatile solids. (Kayhanian and Rich, 1996; Rivard et al., 1990). While digested sludge has a mainly stabilizing effect on the digestion process, primary sludge increases furthermore the methane yield. Besides, it has been described that addition of primary sewage sludge significantly decreases imbalances observed during the start-up of the digesters (Demirekler and Anderson, 1998).

Most co-digestion studies conducted before 1990 were low-solids processes, typically at a total solids concentration of 4–8%. Recently, the high-solids anaerobic digestion process has also been utilized for the co-digestion of OFMSW and sewage sludge. In these high-solids digestion studies, the sludge was mainly used to provide sufficient nutrients for microbial growth and metabolism. The operational total solids concentration was maintained at 25–35%, the feedstock C:N ratio was held between 22 and 30, based on the biodegradable carbon and total nitrogen mass (Kayhanian and Tchobanoglous, 1992; Kayhanian and Rich, 1996).

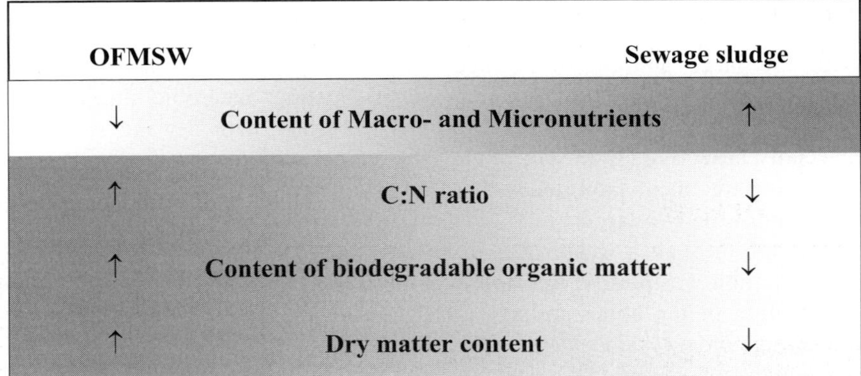

Figure 7.3. The different substrate characteristics of OFMSW and sewage sludge (Symbols: ↑ high, ↓ low).

The optimal mixture of OFMSW and sewage sludge is depending on the specific waste characteristics and the process concept used. For wet digestion systems, several researchers observed the best performance in terms of specific gas production and VS reduction with feedstock having an OFMSW:SS ratio in the range of 80:20 on a TS basis (Demirekler and Anderson, 1998; Diaz et al., 1981), or a volume ratio of 25% OFMSW and 75% sewage sludge, respectively (Hamzawi et al., 1998).

7.2.3. Co-digestion with livestock waste

Livestock waste has been used for a long time as substrate in anaerobic digesters for the purpose of renewable energy production. This is because of its excellent characteristics for the use as basic substrate and because it is available in large amounts since it is the largest agricultural waste stream. In Denmark, positive experiences have been made with the large-scale co-digestion process of livestock waste and industrial organic waste types. The construction of 20 Joint Biogas Plants since the 1980s has given the possibility for combined anaerobic treatment and utilization of livestock waste and several types of organic waste from food processing industries. Nowadays, the organic fraction of municipal solid waste is co-digested together with livestock waste (and industrial organic waste) at five sites (Sinding, Studsgård, Vaarst-Fjellerad, Vegger, Århus).

Livestock waste is an excellent basic substrate for the co-digestion process because of the following reasons (Figure 7.4). Livestock waste has a high buffering capacity originating mainly from the ammonia content and it has a high water content with total solids content typically 3–5% for livestock waste from pigs and 6–9% for livestock waste from cattle and dairy cows. Furthermore, it is rich in a wide variety of nutrients necessary for optimal bacterial growth and it has always been used for fertilizing and thus facilitates an easier application on agricultural soils.

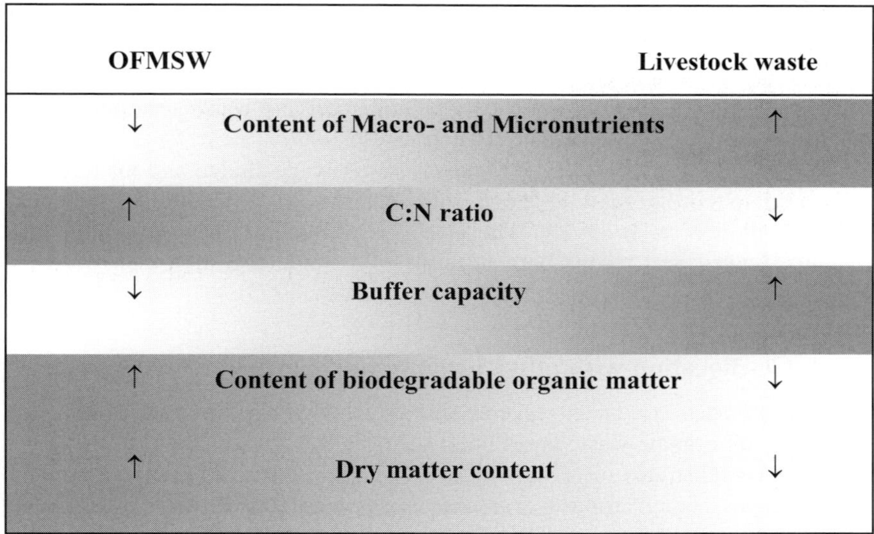

Figure 7.4. The different substrate characteristics of OFMSW and livestock waste (Symbols: ↑ high, ↓ low).

When treating livestock waste alone it has, on the other hand, typically a relatively low methane yield ranging from 10–20 m^3 CH_4/ton. The reason for this is the low solids content and a high content of fibres consisting of lignocellulose. This fibre fraction is highly recalcitrant to anaerobic degradation and will often pass through the reactor mainly undigested. Addition of industrial organic waste will therefore lead to a substrate with a higher biogas yield per m^3 feedstock than livestock waste alone. These industrial organic waste types are characterized by a high content of carbohydrates, proteins and lipids that are often bioavailable to a large extent, leading to a high biogas potential. The high biogas potential of OFMSW makes it like other organic industrial waste types a very attractive substrate for biogas plants (Mathrani et al., 1994). By combining the two different waste types a methane yield higher than 25 m^3 CH_4/ton feedstock can be achieved leading to an economically feasible digestion process (Ahring and Johansen, 1992).

The benefit of using livestock waste as basic medium in the co-digestion process has been shown together with several kinds of waste. The co-digestion process with livestock waste has been profitably applied to, for example, slaughterhouse waste, flotation sludge and fish-oil sludge (Ahring et al., 1992a), fruit and vegetable waste, fish offal, pig manure, dissolved air flotation sludge and brewery sludge (Callaghan et al., 1999). Another investigation showed that the fastest start-up of high-solids anaerobic digestion of a mixture of municipal solid waste, paper sludge and sewage sludge was obtained with reactors using a mixture of equivalent amounts of cow manure, soil and waste activated sludge as inoculum (Poggi-Varaldo et al., 1997).

The co-digestion of OFMSW together with sewage sludge and manure was compared to co-digestion together with water, sewage sludge, or chemical nutrient solution, respectively. The co-digestion together sewage sludge and manure showed best results both in terms of a stable digestion and markedly enhanced gas production rates (Kayhanian and Rich, 1995).

7.2.4. Co-digestion with other organic waste types

There have been conducted several studies where OFMSW was co-digested together with organic waste types other than sewage sludge or livestock waste. As other co-substrates olive mill effluents (OME), macroalgae, waste coming from kitchens, slaughterhouses and meat-processing industries were investigated (Angelidaki and Ahring, 1997; Cecchi et al., 1993; Kübler et al., 2000; Brinkman, 1999; Mata-Alvarez et al., 2000).

The study of the co-digestion together with olive mill effluents (OME) is a good example of determining the right co-substrate for co-digestion of a given waste type (Angelidaki and Ahring, 1997). Olive mill effluents are characterized by high loads of soluble organic matter, low alkalinity, lack of ammonia and inhibitory effects of polyphenols. The main purpose was to find an appropriate co-substrate for the anaerobic digestion of olive mill effluents. Batch experiments were carried out with mixtures of OME together with livestock waste, sewage sludge and OFMSW, respectively, in different dilutions with water. It was shown that the co-digestion of OME together with OFMSW or sewage sludge could only be achieved with high dilution of the waste. OFMSW or sewage sludge had not sufficient buffer capacity to maintain an adequate pH range. The pH dropped severely and the digestion process was inhibited. In co-digestion with livestock waste, however, OME could be degraded without dilution. The positive effect of livestock waste was found to be mainly due to its high alkalinity. Reactor experiments showed that the co-digestion with OFMSW was possible, but more optimal with livestock waste.

7.3. MODELLING THE CO-DIGESTION PROCESS

Organic waste management with co-digestion of the right substrates can be beneficial for the anaerobic digestion process. To exploit co-digestion in full, there is clearly the need to be able to predict the outcome of a certain waste combination strategy. The anaerobic treatment plants using co-digestion operate mainly on an empirical basis. However, successful combination of different types of waste requires careful management. The ability to predict the outcome of the process when mixing new waste is an important tool in optimizing the results. Consequently, the need for accurate modelling of the anaerobic degradation of mixtures of waste has arisen (Angelidaki et al., 1997). There are two main approaches to study the co-digestion of different waste types.

On one hand there is the need to perform experiments and derive a mathematical model for the degradation of mixed wastewaters. As described in the example of co-digestion of olive mill effluents, this can be done by experiments of co-digesting one type of waste together with others in different ratios. On the other hand, from the derived models the biodegradation of given waste types can be modelled in different mixtures, with the objective to obtain a stable process with a high-energy yield to minimize capital and operational costs (Bozinis et al., 1996).

Many models have been developed during the past 20 years. However, only few models include the digestion of complex substrates and only little attention has been directed to co-digestion utilizing different waste types in a mixture (Bozinis et al., 1996; Angelidaki et al., 1997; Gavala et al., 1999; Mata-Alvarez et al., 2000).

A model for the anaerobic degradation of different waste types in co-digestion has been developed which will be described in detail in the following (Angelidaki et al., 1997). The model considers the hydrolysis of complex substrates and accounts the influence of the main characteristics of a substrate, i.e. lipid, protein and carbohydrates concentration. Several inhibition mechanisms are included (Angelidaki et al., 1993; Angelidaki et al., 1999). This model has been verified with several co-digestion experiments, one of them was the co-digestion of olive mill effluent (OME) together with manure. However, the model has not been applied for the co-digestion of OFMSW yet.

According to the anaerobic degradation pathway of complex substrates, the co-digestion model involves one enzymatic process (hydrolysis of undissolved organic matter) and eight bacterial groups (glucose fermenting acidogens, lipolytic bacteria, LCFA- (long-chain fatty acids) degrading acetogens, amino acid-degrading acidogens, propionate, butyrate and valerate-degrading acetogens and aceticlastic methanogens, Figure 7.5).

In the single degradation steps, stoichiometry as previously described (Hill, 1982), has been employed with some minor modifications (Angelidaki et al., 1993). As parameters influencing the equilibrium of the reactions in the single stages the equilibrium relationship of ammonia, carbon dioxide and pH, the gas phase dynamics and temperature effects have been included. Four inhibition mechanisms are considered (free ammonia inhibition of the aceticlastic step, acetate inhibition of the acetogenic steps, volatile fatty acids (VFA) inhibition of the initial enzymatic hydrolytic step, inhibition caused by LCFA on all steps of the degradation process except hydrolysis).

For an accurate application of the model, a thorough analysis of the substrate characteristics is required. The biodegradable organic matter fraction is described by the volatile solids (VS) content together with specific content of insoluble and soluble organic matter. For encountering the inhibition effect, the content of ammonia and nitrogen released as ammonia during the degradation process, the LCFA content, VFA concentration and buffer capacity has to be defined as well. The buffer capacity is described by pH, alkalinity, phosphorus, cation- and CO_2 content.

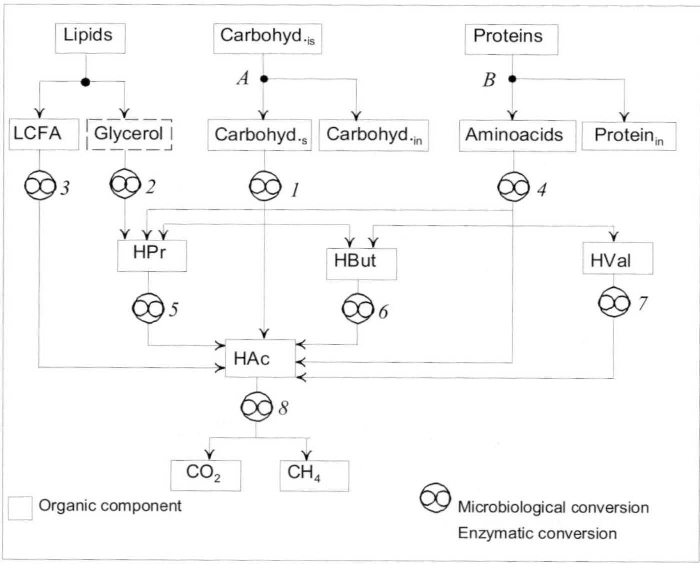

Figure 7.5. Main pathways of the organic component flow used in the model (Angelidaki et al., 1999).

For modelling the co-digestion of OFMSW, further characteristics of OFMSW might be needed for an accurate process description. It has been shown, for example, that factors like the feed particle size play an important role in the digestibility of OFMSW (Hamzawi et al., 1998; Kayhanian and Hardy, 1994).

The model describes well the dynamic response in methane production, pH and VFA concentration after introduction of the co-substrate (OME) to the basic substrate (manure). The simulation shows further the different types of behaviour with different mixture ratios of both substrates (Figure 7.6).

In a simulation a total shift from one substrate to the other (manure to OME) the model predicts a total collapse of the process, indicated by VFA rise, pH drop, drop in methane production and bacterial wash-out (Figure 7.7).

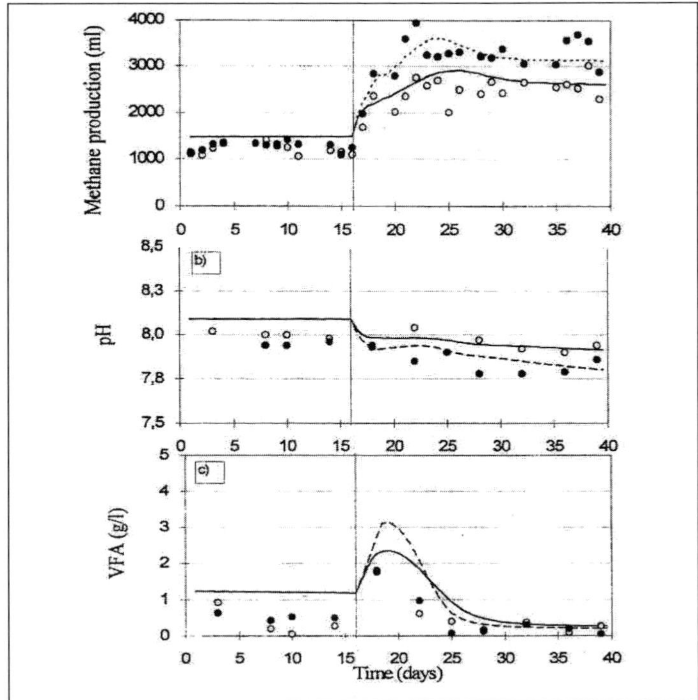

Figure 7.6. Digestion of cattle manure alone and then change to co-digestion of OME and livestock waste at day 16; symbols: o experimental data for the 50:50 mixture of OME:manure; • experimental data for the 75:25 mixture of OME:manure; — simulation data for the 50:50 mixture; - - simulation data for the 75:25 mixture (Angelidaki et al., 1997).

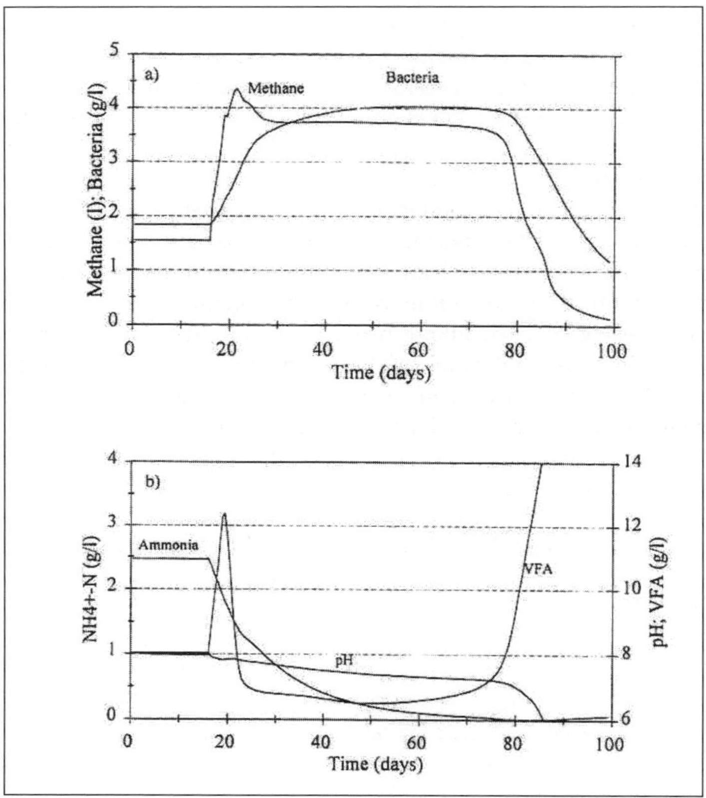

Figure 7.7. Simulation of manure digestion and change of the influent to OME on day 16 (Angelidaki et al., 1997).

7.4. LARGE-SCALE PLANT EXPERIENCES WITH CO-DIGESTION

One of the first large-scale co-digestion plants for OFMSW and sewage sludge was operated in the United States by the Refcom (refuse conversion to methane) project in the 1980s. The Refcom system was based on a conventional low-solids digester design (Kayhanian and Rich, 1996; Cecchi et al., 1988).

One of the first large-scale co-digestion plants for OFMSW in Europe was the Vaasa plant in Finland, which started its operation in 1989 (Rintala and Järvinen, 1996). The process has been tested on several waste types. The digester feed consists of mechanically or source-sorted OFMSW, thickened sewage sludge containing primary and waste activated sludge, slaughterhouse waste, fish waste and manure. The plant uses a wet digestion process with a TS content of approx. 10–15%. Both mesophilic and thermophilic treatment methods are in operation in two parallel reactors. One characteristic of the Vaasa Process is that a small part of the supernatant from the dewatering of the digested material is recirculated to be mixed with the fresh feed. Today, the Vaasa process is also in operation in Kil, Sweden and outside Tokyo, Japan.

The joint biogas plants in Denmark that admix OFMSW (Sinding, Studsgård, Vaarst-Fjellerad, Vegger, Århus) and Sweden's largest biogas plant in Kristianstad work with the same co-digestion concept: household waste is treated together with agricultural waste and industrial waste. The plant at Kristianstad has a capacity of 73,000 tons of biomass per year, of which 15% is OFMSW, 18% is industrial organic waste and 67% is manure. The biogas yield is around 40 m^3 biogas per ton feedstock. The energy production is 1.8–2 MW and the heat requirements of 600 to 800 households is covered. (Hedegaard and Jaensch, 1999). At Studsgård biogas plant in Denmark, 129,000 tons of biomass are treated per year, of which approximately 5% is OFMSW, 7% is industrial organic waste and 88% is manure. This plant achieves a biogas yield of approximately 35 m^3 biogas per m^3. The power production can cover the electricity consumption of about 2700 families and the production of heat is large enough to supply about 700 single-family houses. At all of these plants, the admixing of OFMSW to the feedstock is performed by either direct transport into the digester (for example at Kristianstad, Vegger) or it is first mixed together with manure in a feedstock tank and then pumped into the digester (for example at Studsgård). The effluent quality for the use as fertilizer is very much dependent on the purity of the collected OFMSW. For example, although OFMSW at Studsgård is source-separated, it is necessary to separate out the solid fraction of the digested effluent and incinerate it because of high plastic impurities. The plastic impurities come from the plastic bags used to collect the OFMSW.

An example for the excellent performance of a large-scale co-digestion process of OFMSW and sewage sludge is the sewage treatment plant in Grindsted, Denmark (Skøtt, 1997; Hedegaard and Jaensch, 1999) (Figure 7.8).

The OFMSW is co-digested together with sewage sludge and small amounts of industrial organic waste with the production of biogas for electricity generation and central heating supply. The municipality of Grindsted decided to build this new co-digestion plant to meet the formerly formulated requirements of recirculation of 50% of the municipal solid waste. About 1500 tons of the organic fraction of household waste coming from 6200 households is treated per year. The plant is situated outside inhabited areas of the municipality, to avoid annoyance by waste delivery and to have short distances to nearby farms using the digested effluent as fertilizer.

To achieve an OFMSW supply with a high purity a new waste collection system was introduced into the Grindsted municipality. The OFMSW is source-sorted in paper bags, and because of a good educational campaign the collecting system is accepted well in the community. Thus the problems in the treatment process arising from impurities of plastics clogging pumps and mixing equipment can be avoided. Furthermore, the dewatered effluent is accepted well by farms for use as fertilizer because of its high purity and its good fertilizing qualities.

The OFMSW used in the process is first shredded and metal impurities are removed before it is mixed with the sewage sludge in a pulper. The mixing ratio of OFMSW:sewage sludge is 1:9, leading to a dry matter content of 8–10%. Then, the mixture is macerated and glass and stones are separated before it is lead to a hygienization tank where it is heated at 70 °C for one hour. Together with a minor amount of industrial organic waste (fatty sludge) the subsequent digestion takes place under mesophilic conditions (35 °C). The biogas yield is approximately 25m^3 per m^3 of the influent. The biogas produced is loaded to a 247 kW CPH and/or a 706 kW boiler. The TS content is reduced to 3% in the effluent by the treatment. After dewatering, the solid fraction with a TS content of 21–25% can be easily stored for a whole year until it will be supplied to agricultural land in the spring during growth season.

Figure 7.8. The co-digestion plant for OFMSW and sewage sludge in Grindsted, Denmark. (Shredding and conveyor equipment for OFMSW in the front, reactor in the background.) (Photo: T. Skøtt.)

7.5. CONCLUSION

The anaerobic degradation of OFMSW by co-digestion with other organic substrates shows several advantages in terms of process stability and economical feasibility. The successful application of the co-digestion concept in large-scale has been demonstrated in several existing co-digestion plants. Since appropriate co-substrates (sewage sludge, livestock waste) will often be available, the anaerobic degradation of OFMSW can be realized by co-digestion at most sites. Co-digestion can optimize waste handling of several waste streams in a single treatment step. Modelling of the co-digestion process is a useful tool for determination of the appropriate co-digestion mixture; however, verification of a model that can simulate the co-digestion of the organic fraction of municipal solid waste with other waste is still missing. Furthermore, the benefit of co-digestion in the context of degradation of toxic compounds found in OFMSW needs to be studied in greater detail to ensure safe use of the effluent after anaerobic treatment.

7.6. REFERENCES

Ahring, B.K. (1995). Methanogenesis in thermophilic biogas reactors. *Antonie Van Leeuwenhoek International Journal Of General And Molecular Microbiology* **67**(1), 91–102.

Ahring, B.K., Angelidaki, I. and Johansen, K. (1992a). Anaerobic treatment of manure together with industrial waste. *Water Science And Technology* **25**(7), 311–318.

Ahring, B.K., Angelidaki, I. and Johansen, K. (1992b). Co-digestion of organic solid waste, manure and organic industrial waste. In *Waste Management International* (ed. Thomé-Kozmiensky, K.J.), pp. 661–666. EF-VERLAG für Energie und Umwelttechnik GmbH, Berlin.

Ahring, B.K., Garcia, H., Mathrani, I. and Angelidaki, I. (1996). Codigestion of manure with organic toxic waste in biogas reactors. In *Management of Urban Biodegradable Waste* (ed. J.A. Hansen), pp. 125–132. James & James Science Publishers Ltd., London.

Ahring, B.K. and Johansen, K. (1992). Anaerobic digestion of source-sorted household waste together with manure and organic industrial waste. In *Proceedings of International Symposium on Anaerobic Digestion of Solid Waste, Venice, Italy, April 14–17* (ed. Cecchi, F., Mata-Alvarez, J. and Pohland, F.G.), pp. 203–208. Stamperia di Venezia, Venice.

Angelidaki, I. and Ahring, B.K. (1997). Codigestion of olive oil mill wastewaters with manure, household waste or sewage sludge. *Biodegradation* **8**(4), 221–226.

Angelidaki, I., Ellegaard, L. and Ahring, B.K. (1993). A mathematical model for dynamic simulation of anaerobic digestion of complex substrates – focusing on ammonia inhibition. *Biotechnology and Bioengineering* **42**(2), 159–166.

Angelidaki, I., Ellegaard, L. and Ahring, B.K. (1997). Modeling anaerobic codigestion of manure with olive oil mill effluent. *Water Science and Technology* **36**(6–7), 263–270.

Angelidaki, I., Ellegaard, L. and Ahring, B.K. (1999). A comprehensive model of anaerobic bioconversion of complex substrates to biogas. *Biotechnology and Bioengineering* **63**, 363–372.

Bozinis, N.A., Alexiou, I.E. and Pistikopoulos, E.N. (1996). A mathematical model for the optimal design and operation of an anaerobic co-digestion plant. *Water Science and Technology* **34**(5–6), 383–391.

Brinkman, J. (1999). Anaerobic digestion of mixed slurries from kitchens, slaughterhouses and meat processing industries. In *Proceedings of the Second International Symposium on Anaerobic Digestion of Solid Waste, Barcelona, 15–18 June* (ed. Mata-Alvarez, J., Tilche, A. and Cecchi, F.), pp. 190–195. Gràfiques 92, S.A., Barcelona.

Callaghan, F.J., Wase, D.J., Thayanithy, K. and Forster, C.F. (1999). Co-digestion of waste organic solids: batch studies. *Bioresource Technology* **67**(2), 117–122.

Cecchi, F., Traverso, P.G., Mata-Alvarez, J., Clancy, J. and Zaror, C. (1988). State of the art of R and D in the anaerobic digestion process of municipal solid waste in Europe. *Biomass* **16**(4), 257–284.

Cecchi, F., Vallini, G., Pavan, P., Bassetti, A. and Mata-Alvarez, J. (1993). Management of macroalgae from the Venice lagoon through anaerobic co-digestion and co-composting with municipal solid waste (MSW). *Water Science and Technology* **27**(2), 159–168.

Danish Energy Agency (1995). Progress Report on the Economy of Centralized Biogas Plants. The Biomass Section of the Danish Energy Agency.

De Baere, L. (2000). Anaerobic digestion of solid waste: state-of-the-art. *Water Science and Technology* **41**(3), 283–290.

Demirekler, E. and Anderson, G.K. (1998). Effect of sewage sludge addition on the startup of the anaerobic digestion of OFMSW. *Environmental Technology* **19**(8), 837–843.

Diaz, L.F., Savage, G.M., Trezek, G.J. and Golueke, C.G. (1981). Biogasification of municipal solid wastes. *Transactions of the ASME.Journal of Energy Resources Technology* **103**(2), 180–185.

Gavala, H.N., Skiadas, I.V. and Lyberatos, G. (1999). On the performance of a centralised digestion facility receiving seasonal agroindustrial wastewaters. *Water Science And Technology* **40**(1), 339–346.

Hamzawi, N., Kennedy, K.J. and McLean, D.D. (1998). Technical feasibility of anaerobic co-digestion of sewage sludge and municipal solid waste. *Environmental Technology* **19**(10), 993–1003.

Hedegaard, M. and Jaensch, V. (1999). Anaerobic co-digestion of urban and rural wastes. *Renewable Energy* **16**, 1064–1069.

Hill, D.T. (1982). A comprehensive dynamic model for simulation of animal waste digestion. *Trans ASAE* **25**, 1374–1380.

Kayhanian, M. and Hardy, S. (1994). The impact of 4 design parameters on the performance of a high-solids anaerobic digestion process of municipal solid waste for fuel gas production. *Environmental Technology* **15**(6), 557–567.

Kayhanian, M. and Rich, D. (1995). Pilot-scale high solids thermophilic anaerobic digestion of municipal solid waste with an emphasis on nutrient requirements. *Biomass and Bioenergy* **8**(6), 433–444.

Kayhanian, M. and Rich, D. (1996). Sludge management using the biodegradable organic fraction of municipal solid waste as a primary substrate. *Water Environment Research* **68**(2), 240–252.

Kayhanian, M. and Tchobanoglous, G. (1992). Computation of C:N ratios for various organic fractions. *Biocycle* **33**(5), 58–60.

Kjølholt, J., Thomsen, C. D., and Hansen, E. (1998). Cadmium og DEHP i compost og bioafgasset materiale. Miljøprojekt Nr. 385, Danish Environmental Protection Agency.

Kübler, H., Hoppenheidt, K., Hirsch, P., Kottmair, A., Nimmrichter, R., Nordsieck, H., Mücke, W. and Swerev, M. (2000). Full scale co-digestion of organic waste. *Water Science and Technology* **41**(3), 195–202.

Mata-Alvarez, J., Mace, S. and Llabres, P. (2000). Anaerobic digestion of organic solid wastes. An overview of research achievements and perspectives. *Bioresource Technology* **74**(1), 3–16.

Mathrani, I.M., Johansen, K., and Ahring, B.K. (1994). Experiences with thermophilic anaerobic digestion of manure, organic industrial and household waste at the large scale biogas plant in Vegger, Denmark. In *Proceedings of the 7th International Symposium on Anaerobic Digestion, Cape Town, South Africa, January 23–27* (ed. IAWQ Specialist Group on Anaerobic Digestion, IAWQ Southern African National Committee, Anaerobic Processes Division of the Water Institute of Southern Africa), pp. 365–374. RSA Litho Ltd., Howard Place, South Africa.

Poggi-Varaldo, H.M., Valdés, L., Esparza-García, F. and Fernández-Villagómez, G. (1997). Solid substrate anaerobic co-digestion of paper mill sludge, biosolids, and municipal solid waste. *Water Science and Technology* **35**(2–3), 197–204.

Richard, T.L. and Woodbury, P.B. (1992). The impact of separation on heavy metal contaminants in municipal solid waste composts. *Biomass and Bioenergy* **3**(3–4), 195–211.

Rintala, J.A. and Järvinen, K.T. (1996). Full-scale mesophilic anaerobic co-digestion of municipal solid waste and sewage sludge: methane production characteristics. *Waste Management and Research* **14**(2), 163–170.

Rivard, C.J., Vinzant, T.B., Adney, W.S., Grohmann, K. and Himmel, M.E. (1990). Anaerobic digestibility of 2 processes municipal solid waste materials. *Biomass* **23**(3), 201–214.

Six, W. and De Baere, L. (1992). Dry anaerobic conversion of municipal solid waste by means of the Dranco process. *Water Science and Technology* **25**(7), 295–300.

Skøtt, T. (1997). Husholdningsaffald blandes med slam. *Dansk Bioenergi* Juni 1997. Danish Energy Agency, pp. 12–13.

8

Pretreatments for the enhancement of anaerobic digestion of solid wastes

J.P. Delgenès, V. Penaud and R. Moletta

8.1. INTRODUCTION

Anaerobic digestion is a common process for the treatment of insoluble organic matter or effluents with high COD content. With effluents, the methanogenic reaction is usually considered as the rate-limiting step of the overall process. When considering particulate substrates like solid wastes, both accessibility of hydrolytic microorganisms to the solid matter and hydrolysis of complex polymeric components constitute the rate-limiting step (Eastman and Ferguson, 1981). Therefore, one way of improving performance of digesters treating solid wastes is reduction in the size of the particles: thus, pretreatment of the substrate by mechanical disintegration should have positive effects on the anaerobic biodegradability of the substrate, through an increase of the available specific surface to the medium. The other way of improving performance is to promote

 Biomethanization of the organic fraction of municipal solid wastes. Edited by J. Mata-Alvarez. ISBN: 1 900222 14 0

hydrolysis of organic matter by a pretreatment of the substrate. Such pretreatments, breaking the polymer chains into soluble components, can be mainly biological or physico-chemical.

Whatever the pretreatment may be, the objectives are to obtain an extension and an acceleration of the anaerobic process, an increased amount of biogas as well as a reduction of the amount of anaerobic sludge and of the digestion time.

8.2. MECHANICAL PRETREATMENTS

The performance of digesters operating on solid wastes is dependent on the particle size of influent. Hills and Nakano (1984), working on tomato waste chopped to particles sizes from 1.3 to 20 mm, reported that biogas production rate and solid reduction are inversely proportional to the substrate's average particle diameter. Similar results were obtained by Sharma et al. (1988) with agricultural and forest residues. Therefore, size reduction of the particles, and the resulting increase of the available specific surface, represents an option for increasing degradation yields and accelerating the digestion process.

Several size reduction processes ranging from comminution to cell disintegration were investigated as pretreatments to enhance anaerobic biodegradation of solid materials.

The effect of comminution of organic materials using various machines on anaerobic biodegradability was investigated by Palmowski and Muller (1999) (Table 8.1).

Results demonstrate that both biogas production and reduction of the technical digestion time were increased by comminution, for all substrates, particularly for those of low biodegradability like leaves, seeds and hay stems.

Mechanical size reduction was also found to be efficient for enhancing the biogas potential from fibre-rich wastes like manure. Hartmann et al. (1999) compared the effect on biogas production, of maceration of fibres in manure feedstocks at five different biogas plants. The macerators installed between the storage tank and the digester worked by the same principle : three to four knife blades rotating (rotation speed ranging from 300 to 900 rpm) on one side of a cutter screen inside the manure stream. It was demonstrated that maceration of the whole feed has the effect of increasing the biogas yield to the order of 5–25%, and that the anaerobic biodegradability of the fibres is rather enhanced by shearing which is not necessarily reflected by a change in fibre size. According to the authors, the low costs of pretreatment by maceration (energy requirements of 0.1–1.3 kWh per m^3 manure) make this method attractive for increasing biogas production from manure and probably from other kinds of waste with a high content of particulate matter and with a consistency that means they can be pumped.

Table 8.1. Effect of comminution of several organic materials on their anaerobic biodegradability (from Palmowski and Muller, 1999).

Substrate	State of the samples	Comminution treatment	Improvement of biogas production	Reduction of technical digestion time
Mixture of potatoes, apple and carrots	2.2.2 cm pieces	Rubbed with a grater	—	50 %
Meat	2.2.2 cm pieces	Ground with a kitchen machine by cutting stress	2 %	23 %
Sunflower seeds	5 mm pieces	Ground with a kitchen machine by cutting stress	19 %	45 %
Maple leaves	2.2 cm pieces	Ground with a flour mill by shear stress	14.4 %	59 %
Hay stems	1.5 cm pieces in water suspended	Ground in a stirred ball mill with water	18 %	52 %

Technical digestion time = time to reach 80 % of the maximal biogas production.

Mechanical cell disintegration is a well-known process used in many biotechnological applications to obtain intracellular products, such as proteins or enzymes, (Schwedes and Bunge, 1992). Some authors have proposed using such mechanical treatment as a pre-treatment before biological anaerobic processes treating sludge (Kopp et al., 1997, Müller et al., 1998, Müller and Pelletier, 1998). The objective of such pretreatment is to accelerate the digestion of sludge, to raise the degree of degradation and, thus, to decrease the amount of sludge requiring disposal. The main fraction of sludge consists of cellular materials which often withstand direct anaerobic degradation because cell walls act as a physical barrier to enzymes from hydrolytic microorganisms. The availability of intracellular organic material from cells could be increased by mechanical cell disruption processes. Furthermore, the objective of a cell disruption process is to destroy the floc structure and to fragment the cell walls into low size particles which become more readily available to biological degradation.

Several cell disruption processes were tried out on sewage sludge to improve its anaerobic biodegradability (Muller et al., 1998, Muller and Pelletier, 1998; Baier and Schmidheiny, 1997; Engelhart et al., 1999). In this field, comparative results are reported by Muller et al. (1998) who tested at a laboratory scale four methods of mechanical cell disintegration using sewage sludge with SS content of 1–4% and VSS of 70% of SS. The processes tested were as follows.

- A stirred ball mill which consists of a cylindrical chamber into which the agitator discs projects beads of glass. In this process cell disruption occurs through shear forces and compression loading between the grinding agents. The tested mill operated continuously with a flow of 10 l/hour and an agitation tangential speed ranging from 2 to 10 m/s.
- A high pressure homogenizer in which the key component is the homogenizing valve through which the cell suspension is forced at high pressure (400–900 bars). In this process cell disruption takes place in the narrow homogenizing gap through cavitation and turbulence mechanisms induced by the decrease in pressure.
- An ultrasonic homogenizer where the energy necessary for the disruption is brought into the cell suspension by an oscillator system in the form of acoustic waves. The high-energy eddies resulting from the cavitation mechanism create shear forces which disrupt cell walls.
- A shear gap homogenizer in which a cylindrical rotor turns with high circumferential speed (in the range of 5,000–24,000 rpm) concentrically inside a stator. The resulting shear forces lead to the disruption of cells.

Using the rate of oxygen demand and the COD release as parameters for measuring the degree of cell disruption, it was shown that disintegration rates near 90% were obtained in optimal operating conditions with all tested methods except when using the shear gap homogenizer. Considering the specific energy consumption, the high pressure homogenizer and the stirred ball mill were the most economic processes. Furthermore, the degree of digestion of disintegrated sludges was between 10 and 20% higher in comparison to the untreated sludges (Muller et al., 1998).

The positive effect of mechanical cell disintegration on anaerobic digestion of sludges has been illustrated by other authors. Thiem et al. (1997) investigated the effect of ultrasound pretreatment on sludge biodegradability using ultrasound at a frequency of 31 kHz and high acoustic intensities. They showed in digesters operated with an identical residence time of 22 days that reduction of volatile solid was 45.8% for untreated sludge and 50.3% for disintegrated sludge. Furthermore, reduction of volatile solids was 44.3 % in a digester operated with disintegrated sludge and a residence time of 8 days. Engelhart et al. (1999) showed that mechanical pretreatment of sludge using a high pressure homogenizer lead to an acceleration in the reduction of volatile solid in fixed film digesters. Volatile solids reduction of about 40% was reached with hydraulic residence times down to 5 days without process failure.

Chiu et al. (1997) studied the effect of alkaline and ultrasonic pretreatment of sludge before anaerobic production of volatile fatty acids. It was shown that combination of alkaline treatment and ultrasonic treatment released from 78 to 89% of total COD in the form soluble COD against 36% with alkaline treatment alone. Furthermore, simultaneous ultrasonic and alkaline pretreatment enhanced the production of volatile fatty acids: VFA:COD ratios from 66 to 84% were obtained in these conditions whereas the ratios were 10% and 30% with the raw sludge and the alkaline-pretreated sludge respectively.

In another way, Rivard and Nagle (1996) used a combination of thermal and mechanical pretreatment based on the synergistic of mechanical action with thermal input, to disrupt the macrostructure of dewatered sewage sludge and subsequently to increase its biodegradability. First, they developed a thermal sonication; optimal pretreatment performances were obtained with sludge solids of 1%, treatment times between 4 and 8 min, and a temperature of 55 °C. In these conditions, the maximum enhancement of anaerobic digestion potential for the tested substrate sonication pretreatment was 80–83% as measured by release of soluble COD. Secondly, they proposed a combination of thermal pretreatment and shear using an Ultra-Turrax; The optimum shear pretreatment occurred with sludge solids of 1–2%, treatment times of 6–10 min and a temperature of 87 °C. In these conditions, the enhancement of anaerobic digestion potential from sludge pretreated by shear forces, was 88–90%, as measured by release of soluble COD.

Dohanyos et al. (1997) proposed the mechanical destruction of cells in activated sludge by a specially adapted thickening centrifuge. The aim of this method is the partial destruction of cells during the thickening with the centrifuge, with the exploitation of dissipated kinetic energy generated by the centrifuge. As a consequence, no additional energy is required. A special impact gear was incorporated in the thickening centrifuge and placed at the end of centrifuge where the thickened sludge leaves the bowl. They report that the improvement of methane yield from thickened activated sludge, in comparison with untreated activated sludge, was 84.6%. The improvement of methane yield is strongly influenced by the input sludge quality.

8.3. BIOLOGICAL PRETREATMENTS

8.3.1. Enzymatic

Solid wastes constitute a complex mixture which contains mainly carbohydrates, lipids and proteins, the relative proportions being dependent on the nature of the wastes (Table 8.2).

To use particulate organics as a substrate for biogas production, they must be hydrolysed. This process is carried out by extracellular hydrolytic enzymes. Briefly, hydrolysis of proteins require protease and peptidase which break proteins into peptides and aminoacids. Hydrolysis of lipids leads to the production of glycerol and long-chain fatty acids via the action of lipases.

Cellulose, the most abundant carbohydrate in waste, is hydrolysed by a complex mixture of enzymes including endoglucanase, cellobiohydrolase and β-glucosidase, which act in concert to produce glucose.

Table 8.2. Proportion of organic matter in wastes commonly treated by anaerobic digestion.

Substrate	Proteins	Lipids	Carbohydrates	References
Primary sludge (%MES)	21%	17%	57%	Elefsiniotis and Oldham, 1994
Domestic sludge (%MVS)	28%	28%	33%	Chynoweth and Mah, 1971
Organic fraction of municipal waste (%MES)	12%	44%	44%	Peres et al., 1992
Industrial sludge (%COD)	63%	18%	16%	Penaud et al., 1998
Food waste (%TS)	6.5%	0.6%	78%	Raynal et al., 1998

The rate of anaerobic hydrolysis is a function of various parameters: pH, temperature, hydraulic residence time, the microbial population and especially the source of hydrolytic enzymes. Considering the last parameter, some authors have investigated the addition of hydrolytic enzymes as a pretreatment as designed to increase the yield and the rate of particulate matter solubilisation during anaerobic digestion of solid waste, particularly wastewater sludge or cellulose-rich materials.

In this field, Scheidat et al. (1999) showed the positive effect of the addition of hydrolytic enzymes on the solubilisation of municipal primary sludge. Various commercial enzymes, such as peptidase, carbohydrolase and lipase, were tested in a mixture, the amount of enzymes added being in the range of 1–10% related to the weight of total solids. It was shown that addition of the enzyme mixture at an incubation temperature of 39 °C led to a significant increase in the soluble COD, especially with larger additions of enzyme. In these conditions, measured COD content in the filtrate increased to 18 g/l compared with 11 g/l in the untreated sludge. Furthermore, the structure of enzymatically treated sludge changed from a very viscous material to a more liquified one with better solid / liquid separation properties. Experiments run with separate enzymes showed that carbohydrolases were mainly responsible for the observed structure modifications, through cleavage of cellulose fibres. The positive effect of cellulolytic enzyme addition on the reduction of both cellulose content and viscosity in primary municipal sludge was also reported by Knapp and Howell (1978) and Cinq-Mars and Howell (1977) using a cellulase preparation produced by a submerged culture of *Trichoderma viride* grown on a defined media based on Solka floc. Furthermore, it was shown by the latter authors that the use of primary municipal sludge as a growth medium for

Trichoderma viride is suitable for the production of cellulase. It was possible with nitrogen supplements to obtain an enzyme with a Filter Paper Activity of 2 compared with 4–6 with a growth medium based on Solka floc.

There are other studies on the screening or testing of enzymatic preparations for the solubilisation of solid waste before anaerobic digestion. Aoki and Kawase (1991) developed a process, run during the acidogenic stage of the sludge, consisting of an enzymatic pre-treatment based on the addition of commercial proteases, and obtained 58% total solid reduction compared with 49% without enzyme addition. Another example of enzymatic pretreatment is proposed by Rintala and Ahring (1994) who observed that the biogas production obtained from biological sludge was increased when proteases were added (but not at all with xylanases or lipases). Clanet et al. (1988) tested cellulase from Trichoderma reseii and Penicillium sp. for the saccharification of municipal waste cleared of glass and metal components. With an enzyme / substrate ratio of 10 mg protein / g dry matter, enzymes of T. reseii and Penicillium sp. degraded 45% and 25% of total sugars respectively, after an incubation time of 48 hours at 5 °C. With enzymes/substrate ratio varying from 0.3 to 10, a linear correlation was observed between enzyme activity and sugar yield. Furthermore, many studies have been devoted to researching microorganisms able to produce hydrolytic enzymes useful for the solubilization of solid wastes: cellulolytic activities (Clanet et al., 1988; Knapp and Howell, 1978; Gijzen et al., 1988; Cailliez et al., 1993; Thomas et al., 1993; Haltrich et al., 1994), proteolytic activities (Ensign and Wolfe, 1965; Murao et al., 1976; Arai and Murao, 1978; Uehara and Hagesawa-Iwai, 1979; Mac Kelly et al., 1993). The main drawback of a such approach is that enzyme doses that have to be added to obtain significant improvements are important. Perot (1989), studying sludge hydrolysis, reported that only important doses of cellulases resulted in a significant improvement of biodegradability. Similar observations were reported by Clanet et al. (1988) for the enzymatic pretreatment of municipal wastes. Economic feasibility is an important parameter that has to be considered when such pretreatment is proposed. Interesting results were reported by Radermacher et al. (1999) about a full-scale application of enzyme-supported digestion of municipal sludge. At the wastewater treatment plant of Aachen-Soers (430,000 IE), which comprises two sludge digestion units of 10,000 m3 capacity each, an enzyme complex including mainly cellulase was added continuously to one of the digestion units, at enzyme doses of 500–700 mg/kg. Comparison with the control unit shows that enzyme addition improves the digestion yield, resulting in an additional solid removal of 2 tons/day and a higher biogas production of 840 m3/day. This procedure led to a significant reduction of the sludge treatment costs. Based on sludge disposal costs of 280.19 EU/t dry solid, and a 9,245 EU/m3 biogas benefit for the Aachen-Soers plant, a net annual cost reduction of approximately 175,100 EU/year has been calculated by the authors after one year of experiment.

Another approach tested for enzymatic pretreatment of solid wastes is the use of hydrolytic microorganisms or hydrolytic ecosystems. In this field, Del Borghi et al. (1999) tested the aptitude of a consortium of hydrolytic bacteria to solubilize a mixture of sewage sludge and organic fraction of municipal solid waste. The bacterial consortium was isolated from activated sludge and contained strains of *Bacillus cereus* (38%), *Arthrobacter* sp. (9.5%), *Bacillus polymixa* (52%) and *Micrococcus* (0.15%). During the pretreatment, volatile suspended solids significantly decreased while soluble COD increased as the result of the bacterial hydrolysis of polymeric materials. Bacterial hydrolysis was performed throughout 25 days at 25 °C and pH = 7 by inoculating a 6% feed mixture with 1g/l of selected hydrolytic bacteria. In these conditions, a VSS reduction from 45 to 8 g/l was observed with a soluble COD increase from 18 to 32 g/l. In another study, Gijzen et al. (1988) developed a two-stage anaerobic digestion process based on the utilization during the first phase of a mixed population of rumen bacteria and ciliates for the hydrolysis and the production of volatile fatty acids from cellulose rich substrate. It is known that a cellulolytic ecosystem is present in the rumen of ruminants and that several of the bacteria and ciliates of this system exhibit high levels of cellulase activity, even on crystalline cellulose. The application of rumen microorganisms for an enhanced degradation of cellulose residues was successful: using filter paper cellulose supplemented with ground alfalfa hay as substrate, it was shown that the degradation of the substrate was almost complete at loading rates in the hydrolysing/acidogenic reactor of from 11.9 to 25.8 g volatile solid/l/day. Furthermore, up to a loading rate of 21 gVS/l/day, the acidogenic effluent was almost completely converted into biogas by the action of the UASB methanogenic reactor: total methane production of both reactors yielded up to 98% of the theoretical maximal value after prolonged periods of operation. Another interesting result is that the high rate of cellulose hydrolysis and acidification of cellulose was correlated with an increase of the specific microbial biomass in the reactor, although a dynamic change among the different groups of ciliates was observed.

8.3.2. Aerobic

Aerobic microbial processes can be proposed as a pre-treatment stage before conventional anaerobic digestion. Various aerobic pre-treatment methods are available.

8.3.2.1. Composting

Composting is a bio-oxidative process involving the mineralization and partial humification of the organic matter. The solubilization and transformation of organic matter by the microorganisms initially involves the release in the liquid film around the solid matter of carbohydrates, amino and volatile acids which are then biodegraded, producing mainly CO_2, H_2O and heat. The pretreatment of solid waste in a composting stage is proposed by Capela et al. (1999) to prevent acidification during anaerobic digestion, and particularly during dry anaerobic digestion. The objective of this pretreatment is to obtain a pre-degradation of volatile solids and therefore to decrease the inhibition caused by the build up of volatile fatty acids on the methanogenic activities. With primary sludges from a bleached kraft pulp mill as substrate, it was demonstrated that the composting pretreatment improved overall anaerobic digestion performance. As an illustration, in the anaerobic digestion of un-pretreated substrate, the rate of volatile solid degradation was 34% only after 49 days of operation, whereas the degradation rates were up by to 50% in the anaerobic digestion of pretreated substrate. Furthermore, the effect of composting on the digestion performances was dependent on the degree of composting, i.e. the level of volatile solid reduction by composting. A degree of composting of 35% is efficient for low operation times of anaerobic digestion, leading to the highest volatile solids degradation (volatile solid reduction of 50% for an operation time of 10 days). For high operation times (49 days), a degree of composting of 10% provides the highest rates of both volatile solid reduction (70%) and methane production (0.047 m3/kgVS), although with a slow start-up (Capela et al., 1999). According to the authors, the adequate level of pretreatment for optimization of the combined treatment process depends on a balance between operation time and the objective of the treatment: either high degradation or high biogas production rates for profitable use. They recommended a high degree of composting when the objective of the treatment is mainly reduction of solids with low operation times, while a low level of pre-treatment is more suitable when there are no limitations on anaerobic digester volume and where the objective is to maintain high methane production rates.

8.3.2.2. Aerobic thermophilic digestion

Aerobic thermophilic processes have been found of interest in the treatment of some solid wastes, and particularly of municipal sludges, because they are effective in the reduction of pathogenic organisms, the bio-oxidation of sorbed pollutants removed with solids during primary treatment which are largely recalcitrant to anaerobic degradation, and the solubilization of particulate biodegradable matter. Considering the last aspect, Hamer and Mason (1987) proposed a sequence of events for the solubilization and biodegradation of whole microbial cells by an aerobic thermophilic population, using pressed baker's yeasts as model substrate: (1) initial degradation of cell-wall polymers resulting in the release of soluble cell compounds; (2) accumulation of carboxylic acids; (3) enhancement of thermophilic microbe activity as a result of the utilisation of available soluble substrate; (4) exhaustion of the preferred substrate followed by utilization of low molecular weight carboxylic acids; (5) hydrolysis of remaining cell-wall fragments and utilization of higher molecular weight carboxylic acids.

However, complete aerobic thermophilic treatment is expensive due to the large oxygen demand. It is also capital intensive to build reactors with materials capable of minimizing heat loss. For these reasons, some authors have proposed aerobic thermophilic digestion as a pretreatment stage before conventional anaerobic mesophilic digestion. In this two-stage system, aerobic thermophilic digestion pretreats the sludge to achieve pathogen reduction, hydrolysis and acetogenesis of particulate organic matter into organic acids before the subsequent mesophilic digestion which provides the volatile solid reduction, biogas production and additional pathogen reduction.

In this field several studies have investigated the effect of operating parameters on the solubilization and biodegradation of microbial cells in aerobic thermophilic processes when used for pretreatment. Results of the comparison of oxygen excess and oxygen limitations on the solubilisation/biodegradation of whole yeast cells under thermophilic conditions have demonstrated that aerobic thermophilic pretreatment is most effective in reducing solids and in increasing dissolved organic carbon, when dissolved oxygen concentrations are low (Mason et al., 1987). Their results indicated that enhanced exo-cellular enzyme production occurred under conditions of limited oxygen, and that it was very effective both in terms of the amount of solids removed and in the rate of removal. Furthermore, under oxygen limited conditions, the total suspended solids removal rate was optimal at a temperature of 65 °C and for a residence time of 0.6–1 day. These results are close to those found for municipal sludge by Cheubarn and Pagilla (1999). Using as a substrate, a mixture of thickened waste activated sludge and primary sludge, these authors found that volatile solid reduction by aerobic thermophilic pretreatment increased from 25% to 40% when sludge residence time was increased from 0.6 to 1.5 day and that a

5% increase in volatile solid reduction was obtained at each of the SRT values of 0.6; 1 and 1.5 days when the temperature was increased from 55 °C to 65 °C. Temperature is a key parameter in the aerobic thermophilic digestion of sludge, due to the temperature dependence of enzymatic activities. Bomio et al. (1989) demonstrated that proteolytic activity was the main enzymatic activity in aerobic thermophilic sewage sludge digestion, and was temperature dependent. The proteolytic activity from sludge solubilising thermophilic bacteria was highest at temperatures ranging from 65 °C to 85 °C, with an optimal at 80 °C. In another study, Hasegawa and Katsura (1999) reported that the growth temperature of *Bacillus stearothermophilus*, an active protease-secreting bacteria isolated from sludge-solubilizing thermophilic bacteria, was 55–75 °C.

The ability of aerobic thermophilic pretreatment to produce a digested substrate that can be treated in mesophilic anaerobic digestion with enhanced volatile solid reduction and higher biogas production has been reported by several authors, though with some contradictions. Hasegawa and Katsura (1999) found that the organic sludge that was solubilised under slightly aerobic thermophilic conditions generated 1.5 times as much biogas as the untreated sludge. These results are in accordance with those of Baier and Zwiefelhofer (1991). By contrast, Pagilla et al. (1996) reported that when aerobic thermophilic pretreated mixed sludge is fed to the mesophilic anaerobic digesters (ATP system), the biogas production rate (0.761 m^3/kg VSS destroyed) was lower than the control anaerobic digester sludge biogas production rate (0.918 m^3/kgVSS destroyed). However, the average methane content (66%) of the ATP system biogas was greater than the average methane content of the control biogas (60%). Furthermore, the ATP system produced digested sludge with better dewaterability and a sludge supernatant with lower COD concentration (35% reduction) compared with those from the control. Ward et al. (1998) reported that autothermal treatment of mixed sludge (1 day SRT at 65 °C) did not significantly improve their mesophilic anaerobic digestion at SRTs of 4–14 days. The volatile solid destruction rate was 2.7–2.8% per day with or without pretreatment. Moreover, the effect of autothermal treatment on the yield of methane production was slight at tested SRT of 6–14 days, and was significant at the low SRT of 4 days. According to these authors, the major impact of autothermal pretreatment is to make anaerobic digester operations more stable, in relation with lower volatile fatty acids to alkalinity ratio.

8.4. PHYSICO-CHEMICAL PRETREATMENTS

8.4.1. Oxidative processes

8.4.1.1. Wet oxidation process

Wet air oxidation is a physico-chemical process whose basic principle is to enhance the contact between molecular oxygen and the organic matter to be oxidized. High temperature conditions (from 200 to 370 °C) convert some organic matter to carbon dioxide and water, while the liquid phase is maintained by high pressure (from 50 to 150 bars) which also increases the concentration of dissolved oxygen and thus the oxidation rate. Wet air oxidation provides an efficient method for either partial or total destruction of various classes of bioresistant organic pollutants and micropollutants from wastewater, such as phenol and substituted phenol, nitrogenous compounds and carboxylic acids (Jogeklar et al., 1991; Mantzavinos et al., 1997). Some authors have examined the potential for using wet oxidation processes to treat municipal sludge with the objective of either reducing the quantity to be disposed of through a solid/liquid separation, or of increasing the efficiency of biogas recovery in anaerobic treatment. Khan et al. (1999) reported that wet oxidation was able to greatly reduce the levels of total solids and volatile total solids present in gravity-thickened activated sludge. In solid reduction, temperature was the most significant parameter, more than retention time or oxygen overpressure. In the temperature range 200 to 300 °C, solid reduction increased with temperature: at the lower temperature of 200 °C total solid and volatile total solid destruction was 20 and 40% respectively, compared with 65 and 90% at 300 °C. But maximum soluble COD was generated at lower temperatures. This would be desirable either for the further biological treatment of sludge or for the treatment of wastewater, such as biological nutrient removal. At 200 °C and for retention times of 10 and 30 min, the generation of soluble COD was 900 and 400% respectively, whereas at 300 °C the value was nearly 100%, independent of the retention time. The soluble COD produced was mainly composed of low molecular weight fatty acids and predominantly acetic acid. In addition to volatile fatty acids, ammonia was a major intermediate and, to a lesser extent, formaldehyde.

Song et al. (1992) studied the combination of catalytic wet oxidation and anaerobic digestion for the treatment of sewage sludge. Catalytic wet oxidation was performed in the following conditions: temperature of 270 °C, pressure of 86 bars, stoichiometric air ratio of 1:1 and reaction time of 24 min. The resulting supernatant of the pretreated sludge was 7200 mgCOD/l which contained 53% carbonic acids. Anaerobic digestion of the supernatant by a UASB process achieved approximately 93% removal of COD with a volumetric loading rate from 1.8 to 14.4 kg COD/m^3/day. The biogas produced had a high methane concentration: it consisted of 84–90% CH_4, 10–12% CO_2 and 2–4% N_2. These results suggest that the use of CWPO as a pretreatment for methane recovery may be technically attractive since in conventional anaerobic processes, the COD recovery from sewage sludge is around 50%, with a methane concentration in the biogas around 65%.

8.4.1.2. Ozonation

Ozone, a strong oxidizing agent, can be used for its oxidative properties in the treatment of wastewaters or solid wastes. Industrial ozone production is obtained by the electronic activation of oxygen. Ozonised air is passed through water in the form of fine bubbles. The amount of ozone which transfers into solution depends on the flow of ozonised gas through the water and on the size of the bubbles.

Ozone has two main modes of action on solid matter in water (Brière et al., 1994):

- a direct action of ozone itself on the solids (selective action),
- an indirect action on the solids by radicals formed when ozone is destroyed in water.

Direct or selective ozone action is directed towards unsaturated bonds (dipolar action), high ionic density sites (electrophile action). Radical action is enhanced in the following conditions: presence of hydroxyl ions or H_2O_2, a rise in pH.

Ozonation efficiency is, however, dependent on the nature of the pollutant, on the ozone dose and on the pH reaction.

The ozonation of biorefractory aromatic compounds (o-cresol, benzenesulphonic acid, etc.) leads to biodegradation down to methane. The biodegradibility improves with increasing doses of ozone or with an increase in the extent of ozonation (Wang, 1990). In general, large doses of ozone are needed to obtain biodegradable products. However, low ozone doses may reduce the effect of toxic compounds on methanogenic cultures.

pH has an impact on COD solubilization yields. At very low pH, ozone is the main oxidant (molecular form) and it attacks insoluble fractions. At high pH, radical action is predominant. The fraction of insoluble organic carbon destroyed is higher at very acid pH than at higher pH. Oxidation action of molecular ozone is more aggressive towards suspended than soluble solids. But ozonation products obtained from biorefractory or toxic aromatic compounds in the basic pH range are more biodegradable and less inhibitory than those formed under the same ozone dose in the acid pH range.

Ozone was applied to secondary effluent treatment to decrease the refractory COD content and to limit the effect of bulking and foaming (ozone coagulates the bacterial membrane). Ozone was also effective in the reduction of colour and for disinfecting sewage. Ozone could also be used for particulate matter reduction. Legube et al. (1987) working on secondary and industrial sludges, showed that ozone dosages of 6 to 12 mg/l led to a reduction of faecal bacteria, COD and nitrite ion concentration and to an increase of the concentration of free amino-acids (this increase could represent the attack by ozone on proteins and polypeptides). They found also that a high proportion of aromatics (alkyl toluene, dichlorobenzene) simple unsaturated compounds (furfural) and saturated acids (methylesters) disappeared with ozonation. On the other hand, there were appearances of alcohols, aldehydes and short chain fatty acids (methyl esters). In another study, Watt et al. (1985) observed that ozonation reduced the fraction of organic matter with high molecular weight. In addition, mild ozone doses can provoke cleavage of larger molecules into smaller ones. Such changes led to better biological post-degradation performances for organic matter. Paul et al. (1998) showed that ozone could be very efficient for minimization of sludge production. After a complete ozonation of the sludge, 91% of the particulate carbon was modified, with more than 50% found in a soluble form and 35% mineralized. The VSS/SS ratio decreased from 86% to 35% evidencing sludge mineralization. The initial rate of ozone consumption by the sludge was high (estimated value of 30 $mgO_3/gVSS.min$). More than 50% of the carbon obtained after ozonation was found to be readily biodegradable using a short-term BOD procedure.

Sludge reduction by ozonation could be applied either directly to the activated sludge system, or to pre-treat the sludge before anaerobic digestion. Concerning the first approach, a sludge production minimisation system has been proposed where part of the activated sludge to be recirculated is treated with ozone. A positive effect on sludge production was observed either at both laboratory and industrial scales, with different wastewaters, either domestic or agri-industrial. However, the sludge reduction percentages reported were very different from one case to another: Yasui et al. (1996) claim zero sludge production, while Kamiya and Hirotsuji (1997) using sequenced ozonation, report a 50% reduction with lower O3 dosages. Concerning the second approach, Argeliers (1997) tested ozonation of sludge before anaerobic digestion. With thickened waste activated sludge with total solids of 9% (w/w) it was found that ozonation had a low effect on particulate COD solubilization which remained comparable to the control, whatever the ozone dose (for the tested range from 0 to 3.4 g of O3 /l). However, ozonated sludge showed enhanced anaerobic digestibility. At an ozone dose of 3.4 g /l, biogas production and anaerobic biodegradability were increased by 44% and 60% respectively over the control.

However, ozone pre-treatment can generate toxicity problems caused either by the formation of toxic by-products or by high ozone concentrations (Watt et al., 1985). Toxic compounds originate from ozone reactions with some organic compounds such as organic peroxides, H_2O_2, low molecular-weight alcohol, some carboxylic acids, aldehyde (Gilbert, 1983 ; Moerman et al., 1994). When inhibitory compounds are formed or when large ozone doses have to be used, acclimatization of the sludge is necessary. As an illustration, Narkis and Schneider-Rotel (1980) reported that working with adapted populations allowed to work at higher ozone doses and led to best solubilization results.

8.4.2. Chemical

Mild treatment with chemical agents has been widely used with lignocellulosic substrates for the production of biofuels or chemical (Millet et al., 1975; Datta 1981; Pavlostathis and Gosset, 1985). In this field, chemical pretreatment at ambient temperatures has been proposed using acids or alkalis and it is apparent that the latter are more compatible with the anaerobic digestion process, compared to the former, since this bioconversion generally requires an adjustment of pH by increasing alkalinity (Pavlostathis and Gosset, 1985). Pretreatment of lignocellulosic materials with dilute alkali leads to an important chemical reaction which consists of a saponification of esters of uronic acid associated with xylan chains (Datta, 1981). The effect of saponification is a breaking of cross-linking. Consequently, there occurs a marked increased in the swelling capacity and pore size. This increase not only provides an increased diffusivity for the hydrolytic enzymes but also facilitates improved enzyme-substrate interactions. Hence, acidogenic bacteria can ferment the pretreated lignocellulose even though no delignification or cellulose hydrolysis occurs during the pretreatment (Datta, 1981).

Among alkaline agents, sodium hydroxide is commonly used for the pretreatment of lignocellulose. Pavlostathis and Gosset (1985) reported that pretreatment of wheat straw with sodium hydroxide at ambient temperature resulted in an improvement of its anaerobic biodegradability. The highest anaerobic conversion efficiency found was 80% for the substrate pretreated at 7.5% of total solid, with 50g NaOH / 100 g of total solid and with a treatment time of 24 hours; whereas the untreated wheat straw conversion efficiency was 34.3%. With corn stover as substrate, Datta (1981) studied the effect of mild pretreatment with dilute NaOH solution on the production of volatile fatty acids. It was reported that pretreatment at 25 °C with 1% NaOH solution (at 16 ml/g stover) resulted in a two-fold increase in conversion to volatile fatty acids.

Alkaline pre-treatment based on the addition of sodium hydroxide can be extended to other complex organic substrates, such as domestic sludges. Rajan et al., (1989) studied low-level alkaline solubilization of waste activated sludge at ambient temperature and found that it could solubilise over 45% of the particulate COD. They observed that solubilization occurred in two different phases: a first, rapid one corresponding to a rapid increase in solubilization, and a second one corresponding to a slower increase of solubilization, first order with respect to particulate COD. Furthermore, among the variables tested, it was shown that NaOH concentration and concentration of sludge solids influenced the hydrolysis yield to a significant extent. As an illustration, for a feed sludge of 1% total solid, particulate matter hydrolysis after 12 hours at 20 °C increased from 13% to 45% when NaOH concentration was increased from 10 to 40 meq/l. With a constant alkali level of 4 g NaOH / 100 g total solids, hydrolysis yield increased from 13 to 31% when feed solid concentration was increased from 0.5 to 3%.

It was shown by the same authors that such low level alkaline pretreatment improves performance of the subsequent anaerobic digestion (Ray et al., 1990). Alkaline solubilization of waste-activated sludge at 1% total solids, using 20 meq NaOH / l for 24 hours at room temperature improves volatile solid removal in the range of 25–35%, COD removal in the range of 30–75%, and gas production in the range of 29–115% compared with the un-pretreated sludge. It was reported that the pretreatment showed a more pronounced effect on the improvement in digester performances when the hydraulic residence time was lowered. Thus, the volatile solid and COD removal yields obtained with pretreated sludge at a 7.5 day retention time were similar to those obtained without pretreatment at a 20 day retention time. Moreover, for all runs, the reactors fed with pretreated sludge had a greater content of methane in biogas. These results are in accordance with those obtained by Lin et al. (1997) with the same substrate and using a similar solubilization procedure. Additionally, they showed that increasing the concentration of NaOH (factor two) did not significantly change the digestion performances. By contrast, increasing the concentration of sludge solids from 1 to 2% was more effective for the yield of volatile solid and COD removals and for gas production. Moreover, the dewaterability of digested sludge was improved in reactors fed with pretreated substrate.

Alkaline pretreatment was found to be efficient for enhancing solubilization and methane production from mixed waste-activated sludge or from industrial cell biomass. With combined waste-activated sludge from domestic, commercial and industrial wastewaters as substrate, Tanaka et al. (1997) reported that the volatile suspended solid were solubilized by 15% and the methane production increased by 50% over the control, when the substrate was pretreated at ambient temperature with a dose of 1 g NaOH / gVSS for one hour. With a industrial microbial biomass as substrate, Penaud et al. (1999) reported that maximum COD solubilization of 63% and total solid elimination of 33% were obtained when the substrate was pretreated with 4.6 g/l of NaOH at ambient temperature. The anaerobic biodegradability of the pretreated substrate reached 50%, compared with 17% for the control.

Whatever the nature of the substrate, the major variables involved in chemical pretreatment are solid concentrations, treatment time and alkali concentrations, the later being the most important parameter (Cochaux et al., 1996; Penaud et al., 1998). Efficiency of pretreatment is also dependent on the nature of the alkaline agents used. Datta (1981) reported that pretreatment of corn stover with a mixture of lime and sodium carbonate gives higher conversions yields of substrate into volatile fatty acids than pre-treatment with sodium hydroxide for the same level of alkali used, the beneficial effect being more pronounced when the initial solid concentration increased. The marked improvement in anaerobic digestibility when corn stover was pretreated with the alkali mixture may be correlated with the increased buffering capacity due to calcium salts (Datta, 1981). Rajan et al. (1989) reported that pretreatment of waste- activated sludge was more effective with sodium hydroxide than with lime. At 38 °C, when 30 meq NaOH / l were added to sludge of 1% total solids, 38% and 48% solubilization yields were achieved respectively after a treatment time of 12 hours, compared with 15 and 20% with lime. Furthermore, pretreatment using sodium hydroxide provides better anaerobic digestion performance than pretreatment based on the addition of lime (Ray et al., 1990). During alkaline pretreatment of an industrial cell biomass, Penaud et al. (1999) tested at ambient temperature various alkaline agents; NaOH, KOH, $Mg(OH)_2$ and $Ca(OH)_2$. It was shown that monobasic agents led to better COD solubilization percentages than dibasic ones: NaOH, KOH, $Mg(OH)_2$ and $Ca(OH)_2$ additions resulted in 60.4; 58.2; 29.1 and 30.7% COD solubilization yields, respectively. When considering anaerobic biodegradability of the pretreated substrate, no difference was observed with the four alkaline agents tested.

In every case potential toxicity problems must be considered in the chemical pretreatment of waste, and particularly the inhibition or the toxicity due to high ion concentration. The sodium cation (when NaOH is added) is known to be inhibitory to methanogenic flora (Kugelman and McCarty, 1965; Feijoo et al., 1995). However, the sodium concentration in an anaerobic digester fed with a substrate pretreated at low alkaline level is in general lower than the 0.3 M concentration considered to be the threshold for sodium ion toxicity. Furthermore, Penaud et al. (1999) reported that Cl anions (when KOH, Mg(OH)2, Ca(OH)2 are used for particulate COD solubilization at ambient temperature) have little inhibitory effect on anaerobic acetate degradation, at the level of alkali used in the pretreatment of substrate. These results suggest that some low biodegradability of alkaline-pretreated substrate should be attributed to the formation of refractory compounds under high pH conditions, such as aromatic substances deriving from lignin degradation, in the case of lignocellulosic substrate.

8.4.3. Thermal

Thermal treatment is usually used as a conditioning process for raw or digested sludges and improves the dewaterability properties of such wastes. Dote et al. (1993) when studying liquidisation of dewatered sludges at temperatures above 150–175 °C, considered that the liquidisation phenomenon was due to the rupture of the cells of the microorganisms making up the major part of the sewage sludge. Stuckey and McCarty (1978) also reported that the effect of thermal treatment appeared to be lysis of the cell and partial hydrolysis of organic compounds. However, the increase in biodegradability came from the insoluble as well as from the soluble fraction. Hence heat treatment alters the structure of the insoluble fraction to make it more amenable to biodegradability.

Major disadvantages of such a treatment are: odours produced, corrosion and organic fouling of heat exchange tubes, the high energy requirement, the frequent need of some form of treatment before treated liquor is recycled to the treatment plant. Various authors have proposed using thermal treatment as a pretreatment to anaerobic digestion of wastes, particularly municipal sludges. Potential advantages expected include improved dewaterability, increased biodegradability, a reduction in the strength of liquid streams, odour reduction during the digestion step, substrate sterilization and an improved energy balance. Haug et al. (1978, 1983) examined the effect of thermal pretreatment on the potential for increased methane production from different categories of sludges. They observed that thermal pretreatment of primary sludge at 175 °C had no significant effect on biogas yield. By contrast, thermal pretreatment of activated sludge resulted in an increase of 60% in methane yield with a 36% decrease in effluent volatile suspended solids from anaerobic digestion. For the two tested sludges, dewaterability after thermal pretreatment and anaerobic digestion was increased over the non-thermally pretreated sludges. These authors estimated energy balances for several thermal treatment – anaerobic digestion systems processing waste-activated sludge or mixtures of waste-activated sludge and primary sludge. They concluded that, compared with the conventional system in which anaerobic digestion preceded thermal treatment, thermal pretreatment before anaerobic digestion may result in an increase in net energy production from the system because of increased anaerobic biodegradability and reduced digester heating requirements. Another study reported the beneficial effect of the combined thermal pretreatment/anaerobic digestion mode. Li and Noike (1992) showed that the retention time for anaerobic digestion of waste-activated sludge thermally pretreated at 170 °C for 60 min, could be reduced by 5 days. In these conditions, the COD removal yield was 60%, with a biogas production yield of 223 ml / g COD, these values being two times higher than those of the control. According to these authors, the positive effect of thermal pretreatment on the anaerobic biodegradability of activated sludge was due to: (i) thermal pretreatment hydrolysed a large part of the particulate fraction of WAS; (ii) thermal pretreatment led to the production of volatile fatty acids which are easily converted into biogas during the subsequent biological step; (iii) the remaining particulate fraction contained in the thermally pretreated waste-activated sludge was easily hydrolysed by the anaerobic consortium. In another study, Sawayama et al., (1995) reported that thermochemical liquidization of a mixture of primary and waste- activated sludges (moisture 83.8%; volatile solids 12%), at 175 °C with holding time of 1 hour, increased biogas from 250 ml/g VS to 380 ml/g VS, with a methane concentration around 75% in the biogas. Furthermore, the supernatant of the liquidized sludge was successfully digested. The biogas yield from the supernatant during 8 days of incubation was 339 ml / g-added VS and the digestion ratio was 61% (w/w). From this data, the authors suggest treating sewage sludge according to the following procedure: liquidisation of dewatered sludge followed by the anaerobic digestion of the supernatant and the incineration of the precipitate separated from the liquidised sludge, the precipitate having a positive heating value compared with the dewatered sewage sludge (1.2 MJ/kg against −1.0 MJ/kg). Tanaka et al. (1997) reported that in the thermal pretreatment of combined waste-activated sludge from domestic, commercial and industrial wastewaters at 180 °C for 1 hour, the VSS solubilisation rate attained a 30% increase and the methane was produced at a rate 1.9 times higher than the control. When considering a process combining thermal hydrolysis and anaerobic digestion of canteen and fine food scraps, Schieder et al., (1999) demonstrated the many advantages compared to a conventional fermentation process: shortened treatment time and, thus, higher mass flow, reduction of the methane reactor volumes during the methanization phase, reduction of the total investment costs in spite of the additional cost of thermal treatment, reduction of dry matters by up to 70% and, at the same time, higher biogas recovery, no separate hygienic steps necessary since the hydrolysis process already guarantees the requisite hygiene.

Waste liquidisation is dependent on the thermal treatment conditions (temperature, operating time). Dote et al., (1993) showed that sewage sludge could be liquidised at temperatures above 150–175 °C. The viscosity of the sludge decreased by increasing temperature and holding time. However, too severe conditions seem to have a negative effect on the subsequent digestion stage. As an illustration, Pinnekamp (1989) observed that a reverse effect occurred at temperatures over about 180 °C. Despite increased thermal dissolution, gas production obtained from a thermally treated sludge decreased sharply, in some case to a value below that of the control sludge. In the same way, Stuckey and McCarty (1978) observed that increased temperature of treatment increased the solubilization up to a maximum of 51% at 225 °C. At a higher temperature, the amount of soluble organics decreased, suggesting the formation of larger molecules through polymerization. Such a phenomenon was ascribed to reactions of low sugars contained in the sludge with amino acids forming compounds difficult to degrade (such reactions will be further detailed in the consideration of pretreatment combining thermal and chemical action). At temperatures above 175 °C, they observed a decreased anaerobic biodegradability that could be due either to the formation of refractory compounds during heat treatment or to the inhibition of anaerobic microorganisms by the treated waste-activated sludge. Biological acclimatization occurred at both 225 and 250 °C after only 8 days of incubation, which would also suggest that refractory materials were the main cause for lower biodegradability at higher temperature. In the same way, Haug et al. (1978) observed that the temperature at which thermal pretreatment was conducted had a pronounced effect on biodegradability. With activated sludge, biodegradability increased with temperature to an optimum near 175 °C, beyond which gas production decreased.

8.4.4. Combination

Yet another way to enhance biogas production from organic wastes is to combine thermal and chemical pretreatments. Thermo-chemical hydrolysis of particulate COD is commonly carried out with alkaline agents, although some data about thermal acid hydrolysis of wastes are reported in the literature (Bouthilet and Dean, 1970; Barlindhaug and Ødegaard, 1996; Rocher et al., 1999). In this field, the effect of combining thermal and alkaline pretreatment on anaerobic biodegradability of waste- activated sludge was studied by Stuckey and McCarty (1978) and by Haug et al. (1978). Stuckey and McCarty (1978), working with waste-activated sludge, reported that under thermo-chemical pretreatment, waste-activated sludge would react in the following ways: lipids are hydrolysed under acid or alkaline conditions to glycerol and fatty acids; carbohydrates, and more particularly bacterial polysaccharides, are hydrolysed to simpler polysaccharides or sugars; proteins are hydrolysed by acid solutions to amino acid monomers (some peptide bonds – those of valine, isoleucine and leucine for example – are more stable than others and require longer hydrolysis times and stronger acids. Peptide bond cleavage is noticeably faster in hydrochloric acid than in sulphuric). Amino acids can be further degraded to ammonia and organic acids. Under alkaline conditions, proteins can also be hydrolysed; however, the rate and extent are generally less than with acid. The nucleic acids RNA and DNA will be hydrolysed to produce constituent bases, sugars and orthophosphate. Various intermolecular reactions can also be expected, such as the browning reaction which involves the polymerization of carboxyl groups with amino groups to form brown nitrogenous polymers and co-polymers known as melanoidins. High temperature and extremes in pH increase the rate of this polymerization. Such compounds are known to be difficult to degrade. Many authors have compared the efficiency of thermal, chemical or thermo-chemical pretreatments (Stuckey and McCarty, 1978; Tanaka et al., 1997; Penaud et al., 1998) and observed that best performances in terms of COD solubilization and anaerobic biodegradability were obtained when thermo-chemical pretreatment was used. Optimal conditions defined depend, of course, on the substrate used (Table 8.3).

Tanaka et al. (1997) worked with domestic activated sludge and combined sludges from domestic and industrial wastewaters. With domestic waste-activated sludge, VSS solubilization ratio was 70–80% but the increase of the CH_4 production was about 30%. With combined waste-activated sludge, VSS solubilization was 40–50% and the CH_4 production increased by more than 200% over the control (Table 8.1). Although the effect of the thermo-chemical pretreatment was greater on the combined than on the domestic waste-activated sludge, CH_4 production rate was 21.9 mlCH_4/gVSS/Day with domestic waste-activated sludge and only 12.8 mlCH_4/gVSS/day with combined waste-activated sludge. Patel et al. (1993), working with water hyacinth (Table 8.1), concluded that the alkali effect was probably a result of the solubilization of lignin. It constitutes an advantage in the sense that it liberates the remaining solid phase carbohydrates and increases its porosity. Penaud et al. (1998) (Table 8.1) observed that alkali pretreatment of an industrial sludge led to enhanced COD solubilization which was even more pronounced when heating was applied. When more than 5 gNaOH/l were added, 75–80% COD solubilization was achieved when heating at 140 °C for 30 minutes, instead of 65% at ambient temperature. Nevertheless, too high an addition of NaOH (more than 5 g NaOH / l) did not lead to any further significant increase of COD solubilization.

Table 8.3. Effect of thermo-chemical pretreatment on solubilization and anaerobic biodegradability of various wastes.

Reference	Substrate	Optimal conditions defined	Effect on solubilization	Effect on biodegradability
Stuckey and McCarty (1978)	Waste-activated sludge	175 °C, 30 meqNaOH/l, 1h	55% COD solubilization	78% conversion of COD to CH_4
Haug et al. (1978)	Organic sludge	175 °C pH=12, 30 min	68% COD solubilization	Increase of 57% of CH_4 production
Patel et al. (1993)	Water hyacinth	121 °C pH=11, 1h	58,48% COD solubilization	Increase in CH_4 production
Tanaka et al. (1997)	Combined sludge	130 °C 0.3 g NaOH / gVSS 5 min	45% VSS solubilization	Increase of 220% of CH_4 production
Penaud et al. (1998)	Industrial sludge	140 °C pH=12 30 min	75% COD solubilization	40% biodegradability

In any case, it has been reported that even if thermo-chemical pretreatment facilitates considerable solubilization of organic wastes, the solubilization obtained can be of no advantage for anaerobic biodegradability if :

- inhibitory molecules are formed;
- solubilised molecules rearrange to form polymers that remain difficult to degrade;
- chemical agents induce toxicity problems.

First, as concerns the potential formation of inhibitory compounds, Haug et al. (1978), who treated organic sludges at 175 °C, pH=12, observed that such compounds were produced and that their production adversely affected digester performances soon after feeding began. No acclimatization occurred to this inhibition after 43 days of feeding (gas production was less than 15% of the control and VFA had risen up to 4500 mg/l). In another study, Patel et al. (1993) underlined that soluble lignin derivatives obtained from an alkali-thermal treatment exert toxic effects (principally as a consequence of aromatic substances produced as a result of lignin degradation). In the same way, Penaud et al. (1998) did not observe any acclimatization of anaerobic bacteria to a thermo-chemically pretreated industrial biomass, suggesting that inhibitory materials were formed.

Secondly, various intermolecular reactions can be expected. Stuckey and McCarty (1978) reported that decreased biodegradability was due primarily to the formation of refractory compounds, both soluble and insoluble. The hypothesis also proposed as an explanation of the refractory nature of the molecules formed that some intramolecular reactions occurred between solubilised compounds, leading to the formation of complex substances. Such reactions, commonly named Maillard reactions, are frequently observed in the food industry. Coloured compounds formed are complex and very difficult to degrade, even by rumen microorganisms (see Marounek et al., 1995; Kostyukovsky and Marounek, 1995). Penaud et al. (2000) characterized the soluble molecules generated during the thermochemical pretreatment of an industrial sludge (NaOH addition for pH = 12; temperature = 140 °C). Fractionation of the soluble phase of the pretreated substrate was carried out by two methods, (i) treatment with adsorbent resins; (ii) precipitation by pH adjustment. Both methods demonstrated that compounds of high molecular weight (up to 100 kDa) are involved in the poor anaerobic biodegradability and the biotoxicity observed.

Finally, concerning the hypothesis of an inhibition caused by concentrations of alkali agents, no author concluded that such an inhibition exists. Pavlostathis and Gosset (1985), who treated wheat straw with 10 gNaOH / 100gTS in a continuous flow system, demonstrated that for 5% influent solids concentration to the digester, the sodium concentration in the anaerobic digester at steady state would be approximately 0.125M, which is lower than the 0.3M considered to be the threshold for sodium ion toxicity (Kugelman and McCarty, 1965; McCarty and McKinney, 1971). In another study, Penaud et al. (1998) observed no differences in the biodegradability performances obtained when the pretreatment was done at a temperature of 140 °C, with various alkaline agents (NaOH, KOH, $Mg(OH)_2$ or $Ca(OH)_2$). It suggested that the low biodegradability performances observed were not a consequence of sodium cation concentration. This was also confirmed by biodegradability tests run with increasing sodium cation concentrations.

8.4.5. Other physical pretreatments

Other physical pretreatment methods widely used for lignocellulosic materials can be proposed as a pre-treatment step before the anaerobic digestion of solid wastes. Millet et al. (1975) and Grethlein (1984) propose the description of physical pre-treatment, such as irradiation or steam explosion (autohydrolysis), for the enhancement of cellulosic biomass utilization. Concerning irradiation, Grethlein (1984) reported that irradiation markedly reduced the degree of polymerization of cellulose. It also oxidized a part of the substrate and organic acids. Millet et al. (1975) concluded that such a technique of irradiation of wood by gamma rays or by high-velocity electrons substantially improved the digestibility of these materials by rumen microorganisms. Steam explosion is also a pre-treatment method that has received much attention. Steam treatment is a combination of two effects: autohydrolysis of the polymeric components caused by the organic acids generated in situ, and physical disintegration of lignocellulosic matrix when it explodes on decompression. Most research experiments have used steam or steam with CO_2 or N_2. Such physical pre-treatments were also proposed and evaluated for the pretreatment of solid wastes before anaerobic digestion but they exhibited a strong species preference that limited their broad applicability.

8.5. REFERENCES

Aoki, N. and Kawase, M. (1991). Development of high-performance thermophilic two-phase digestion process. *Water Science and Technology* **23**(7–9) , 1147–1150.

Arai, M. and Murao, S. (1978). Red yeast cell wall lytic enzyme and protease. *Agricultural Biology Chemistry* **42**(8), 1461–1467.

Argeliers, S (1997). Personal communication

Barlindhaug, J. and Ødegaard, H. (1996). Thermal hydrolysate as a carbon source for denitrification. *Water Science and Technology* **33**(12), 99–108.

Baier, U. and Zwiefelhofer, H.P. (1991). Effects of aerobic thermophilic pretreatment. *Water Science and Technology* **23**, 56–63.

Baier, U. and Schmidheiny, P. (1997). Enhanced anaerobic degradation of mechanically disintegrated sludge. *Water Science and Technology* **36**(11), 137–143.

Bomio, M., Sonnleitner, B. and Fiechter, A. (1989). Growth and biocatalytic activities of aerobic thermophilic populations in sewage sludge. *Applied Microbiology and Biotechnology* **32**, 356–362.

Bouthilet, R.J and Dean, R.B (1970). Hydrolysis of activated sludge. Advances in Water Pollution Research, Proceedings of the 5th international Conference, III (31), 1–13.

Brière, F.G., Béron, P. and Hausler, R (1994). Influence of suspended solids in ozonation process. IOA regional Conference, Zurich (Suisse), S7.43–S7.52.

Cailliez, C., Benoit, L., Gelhaye, E., Petitdemange, H. and Raval, G. (1993). Solubilization of cellulose by mesophilic cellulolytic clostridia isolated from municipal solid wastes. *Bioresource Technology* **43**, 77–83.

Capela, I.F., Azeiteiro, C., Arroja, L. and Duarte, A.C. (1999). Effects of pre-treatment (composting) on the anaerobic digestion of primary sludges from a bleached kraft pulp mill. II International Symposium on Anaerobic Digestion of Solid Waste. Barcelona 15–17 June, pp. 113–120.

Cheubarn, T and Pagilla, K.R (1999). Temperature and SRT effects on aerobic thermophilic sludge treatment. *Journal of Environmental Engineering*, July, 626–629.

Chiu, Y.C., Chang, C.N., Lin, J.G. and Huang, S.J. (1997). Alkaline and ultrasonic pretreatment of sludge before anaerobic digestion. *Water Science and Technology* **36**(11), 155–162 .

Chynoweth, D.P. and Mah, R.A (1971). Volatile acid formation in sludge digestion. *Adv Chem Se*, **105**, 41–54.

Cinq-Mars, G.V. and Howell, J. (1977). Enzymatic treatment of primary municipal sludge with Trichoderma viride cellulase. *Biotechnology and Bioengineering* **19**, 377–385.

Clanet, M., Durand, H. and Tiraby, G. (1988). Enzymatic saccharification of municipal wastes. *Biotechnology and Bioengineering* **32**, 930–932.

Cochaux, A., Robert, A., Marotte, F., Pla, F. and D'Aveni, A. (1996). Alkaline hydrolysis of cellulose. *Revue ATIP* **50**(4), 148–156.

Datta, R. (1981). Acidogenic fermentation of corn stover. *Biotechnology and Bioengineering* **23**, 61–77.

Del Borghi, A., Converti, A., Palazzi, E. and Del Borghi, M (1999). Hydrolysis and thermophilic anaerobic digestion of sewage sludge and organic fraction of municipal solid waste. *Bioprocess Engineering* **20**, 553–560.

Dohanyos, M., Zabranska, J. and Jenicek, P. (1997). Enhancement of sludge anaerobic digestion by using of a special thickening centrifuge. *Water Science and Technology* **36**(11), 145–153.

Dote, Y., Yokoyamasy, S.Y., Minowa, T., Masuta, T., Sato, K., Itoh, S. and Suzuki, A. (1993). Thermochemical liquidization of dewatered sewage sludge. *Biomass and Bioenergy* **4**(4), 243–248.

Eastman, J.A. and Ferguson, J.F. (1981). Solubilization of particulate organic carbon during the acid phase of anaerobic digestion. *Journal of the Water Pollution Control Federation* **53**(3), 352–366.

Elefsiniotis, P and Oldham, W.K (1994). Substrate degradation patterns in acid phase anaerobic digestion of municipal primary sludge. *Environmental Technology* **15**, 741–751.

Engelhart, M., Krüger, M., Kopp, J. and Dichtl, N. (1999). Effects of disintegration on anaerobic degradation of sewage sludge in downflow stationary fixed film digester. II International Symposium on Anaerobic Digestion of Solid Waste. Barcelona 15–17 June, pp. 153–160.

Ensign, J.C. and Wolfe, R.S. (1965). Lysis of bacterial cell walls by an enzyme isolated from a myxobacter. *Journal of Bacteriology* **90**(2), 395–402.

Feijoo, G., Soto, M., Mendez, R. and Lema, J.M (1995). Sodium inhibition in the anaerobic digestion process: antagonism and adaptation phenomena. *Enzyme and Microbial Technology* **17**, 180–188.

Gijzen, H.J., Zwart, K.B., Verhagen, F.G.M. and Vogels, O.D. (1988). High rate two-phase process for the anaerobic degradation of cellulose employing rumen microorganism for an efficient acidogenesis. *Biotechnology and Bioengineering* **31**, 418–425.

Gilbert, E. (1983). Investigations on the changes of biological degradability of single substances induced by ozonation. *Ozone: Science and Engineering* **5**, 137–149.

Grethlein, H.E. (1984). Pre-treatment for enhanced hydrolysis of cellulosic biomass. *Biotechnologies Advances* **2**, 43–62.

Haltrich, D., Laussmayer, B., Steiner, W., Nidetzky, B. and Kulbe, K.D. (1994). Cellulolytic and hemicellulolytic enzymes of *Sclerotium rolfsii*: optimization of the culture medium and enzymatic hydrolysis of lignocellulosic material. *Bioresource Technology* **50**, 43–50.

Hamer, G. and Mason, C.A. (1987). Fundamental aspects of waste sewage treatment : microbial solids biodegradation in an aerobic thermophilic semi-continuous system. *Bioprocess Engineering*, **2**, 69–77.

Hartmann, H., Angelidaki, I. and Ahring, K. (1999). Increase of anaerobic degradation of particulate organic matter in full-scale biogas plants by mechanical maceration. II International Symposium on Anaerobic Digestion of Solid Waste. Barcelona15–17 June, pp. 129–136.

Hasegawa, S and Katsura, K (1999). Solubilization of organic sludge by thermophilic aerobic bacteria as a pretreatment for anaerobic digestion. II International Symposium on Anaerobic Digestion of Solid Waste. Barcelona 15–17 June, 145–152.

Haug, R.T., Stuckey, D.C., Gossett, J.M. and McCarty, P.L. (1978). Effect of thermal pretreatment on digestibility and dewaterability of organic sludges. *Journal of the Water Pollution Control Federation* January 1978, 73–85.

Haug, R.T., Le Brun, T.J. and Tortorici, D. (1983). Thermal pretreatment of sludges: a field demonstration. *Journal of the Water Pollution Control Federation* **55**(1), 23–34.

Hills, D.J and Nakano, K. (1984). Effects of particle size on anaerobic digestion of tomato solid wastes. *Agricultural Wastes* **10**, 285–295.

Jogeklar, H.S., Samant, S.D. and Joshi, J.B. (1991). Kinetics of wet air oxidation of phenols and substituted phenols. *Water Research* **25**(2), 135–145.

Kamiya, T and Hirotsuji, J (1997). Combined system of activated sludge and ozonation treatment for improving wastewater treatment. In *13th Ozone World Congress, Kyoto (Japan),* vol. 1, 199–204.

Khan, Y., Anderson, G.K. and Elliott, D.J. (1999). Wet oxidation of activated sludge. *Water Research* **33**(7), 1681–1687 ;

Knapp. J.S. and Howell, J.A. (1978). Treatment of primary sewage sludge with enzymes. *Biotechnology and Bioengineering,* **20**, 1221–1234.

Kopp, J., Müller, J., Dichtl, N. and Schwedes, J. (1997). Anaerobic digestion and dewatering characteristics of mechanically disintegrated excess sludge. *Water Science and Technology,* **36**(11), 129–136.

Kostyukovsky, V. and Marounek, M. (1995). Maillard reaction products as a substrate in in vitro rumen fermentation. *Animal Feed Science Technology* **55**, 201–206.

Kugelman, I.J. and McCarty, P.L. (1965). Cation toxicity and stimulation in anaerobic waste treatment. *Journal of the Water Pollution Control Federation* **37**(1), 97–116.

Legube, R., Dore, M., Langlais, B., Bourbigot, M.M. and Gouesbet, G. (1987). Changes in the chemical nature of a biologically treated wastewater during disinfection by ozone. *Ozone: Science and Engineering* **9**, 63–94.

Li, Y.Y. and Noike, T. (1992). Upgrading of anaerobic digestion of waste activated sludge by thermal pretreatment. *Water Science and Technology* **26**(3–4), 857–866.

Lin, J.G., Chang, C.N. and Chang, S.C (1997). Enhancement of anaerobic digestion of waste activated sludge by alkaline solubilization. *Bioresource Technology* **62**, 85–90.

McCarty, P.L. and McKinney, R.E. (1971). Salt toxicity in anerobic digestion. *Journal of the Water Pollution Control Federation* **33**(4), 399–415.

McKelly, R., Robinson, A.K., Blumentals, I. I., Brown, S.H. and Anfinsen, C.B. (1993). Proteolytic enzymes from hyperthermophilic bacteria and processes for their production. US Patent, US005391489A [5,391,489], 21 February 1995.

Mantzavinos, D., Hellenbrand, R., Livingston, A.G. and Metcalfe, I.S. (1997). Reaction mechanisms and kinetics of chemical pretreatment of bioresistant organic molecules by wet air oxidation. *Water Science and Technology* **35**(4) 119–127.

Marounek, M., Kostyukovsky, V.A., Hodrova, B. and Fliegerova, K. (1995). Effect of soluble Maillard reaction products on rumen microorganisms. *Folia Microbios.* **40**(3), 267–270.

Mason, C.A., Hamer, G., Fleischmann, T. and Lang, C. (1987). Aerobic thermophilic biodegradation of microbial cells. *Applied Microbiology and Biotechnology* **25**, 568–576.

Millet, M.A., Baker, A.J. and Satter, L.D. (1975). Pretreatments to enhance chemical, enzymatic and microbiological attack of cellulosic materials. *Biotechnology Bioengineering Symposium.* **5**, 193–219.

Moerman, W.H., Bamelis, D.R., Vergote, H.L., Van Holle, P.M., Houwen, F.P. and Verstraete, W.H. (1994). Ozonation of activated sludge treated carbonization wastewater. *Water Research.* **28**(8), 1791–1798.

Müller, J., Lehne, G., Schwedes, J., Battenberg, S., Näveke, R., Dichtl, N., Scheminski, A., Krull, R. and Hempel, D.C. (1998). Disintegration of sewage sludges and influence on anaerobic digestion. *Water Science and Technology* **18** (8), 425–433.

Müller, J. and Pelletier, L. (1998). Désintégration mécanique des boues activées. *L'eau, l'industrie, les nuisances* **217**, 61–66.

Murao, S., Yamamoto, R. and Arai, M. (1976). Isolation and purification of red yeast cell wall lytic enzyme producing microorganism. *Agricultural and Biological Chemistry* **40**, (1), 23–26.

Narkis, N. and Schneider-Rotel, M. (1980). Evaluation of ozone induced biodegradability of wastewater treatment plant effluent. *Water Research* **14**, 929–939.

Pagilla, K.R., Craney, K.C. and Kido, W.H. (1996). Aerobic thermophilic pretreatment of mixed sludge for pathogen reduction and *Nocardia* control. *Water Environ Res.* **68**, 1093–1098.

Palmowski, L and Muller, J. (1999). Influence of the size reduction of organic waste on their anaerobic digestion. II International Symposium on Anaerobic Digestion of Solid Waste. Barcelona 15–17 June, 137–144.

Patel, V., Desai, M. and Madawar, D. (1993). Thermochemical pretreatment of water hyacinth for improved biomethanation. *Applied Biochemistry Biotechnology.* **42**, 67–74.

Paul, E., Deleris, S., Cantet, J., Audic, J.M., Roustan, M. and Debellefontaine, H (1998). Effect of ozonation on activated sludge solubilization and mineralization. IOA International Regional Conference "Ozonation and AOPs in water treatment: applications and research", Poitiers, 23–25 September, 39/1–39/9.

Pavlostathis, S.G and Gossett, J.M. (1985). Alkaline treatment of wheat straw for increasing anaerobic digestion. *Biotechnology and Bioengineering.* **27**, 334–344.

Penaud, V., Delgenes, J.P., Moletta, R. (1998). Influence of thermo-chemical pretreatment conditions on solubilization and anaerobic biodegradability of a microbial biomass. *Environmental Technology*.

Penaud, V., Delgenes, J.P. and Moletta, R. (1999). Thermo-chemical pretreatment of a microbial biomass: influence of sodium addition on solubilization and anaerobic biodegradability. *Enzyme and Microbiol Technology* **25**, 258–263.

Penaud, V., Delgenès, J.P. and Moletta, R (2000). Characterization of soluble molecules generated during the thermochemical pretreatment of an industrial microbial biomass. *Journal of Environmental Engineering* **126** (5), 397–402.

Peres, C.S., Sanchez, C.R., Matsumoto, C. and Schmidell, W (1992). Anaerobic biodegradability of the organic components of municipal solid wastes. *Water Science and Technology* **25**(7), 285–293.

Perot, C. (1989). Optimisation de la digestion anaerobie en deux étapes des boues de station d'épuration: étude de l'étape d'hydrolyse de la matière organique. Thèse. Université de Clermont-Ferrand II, France.

Pinnekamp, J. (1989). Effects of thermal pretreatments of sewage sludge on anaerobic digestion. *Water Science and Technology*. **21**, 97–108.

Radermacher, H., Zobel, T., Pascik, I. and Kery, K (1999). Enzyme supported digestion of municipal sludges at the wastewater treatment plant Aachen-Soers. II International Symposium on Anaerobic Digestion of Solid Waste. Barcelona, 15–17 June, 356–360.

Rajan, R.V., Lin, J.G. and Ray, B.T. (1989). Low-level chemical pretreatment for enhanced sludge solubilization. *Journal of the Water Pollution Control Federation* **61**(11/12), 1678–1683.

Ray, B.T., Lin, J.G. and Rajan, R.V. (1990). Low-level alkaline solubilization for enhanced anaerobic digestion. *Journal of the Water Pollution Control Federation* **62**(1), 81–87.

Raynal, J., Delgenès, J.P. and Moletta, R (1998). Two phase anaerobic digestion of solid wastes by a multiple liquefaction reactors process. *Bioresource Technology* **65**, 97–103.

Rintala, J.A. and Ahring, B.K. (1994). Thermophilic anaerobic digestion of source shorted household solid waste: the effects of enzyme additions. *Applied Microbiology Biotechnology* **40**, 916–919.

Rivard, C.J. and Nagle, N.J. (1996). Pretreatment technology for the beneficial biological reuse of municipal sewage sludges. *Applied Biochemistry Biotechnology* **57/58**, 983–991.

Rocher, M., Goma, G., Pilas Begue, A., Louvel, L. and Rols, J.L (1999). Towards a reduction in excess sludge production in activated sludge processes: biomass physicochemical treatment and biodegradation. *Applied Microbiol, Biotechnol*. **51**, 883–890.

Sawayama, S., Inoue, S., Yagishita, T., Ogi, T. and Yokoyama. (1995). Thermochemical liquidization and anaerobic treatment of dewatered sewage sludge. *Journal of Fermentation and Bioengineering* **79**(3), 300–302.

Scheidat, B., Kasche, V. and Sekoulov, I. (1999). Primary sludge hydrolysis under addition of hydrolytic enzymes. II International Symposium on Anaerobic Digestion of Solid Waste. Barcelona, 15–17 June, pp. 161–168.

Schieder, D. and Schneider, R. and Bischof, F. (1999). Thermal hydrolysis as a pretreatment method for the digestion of roganic waste. II International Symposium on Anaerobic Digestion of Solid Waste. Barcelona15–17 June, 169–174.

Schwedes, J. and Bunge, F. (1992). Mechanical cell disruption processes. *Biotechnology Focus*, Hanver verlag, **3**, 185–205.

Sharma, S.K., Mishra, L.M., Sharma, M.P. and Saini, J.S. (1988). Effect of particle size on biogas generation from biomass residues. *Biomass* **17**, 251–263 ;

Song, J.J., Takeda, N. and Hiraoka, M. (1992). Anaerobic treatment of sewage treated by catalytic wet oxidation process in upflow anaerobic blanket reactors. *Water Science and Technol*ogy **26**(3–4), 867–875.

Stuckey, D.C. and McCarty, P.L. (1978). Thermochemical pretreatment of nitrogenous materials to increase methane yield. *Biotechnology and Bioengineering Symposium*. **8**, 219–233.

Tanaka, S., Kobayashi, T., Kamiyama, K. I. and Bildan, L. N. S. (1997). Effects of thermochemical pretreatment on the anaerobic digestion of waste activated sludge. *Water Science and Technol*ogy **35**(8) 209–215.

Thiem, A., Nickel, K. and Neis, U. (1997). The use of ultrasound to accelerate the anaerobic digestion of sewage sludge. *Water Science and Technology* **36**(11), 121–128.

Thomas, L., Jungschaffer, G. and Sprössler, B. (1993). Improved sludge dewatering by enzymatic treatment. *Water Science and Technol*ogy **28**(1), 189–192

Uehara, S. and Hasegawa-Iwai, K. (1979). On cell wall lytic enzymes produced by *C. centifugum*. *Agricultural and Biological Chemistry* **43**(9), 1991–1992.

Wang, Y.T. (1990). Methanogenic degradation of ozonation products of biorefractory or toxic aromatic compounds. *Water Research*. **24**(2), 185–190.

Ward, A., Stensel, H.D., Ferguson, J.F., Ma, G. and Hummel, S. (1998). Effect of autothermal treatment on anaerobic digestion in the dual digestion process. *Water Science and Technology* **38**(8–9), 435–442.

Watt, R.D., Kirsch, E.J. and Leslie Grady, C.P. (1985). Characteristics of activated sludge effluents after and before ozonation. *Journal of the Water Pollution Control Federation*. **57**(2), 157–166.

Yasui, H., Nakamura, K., Sakuma, S., Iwasaki, M. and Sakai, Y (1996). A full-scale operation of a novel activated sludge process without excess sludge production. *Water Science and Technology* **34**, 395–404.

9

Use of hydrolysis products of the OFMSW for biological nutrient removal in wastewater treatment plants

F. Cecchi and P. Battistoni

9.1. INTRODUCTION

Wastewater treatment facilities have gone through massive development from the early 1900s until now and they are still subject to continuous improvement and optimization in order to answer for new treatment objectives. From irrigation and intermittent filtration methods commonly used at the beginning of the 20th century, the development of completely stirred or continuous fed reactors have turned out to be significant to accelerate natural processes under

 Biomethanization of the organic fraction of municipal solid wastes. Edited by J. Mata-Alvarez. ISBN: 1 900222 14 0

controlled conditions in treatment facilities of comparatively smaller size. In general, from about 1900 to the early 1970s, the various forms of organic carbon were thought to be the main pollutants and the treatment objectives were the removal of suspended and floatable material, the treatment of biodegradable organics and the elimination of pathogenic organisms. The activated sludge process was traditionally developed for the oxidation of carbonaceous materials to carbon dioxide and biomass. Next, in recent decades, further attention was paid, for the first time, to nutrient removal (i.e. nitrogen and phosphorus), particularly in relation to the discharge in unsalted waters (i.e. streams, lakes): in this sense, a major effort was undertaken by states and environmental agencies to achieve more effective and widespread treatment of wastewater. Since 1980 wastewater treatment has begun to focus on the health concerns related to toxic and potentially toxic chemicals released to the environment which may cause long-term health effects. As a consequence, while the early treatment objectives remained valid, the required degree of treatment increased significantly, and additional treatment objectives and goals have been added.

Under mounting pressure of problems caused by eutrophication, at the beginning of 1990s design and construction of full-scale biological nutrient removal (BNR) activated sludge processes have moved ahead, despite the lack of mechanistic understanding of the process, in particular biological phosphorus removal and the related scientific research. In the European feature this was mainly due to CEC Directive 91/271, which has set more stringent environmental standards for nutrient discharge in sensitive surface waters (Table 9.1), imposing, thus, new treatment objectives hand-in-hand with the new water quality established standards. Importance has been given to micropollutants too like metals, surfactants, phenol compounds, organic solvents, pesticides, etc. since quality standards have been imposed.

Table 9.1. CEC Directive 91/271 nutrient limits.

Population	**N_{tot} (mg/l)**	**P_{tot} (mg/l)**
10^4–10^5 P.E.	15	2
> 10^5 P.E.	10	1

The aim of this chapter is to summarily explain the interactions between wastewater treatment and fermentation of the organic fraction of municipal solid waste (OFMSW) and to underline the importance of the latter in promoting biological nutrient removal from wastewaters. This issue is of primary importance considering that carbon addition, performed to allow more efficient nutrient removal, accounts to a significant extent for the costs in wastewater treatment plants. Furthermore, it must be considered that with this approach the organic fraction of municipal solid wastes is diverted from landfill disposal and is used twice: firstly as a carbon source for biological nutrient removal and then, after anaerobic digestion of the waste-activated sludge (WAS) as an energy source for biogas production. The implementation of the availability of organic carbon (both internal and external) in wastewater biological treatment plants can promote a better recovery of non-renewable resources like phosphorus itself, which, as we will see later in the text, can be collected and converted to final products usable as fertilizers: otherwise, because phosphorus is a not renewable resource, a loss from the global system would occur.

9.1.1. Biological nutrient removal (BNR)

9.1.1.1. Biological nitrogen removal

Total nitrogen content in domestic sewage is about 12 g/PEd and comprises nitrogen either organically bound as protein and nucleic acids, as urea ($OC(NH_2)_2$) or as the ammonium ion (NH_4^+) while nitrates and nitrogen are rarely present. Conventional wastewater treatment processes are capable of removing 1–2 gN/PEd through primary sedimentation whereas about 35 g of dry sludge per PE is daily produced with 5% of nitrogen (on a dry weight basis) and total nitrogen removed for cell synthesis is about 15%. To remove the remaining ammonia, nitrification is needed (Figure 9.1).

In many cases, conversion of ammonium to nitrate provides adequate treatment though, if eutrophication is likely to be a problem in the receiving water, it is essential to remove nitrates as well by encouraging denitrification within the reactor.

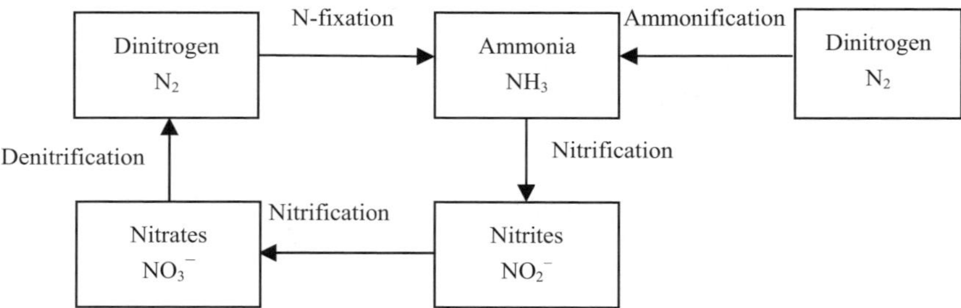

Figure 9.1. Simplified nitrogen cycle showing the nitrogen interconversions that can occur in an aquatic environment (Horan, 1990).

(a) Nitrification

Nitrification is the initial step in the removal of nitrogenous compounds from wastewaters. It involves the two-step conversion of ammonia to nitrite (ammonia oxidation) and nitrite to nitrate (nitrite oxidation). There are a range of autotrophic and heterotrophic bacteria capable of nitrification. Unlike heterotrophic bacteria, autotrophs are dependent on this reaction to generate energy for cell maintenance and growth. In wastewater treatment systems, autotrophs comprise only a small percentage of the mixed liquor microbial community but they are responsible for the bulk of nitrification (Burrel et al., 1999). They are essentially divided into two groups: *Nitrosomonas* that catalyse oxidation of ammonia to nitrite using molecular oxygen (equation 9.1) and *Nitrobacter* which further oxidize nitrite to nitrate using oxygen derived from the water molecule (equation 9.2).

$$NH_3 + 1.5O_2 \Rightarrow NO_2^- + H^+ + H_2O \quad (9.1)$$

$$NO_2^- + H_2O + 0.5O_2 \Rightarrow NO_3^- + H_2O \quad (9.2)$$

These reactions provide the nitrifiers with their energy and carbon is assimilated via the Calvin cycle in the form of carbonate and bicarbonate: thus no organic carbon source is required for nitrification.

Considering also the cell synthesis process (assuming a cell composition for *Nitrosomonas* and *Nitrobacter* equal to $C_5H_7O_2N$) the whole process can be summarised as follows:

$$NH_4^+ + 1.83O_2 + 1.98HCO_3^- \Rightarrow$$
$$0.02C_5H_7O_2N + 0.98NO_3^- + 1.88H_2CO_3 + 1.04H_2O \quad (9.3)$$

From the equation 9.3 it can be seen that:

- 4.2 g of oxygen are necessary for each gram of ammonia nitrogen removed (roundabout 4.3 g of oxygen per gram of nitric nitrogen produced);
- the biomass production is 0.16g/g ammonia nitrogen removed (that is to say 0.165 g of biomass per gram of nitric nitrogen produced);
- the required alkalinity to buff the produced acidity is nearly 7 $gCaCO_3/gN{-}NH_4$ removed.

(b) Denitrification

In the absence of a supply of dissolved oxygen, the utilization of oxygen as the terminal electron acceptor for respiration is inhibited. Under these conditions, most facultative microorganisms have to rely on fermentation to regenerate NAD^+. However, some chemoorganotrophs are capable of replacing O_2 with NO_3^- as the terminal electron acceptor and respiration can proceed with the reduction of nitrate to nitrite, nitric oxide, nitrous oxide or nitrogen (equation 9.4).

Oxidation state:	+5		+3		+2		+1		0	
Species:	NO_3^-	$\Rightarrow$	NO_2^-	$\Rightarrow$	NO	$\Rightarrow$	N_2O	$\Rightarrow$	N_2	(9.4)
	nitrate		nitrite		nitric oxide (gas)		nitrous oxide (gas)		nitrogen (gas)	

This process of anaerobic respiration is carried out by a variety of bacteria such as *Alcaligenes*, *Achromobacter*, *Micrococcus* and *Pseudomonas*. The redox state of the intermediates in denitrification shows that the reaction proceeds in a series of steps, each one associated with the gain of one or two electrons. An electron donor is therefore required as a source of these electrons. In sewage treatment, this reaction is carried out primarily by heterotrophic bacteria and only organic carbon sources can be used. Although the wastewater itself contains a suitable source of organic carbon, this could be insufficient for N-removal, thus requiring the provision of a supplemental source of carbon. Depending on the adopted process scheme, carbon sources used as electron donor can be represented by:

- chemical compounds added from external sources (e.g. methanol, acetic acid, etc.);
- organics contained in sewage (i.e. internal carbon sources; general composition $C_{10}H_{19}O_3N$);
- residual organic fraction of cellular mass deriving from endogenous sludge respiration (general composition $C_5H_7O_2N$).

Many validations exist in the literature that account for the fact that the reaction kinetics decrease passing from pure methanol to endogenous carbon. The maximum denitrification rate (at 20 °C) is usually 8.3–12.5 mgN$-NO_3$/gVSSh when using external carbon sources and 4.2 mgN$-NO_3$/gVSSh with internal carbon sources: remarkably lower values are found when using endogenous carbon (0.8–2.1 mgN-NO_3/gVSSh) (Battistoni et al., 1999).

9.1.1.2. Biological phosphorus removal

Phosphorus load control has been demonstrated as one of the most effective ways of dealing with man-made eutrophication; for this reason several countries apply a phosphorus standard for sewage effluent discharges, as well as one for nitrogen. The phosphorus content in domestic sewage is about 1–1.4 gPE/d:

nearly 50% is represented by orthophosphate, with 40% as polyphosphate and 10% as organic phosphate (these percentages can vary depending on sewer length).

Activated sludge systems can be operated for the biological removal of phosphorus from wastewater and the process is known as enhanced biological phosphorus removal (EBPR). It is a relatively new development which followed from the observation that, if an activated sludge is allowed to become anaerobic in the presence of volatile fatty acids, the amount of phosphorus (as orthophosphate, PO_4^{3-}) in the supernatant increases. Upon resumption of aeration, however, there is a rapid uptake of phosphate by the sludge which is in excess of that released during anaerobiosis. This 'luxury uptake' of phosphate results in a phosphate depleted mixed liquor and a phosphate rich sludge.

Although many different process configurations have been proposed and implemented, any EBPR process is based on the enrichment in the activated sludge of polyphosphate-accumulating microorganisms (PAOs), owing to the presence of an anaerobic zone in the plant before the aerobic one. According to some authors (Knight et al., 1995; Beacham et al., 1992) the removal and release of phosphorus within a sludge is the result of a single genus of bacteria known as *Acinetobacter* spp. and, more specifically, a single species, *Acinetobacter calcoaceticus* is implicated (up to 40% of the viable fraction of bacteria in a conventionally activated sludge plant have been identified as *Acinetobacter* spp; this rises to 80% for enhanced biological phosphorus removal plants). However, other studies have shown that microorganisms involved in biological phosphorus removal can belong to a wider spectrum of bacteria. In the anaerobic zone the PAOs utilize their internal reserve of polyphosphates to provide the energy to sequester readily biodegradable COD (RBCOD) and store it as polyhydroxyalkanoates (PHAs), mainly poly 2-hydroxy butyrate (PHB). In the aerobic zone the storage products are metabolised to provide carbon (for biosynthesis) and energy (for cell requirements and for regeneration of the phosphate pool) and phosphates are stored as volutin or metachromate (granules of polyphosphate formed from phosphate according to equation 9.5).

$$ATP + (PO_4)_n \Leftrightarrow ADP + (PO_4)_{n+1} \tag{9.5}$$

The ability to store the available VFA under anaerobic conditions has been considered a very particular feature of PAOs, owing to the availability of polyphosphate for the storage energy requirement, and thus to be the key-factor for their enrichment in the sludge, which causes an increase of P-content in the excess sludge.

Two main biochemical models describing EBPR have been proposed: the Comeau-Wentzel model (Comeau et al., 1986) and the Mino model (Mino et al., 1987). Common features of these models are that acetic acid (or other volatile fatty acids, VFAs) is the sole external substrate utilised in the anaerobic zone and that PHBs (or other PHAs) is the sole storage compound formed. The main difference between the models is the way in which the reducing power needed for PHB formation from acetate is provided. In the Comeau-Wentzel model, the reducing power is provided from a parallel metabolism of acetate (anaerobic TCA cycle), whereas in the Mino model, the anaerobic consumption of intracellular glycogen is also involved (which also gives more energy). Such a difference can cause different ratios among acetate removed, PHBs produced and phosphorus released between the two models.

Biological nutrient (N and P) removal (BNR) by nitrification and denitrification (ND) and biological excess P removal (BEPR) in single sludge activated sludge systems (NDBEPR) have provided significant benefits to ameliorate eutrophication of surface water. Moreover, the system can provide BNR at considerably reduced cost if some significant difficulties can be overcome, that is to say (Ekama & Wentzel, 1999):

- the limitation of the N and P removal placed on the system by the influent wastewater characteristics, in particular the readily biodegradable COD (RBCOD) fraction and the TKN/COD and P/COD ratios;
- the problems that arise in the treatment of the P-rich waste sludge.

As for the first problem, it is known that the efficiency of biological nutrient removal (BNR) systems for the treatment of civil wastewaters is directly dependent on the availability of readily biodegradable carbon (RBCOD) sources. The rates of denitrification are affected by the specific organic compounds available and from a stoichiometric point of view it has been determined that 8.6 mg COD/l of municipal wastewater is needed to reduce 1 mg/l of nitrate nitrogen to nitrogen gas (Ekama & Marais, 1984). On the other hand, the COD consumption during biological phosphorus removal has been estimated to be 50–59 mgCOD/l per mg/l of phosphorus removed from municipal wastewater. Actually, the efficiency varies with the specific organic compound available in the anaerobic zone and experimental data show that acetic acid is the most efficient organic compound for biological phosphorus removal (Abu-Ghararah & Randall, 1991). Full-scale and pilot-scale experiences indicate that a BOD_5/P_{tot} ratio of at least 20:1 is needed to achieve an effluent P_{tot} concentration of 1.0 mg/l or less , with secondary clarification as the last treatment step (Randall et al., 1992). In any case, the phosphorus removal actually obtained is determined by the amount of stored substrate contained by the poly-P bacteria when they enter the aerobic zone which is a function of the quantity and the types of VFA in the anaerobic/anoxic zone.

9.2. STATE OF THE ART

9.2.1. Production and utilization of internal and/or external carbon sources

The enhancement of N and P removal by addition of readily biodegradable COD is, by now, a common practice when wastewater organic content is not enough: Randall et al. (1992) recommend the use of external carbon sources when the ration COD/TKN in a given wastewater is under 9.

The various substrates that are mainly used are differentiated on their provenance, that is to say they are divided into internal carbon sources and external carbon sources referring to the wastewater treatment process itself.

Internal carbon sources are mainly represented by primary sludge elutriation and/or fermentation (Lotter & Pitman, 1992) and by hydrolysis and fermentation of primary and chemical sludge (Aesoy and Ødegaard, 1994; Isaacs & Henze, 1995; Kristensen et al., 1992).

External carbon sources are essentially ethanol, methanol, acetic acid and the anaerobically fermented organic fraction of municipal solid waste (OFMSW) (Cecchi et al., 1994).

9.2.1.1. Internal carbon sources

Internal carbon sources are preferred for economic matters, the lower waste sludge production and the optimal usage of organic substances within the wastewater plant (Isaacs & Henze, 1995). In this field elutriation and fermentation of primary sludge is a widespread solution in real biological nutrient removal plants (Lotter & Pitman, 1992). By fermenting the primary sludge, readily biodegradable carbon substrates can be produced in the wastewater treatment plant itself. The products are mainly short chain fatty acids (SCFAs) with two to five carbon atoms (Pitman et al., 1992). The use of dissolved fermentation products results in higher denitrification rates compared with other substrates such as acetate or methanol. To enhance the amount of RBCOD for nutrient removal in the activated sludge system, several configurations for separate fermentation and elutriation exist, e.g. sludge fermentation in a separate reactor and recycling it to the primary clarifier for elutriation.

The use of fermented sludge to improve nutrients removal in wastewater treatment plants was considered in many scientific experiences where VFA increase was in the range 30–850% with the best performances corresponding to mesophilic conditions and HRT = 1–10 days (Pavan et al., 1994).

Pitman et al. (1983) first proposed to dose a VFA-rich effluent coming from a digester and to install primary settlers before the BNR phase for concentrating solids to produce VFA.

In 1985 Rabinowitz & Oldham too experimented the psycrophilic pre-fermentation of primary sludge in a UCT system for biological phosphorus removal achieving an RBCOD increase of 23% while phosphorus uptake increased by 20%. The same process was used by Rabinowitz et al. (1987) who experimented at pilot-scale lower HRT (0.5 d) and mesophilic conditions to obtain a P removal efficiency increase of 60% and a VFA increase in the range 28–46%.

In 1990 Lilley et al. found that optimal acidogenic primary sludge fermentation in mesophilic conditions with an HRT = 3 d gave a VFA increase of 38% in the BNR plant.

Primary sludge fermentation has been tested both in psychrophilic and mesophilic conditions by Rozzi et al. (1995), showing a production of 20–50 mg SCOD/g of primary sludge and 1.1–8.8 mg VFA/g. Also hydrolysis and fermentation of primary and chemical sludge were applied as successful operations to assure high performances in nitrogen removal (Kristensen et al., 1992; Aesoy and Ødegaard, 1994): the first-order kinetics of the hydrolysis of particular SCOD allows a high performance for solid residence times of a few days (Moser-Engeler et al., 1998). Other full-scale plant applications of the acid fermentation of primary sludge for the RBCOD production are reported by Skalsky and Daigger (1995).

Pilot-scale tests have been recently done on the possibility of fermenting wastewaters to be treated with a sequencing batch reactor (SBR) for biological nutrient removal: in these conditions, VFA concentrations up to 223 mg/l can be achieved, differentiated in acetic (63%), propionic (25%) and butyric (12%) acids (Cuevas-Rodriguez et al., 1998) which have proved to increase the capacity of bacteria to accumulate phosphates.

As for the quantitative aspects, it must be emphasised that the yield of the primary sludge fermentation process is nearly 4 gC/PEd (Battistoni et al., 1998).

9.2.1.2. External carbon sources

Supplementation of carbon sources with external addition is frequently achieved by using industrial and agricultural wastes such as brewery waste, molasses or corn-silage which are readily biodegradable by microorganisms. In the absence of such an alternative, methanol is generally accepted as the most appropriate commercially available carbon source, though other organic compounds, as acetic acid, may be used.

In a classification of different substrates, denitrification rates for acetate, propionate and butyrate were found to be as much as four times higher than for ethanol, methanol or citrate (Moser-Engeler et al., 1998).

Methanol has been used, in the past, as an external carbon source for denitrification in separate activated sludge systems following nitrification and, also, to post-anoxic zones following nitrification to provide necessary organic substrate or to increase the denitrification rate. In these conditions, the specific denitrification rate (SDNR) ranges from 0.10 to 1.2 g NO_3–N/g TSS d due to different organic substrate levels during the denitrification reactions (U.S: EPA, 1975).

As concerns the external addition of non-pure substrates, Choi et al. (1996) suggested the introduction of nightsoil to reduce nitrate presence in the returning activated sludge to favour phosphorus removal.

The use, as external carbon source, of anaerobically fermented organic fraction of municipal solid waste (OFMSW) represents, at the same time, a good way for OFMSW final disposal and the promoting of BNR processes in wastewater treatment; furthermore it is characterised by negative operational cost for the other external carbon sources (methanol, acetic acid, etc.). In fact, the large amount of volatile fatty acids (VFAs) which characterize the fermented OFMSW can be used as a source of carbonaceous compounds. In particular, the linear short-chain fatty acids (SCFAs) such as acetate, propionate, n-butyrate and n-valeriate are preferentially consumed though acetate has been mentioned as the most efficient carbon source for nutrient removal (Moser-Engeler et al., 1998).

Experiences by Sans-Mazon et al. (1992) and Virtutia et al. (1992) concerning anaerobic fermentation of organic wastes comprising mainly vegetal wastes by municipal markets in a pilot-scale controlled reactor have demonstrated the possibility of obtaining up to 40 g VFA/l by fermentation of the separately collected organic fraction of municipal solid waste (SC-OFMSW).

This strategy allows the production of 16 gC/PEd (Battistoni et al., 1998) and to integrate wastewater and waste treatments, aiming for economic and environmental benefits as confirmed by the experiments carried out by Pavan et al. (1994) and Cecchi et al. (1994).

9.2.1.3. The organic fraction of municipal solid waste (OFMSW) as external carbon source

When choosing a carbon source suitable for BNR process, two characteristics must be taken into account: the source availability to assure a continuous enhancement of nitrogen removal and the COD/N_{tot} ratio for a complete use of the carbon source to be guaranteed. As for this latter aspect, it can been stated that the difference between the COD/N_{tot} ratio required by the plant and the one deriving from the fermented OFMSW addition gives the real availability of carbon for nitrogen and P removal. On this basis, comparing these properties in different fermented or hydrolysed substrates, it turns out that OFMSW can be considered the best substrate (Table 9.2).

Table 9.2. Processes and characteristics of external carbon source production.

		Characteristics of products					
Substrate	**Process**	**COD_{tot}/ N-NH_4 %**	**SCOD/ N-H_4 %**	**COD_{tot}/TKN %**	**VFA/SCOD%***	**$COD_{conv.}$ %****	**Reference**
PC	HF		11.3		67	11	Aesoy and Ødegaard (1994)
PC	HF		18		60–70		Isaac & Henze (1995)
PC	HF		18	35	60–70	10–13	Kristensen et al., (1992)
P	FE	16–26					Lotter & Pitman, (1992)
OFMSW	F	250	167		30	66	Pavan et al. (1994)
Blend	F		12–32		55–73	12.5	Ghosh et al. (1975)
P	F					3.8–3.9	
Blend	F				27–40	2.5–10.5	Bhattacharya et al., (1996)
Vegetable and fruit mixtures	F	313	147	94	28	40	Traverso et al. (2000)

PC = primary and chemical sludge; P = primary sludge; OFMSW = organic fraction of municipal solid waste. HF = hydrolysis and fermentation; FE = fermentation and elutriation; F = fermentation.
* VFA measured in terms of COD; ** ratio between soluble COD in the effluent and total COD in the feed.

The chemical and physical characteristics of the fermented OFMSW (Table 9.3) allow high phosphate release and high denitrification rates that can be attributed to the high presence of VFA which represent the 30% of soluble COD (SCOD) and to the large portion of acetic acid in the VFA mix (70–85%).

Table 9.3. Specific production and properties of fermentate.

Parameter	Unit	Mean value
Fermentate		
Specific production	l/PE d	0.23
TCOD	mg/l	75,000
RBCOD	mg/l	34,000
VFA	mg/l	15,000
N_{tot}	mg/l	1,000
N–NH_4	mg/l	300
P_{tot}	mg/l	150
TSS	mg/l	25,000
Residual solid		
Specific production	Kg/PE d	0.07
TS	%	18.4
Yields		
Maximum denitrification rate (NUR_{max})	mg NO_x–N/gTVS h	11.7
Maximum phosphate release rate	mg PO_4–P/gTVS h	12.5

OFMSW fermentation tests (Pavan et al., 2000) in mesophilic conditions show a quite constant total VFA production in the HRT range 3–6 d with small increases in VFA production for increasing HRT. In this conditions acetic and lactic acid are the main components (44% and 51% of total soluble COD produced respectively).

In general, the effectiveness of the fermented effluent in the denitrification step leads to a maximum nitrate utilization rate (NUR_{max}) of 11.7 mg NO_3–N/gVSS h (Bolzonella et al., 2001). This value can be compared with the one obtainable using pure methanol and other easily biodegradable substrates as RBCOD sources (12.5 mg NO_3–N/gVSS h; Beccari et al., 1993) and is double if compared with the one obtained by elutriation of primary sludge (5.8 mg NO_3–N/gVSS h; Kristensen et al., 1992).

In particular, with a RBCOD excess availability of 54% of TCOD and a satisfactory carbon:nutrients ratio, it is possible to achieve a phosphorus content in waste activated sludge of 4.9%: this percentage is in good agreement with the ones reported in Pöpel and Jardin (1993) for good biological phosphorus removal in enhanced biological phosphorus removal (EBPR) plants.

Valuable results have been obtained also by mesophilic fermentation of source separated mixtures of vegetables and fruits wasted by supermarkets (Traverso et al., 2000). The treatment of these mixtures allows the transformation of about 43% of total COD into soluble COD and total VFA production yields 40% of total influent COD. Aside from HRT values, the main products in the liquid phase are lactic and acetic acids (Table 9.4).

Table 9.4. Average values and standard deviations of the contributions to SCOD coming from the fermentation products when HRT > 6 d (Traverso et al., 2000).

Product	Lactic acid	C2	C3	i-C4	C4	i-C5	C5	i-C6	C6	C7	Ethanol
Average (%)	41	33	7	0.7	8	0.3	1	0.4	0.7	0.5	7
Std. Dev.	4	3	0.8	0.2	3.3	0.1	0.5	0.3	0.3	0.4	3

The conversion factors (COD_{conv}) express the amount of total COD in the feed (COD_{tot}) that is converted to soluble COD in the effluent; comparing the different values from the various substrates (Table 9.2), it can be seen that the OFMSW shows a higher value than the sludge (66% vs. ≅ 13%). This approach also involves a smaller nutrients addition, as it can be seen comparing the $COD_{tot}/N{-}NH_4$ ratio in Table 9.2 for the sludge and for OFMSW. Considering the $COD_{tot}/N{-}NH_4$ ratio for the OFMSW fractions, the range is from the lowest 17 (meat) to the highest 155 (fruits and vegetables), which, furthermore, represent the largest part of OFMSW (Pavan et al., 1998). The hydrolysed primary sludge, or chemical and primary sludge mixtures, is characterised by a high nitrogen content, since the COD_{tot}/TKN ratio goes down to values of 18, as reported in several works (Isaacs & Henze, 1995; Kristensen et al., 1992): this leads to a higher nitrogen load to the wastewater treatment plant and reduces the carbon availability for nitrogen removal to the 61% of COD_{tot} fed (assuming a request of 7 mg COD / mg N according to Isaac & Henze, 1995). In terms of qualitative effect, the addition of hydrolysed primary sludge only leads to a change in the process kinetics while the addition of the OFMSW to the wastewater also positively affects the C:N ratio. As regards the quantitative aspects, the magnitude of these effects depends on the extent to which the carbon addition has been performed.

9.3. IMPACT OF THE FERMENTED OFMSW ADDITION ON WASTEWATER TREATMENT PROCESSES

The impact of fermented OFMSW addition has been studied by means of the application of Activated Sludge Model No. 2 (IAWQ Task Group, 1995) by Battistoni et al. (1998). The population equivalent on which the model has been tried is 100,000 PE and the fermented OFMSW addition was hypothesised equal to 23 m^3/d. The main characteristics for the inlet wastewater are 40 mg/l of total nitrogen and 400 mg/l for total COD. The effect has been observed both for a process only for carbon and nitrogen removal and for a BNR process. Figures 9.2 and 9.3 show the effect of this relationship. The working temperatures and the sludge retention time values (SRT) are reported in Table 9.5.

Table 9.5. SRT and T variation applied during Activated Sludge Model No. 2 simulation (Battistoni et al., 1998).

T (°C)	12	14	16	18	20	22
SRT_{CN} (d)	15	15	15	15	10	10
SRT_{BNR} (d)	20	20	15	15	10	10

In Figures 9.2 and 9.3 two main parameters are presented: the mixed liquor suspended solid concentration (MLSS) and the safety coefficient (Cs) which can be defined as the ratio between the standard limits imposed by CEC Dir. 91/271 and the effluent nitrogen concentration ($N_{tot} = 15$). If the Cs is larger than 1.5 the effluent fits the >100,000 PE standards, if Cs is lower than 1.5 it fits the < 100,000 PE standard. If Cs is lower than 1 there is no way to fit the standards.

As for the phosphorus, the limit value was 1 mgP–PO_4/l when Cs < 1.5 and 0.5 mgP–PO_4/l when Cs > 1.5 to maintain the consistency with 10 mgN_{tot}/l. Their trend was observed with varying temperatures and in various combinations, that is to say with primary sedimentation (PS), with primary sedimentation together with fermentate addition (PS-F), without primary sedimentation (WPS) and without primary sedimentation and with the fermentate addition (WPS-F).

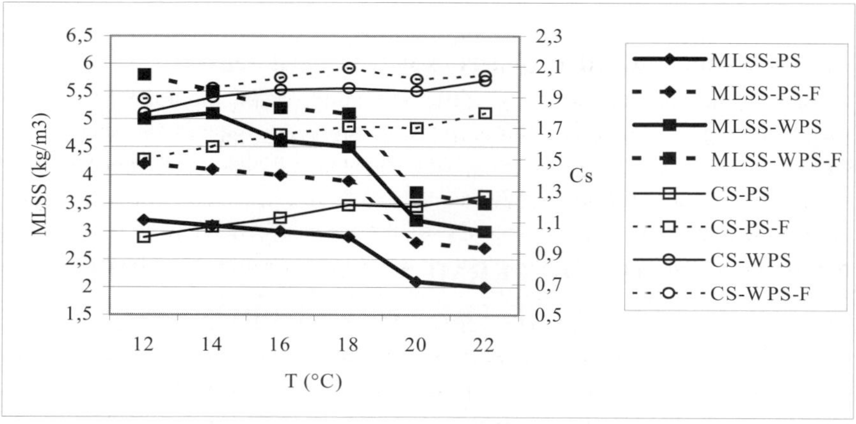

Figure 9.2. Process for only carbon and nitrogen removal (CN process). Effect of fermentate addition and primary sedimentation at various temperatures (Battistoni et al., 1998). MLSS = mixed liquor suspended solids; PS = primary sedimentation; WPS = without primary sedimentation; F = with fermentate addition; CS = safety coefficient.

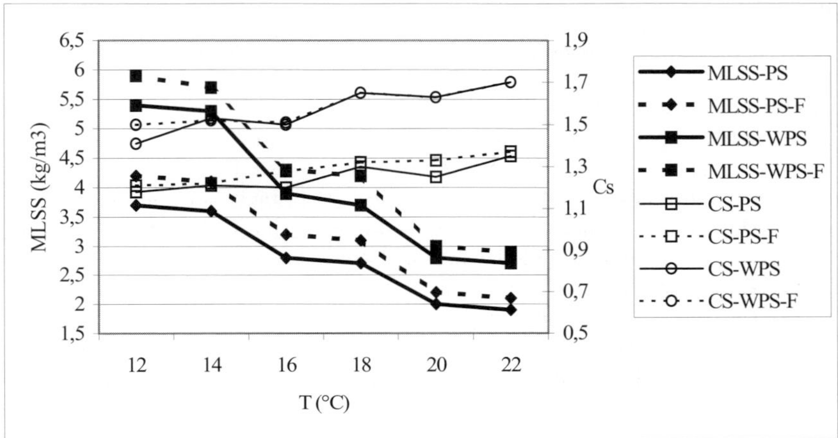

Figure 9.3. BNR process. Effect of fermentate addition and primary sedimentation impact at various temperatures (Battistoni et al., 1998). MLSS = mixed liquor suspended solids; PS = primary sedimentation; WPS = without primary sedimentation; F = with fermentate addition; CS = safety coefficient.

9.3.1. Carbon and nitrogen removal (CN process)

In the CN process an important increase of the safety coefficient Cs with temperature can be observed, coupled with a decrease of the MLSS (Figure 9.2); furthermore a shift is observed when SRT has changed (see also Table 9.5).

The fermentate addition seems to have a double effect: it allows a higher MLSS concentration and a higher Cs for the process. The first increase is due both to the fermentate total solid content and to carbon increase which enhances biomass growth; the second effect is due to the enhancement of denitrification rates thanks to external carbon addition. More precisely, the increase in Cs is smaller in processes without primary sedimentation if compared with processes with primary sedimentation. In fact primary sedimentation is an operation which causes a reduction in the influent's constituents: the primary treatment can provide the 27% reduction of total influent COD and only 4.3% of influent total nitrogen (Battistoni et al., 1998). This can lead to an inadequate COD:TKN ratio coming into the plant and makes the implementation of COD with external carbon sources the only possibility to fit the effluent to the 10 mgNtot/l limit. On the contrary, if the plant is not provided with a primary settling tank, the internal RBCOD, obtained by hydrolysis of SBCOD, allows the high nutrient removal yields and the OFMSW-fermentate addition can be substantially cut. In this case the OFMSW fermentate addition is necessary first of all for the final disposal with energy recovery and for avoiding high nitrogen effluent concentration due to seasonal variations of influent characteristics.

The global impact of fermentate addition is compared with the 'no addition' situation in Table 9.6 in terms of Fe^{2+} to be added for excess phosphorus coagulation and precipitation, oxygen required for organic substances oxidation, biogas obtainable by anaerobic digestion of excess sludge and/or mixed sludge (primary plus secondary) and amount of excess sludge produced during the treatment process.

Table 9.6. Comparison between fermentate addition situation and no addition situation in a process for carbon and nitrogen (CN) removal (Battistoni et al., 1998).

Process	**Fe^{2+} (mg/l)**	**O_2 (kg/h)**	**Biogas (m^3/d)**	**Sludge disposal (kgTS/d)**
With primary sedimentation	8	300	3320	4530
With primary sedimentation and fermentate addition	16	320	3930	5150
Without primary sedimentation	16	360	990*	3500
Without primary sedimentation and with fermentate addition	18	400	1730*	3900

* Calculated on the basis of 0.2 m^3 biogas production / Kg TVS of waste-activated sludge.

The fermentate addition involves a higher oxygen demand since more carbon needs to be oxidised while it determines a reduction in ferrous ions needed for excess phosphorus precipitation; at the same time, it generates an over-production of biogas if co-digestion technology is applied which means digestion of the mixture obtained by waste-activated sludge and the solid residue from the OFMSW acidogenic fermentation.

9.3.2. Biological nutrients removal (BNR process)

In Figure 9.3 MLSS and Cs trends are reported at various temperatures and in different operational conditions (see also Table 9.5). A remarkable increase of the safety coefficient can be seen, coupled with a decrease in MLSS concentration when passing from plants with primary settling tank to situations without primary treatment. Furthermore a significant shift can be seen in correspondence with a change in SRT.

As regards the fermentate addition, the difference is not as remarkable as it was in the CN process previously considered mainly because the BNR process is already characterised by a significant safety coefficient. A further consideration is to be made about the phosphorus removal process: the external carbon addition is vital to support the poly-P accumulating biomass activity and to increase the process effectiveness. BNR tests, carried out to evaluate the differences in process behaviour with and without the addition of the liquid phase of the fermented OFMSW effluent with particular concern for phosphorus removal (Pavan et al., 2000), show that, with an RBCOD equal to 50% of total influent COD, P removal has been enhanced from 43% to 76%.

Thus the external carbon addition exerts a double role by enhancing both phosphorus removal and nitrates removal and, at the same time, it involves the effects on oxygen requirements, biogas production and solid residue production discussed before when talking about the CN process.

Regarding Table 9.7 some considerations arise: first of all, the fermentate addition substantially reduces the ferrous ion implementation for phosphorus precipitation of 63% with primary sedimentation and 75% without. The oxygen requirements are obviously larger for the increased availability of degradable matter and this reflects also on the biogas production by sludge co-digestion.

Table 9.7. Comparison between fermentate addition situation and no addition situation in a BNR process (Battistoni et al., 1998).

Process	**Fe^{2+} (mg/l)**	**O_2 (kg/h)**	**Biogas (m^3/d)**	**Sludge disposal (kgTS/d)**
With primary sedimentation	8	290	3320	4530
With primary sedimentation and fermentate addition	3	320	3930	5150
Without primary sedimentation	4	350	810*	2900
Without primary sedimentation and with fermentate addition	1	400	1510*	3200

* Calculated on the basis of 0.2 m^3 biogas production / Kg TVS of waste-activated sludge.

9.3.3. Impact of fermented OFMSW addition on nitrogen overloading

Nitrogen overloading is frequently encountered when considering civil wastewater treatment plants that receive external discharge wastewater such as landfill leachate or septic tank effluents exerting the role of environmental service for the drainage basin. In this case the high nitrogen load treatment requires the use of external carbon sources which is usually performed by the addition of methanol or acetic acid. The choice of the addition of OFMSW fermentate can lead to the same final effect with considerable economic advantages. The effect of fermentate addition at various percentages of ammonia nitrogen can be observed in Figure 9.4 where the excess nitrogen in the influent (expressed as a percentage of ammonia nitrogen on the average ammonia nitrogen daily load) is plotted vs. the safety coefficient for nitrogen.

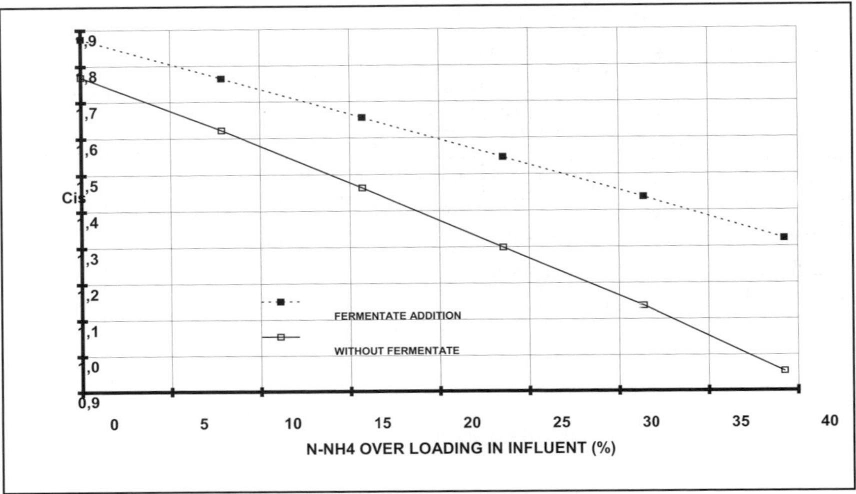

Figure 9.4. Effect of fermentate addition on nitrogen overloading in plants without primary sedimentation (Battistoni et al., 1998).

As can be seen from the graph, the slope of the fermentate line is smaller than the one 'without fermentate': this underlines the fact that the nitrogen overloading effect is significantly smoothed and never goes below the value of 1.3. In these terms, the plants can guarantee the accomplishment of the safety standards.

9.4. THE INTEGRATED PROCESS: BASIC BALANCES AND DESIGN

9.4.1. Process description

The integrated wastewater and OFMSW treatment system proposed by Cecchi et al. (1994) involves a combination of anaerobic fermentation (AF), biological nutrient removal (BNR) and struvite crystallization processes (SCP) (AF-BNR-SCP) as shown in Figure 9.5.

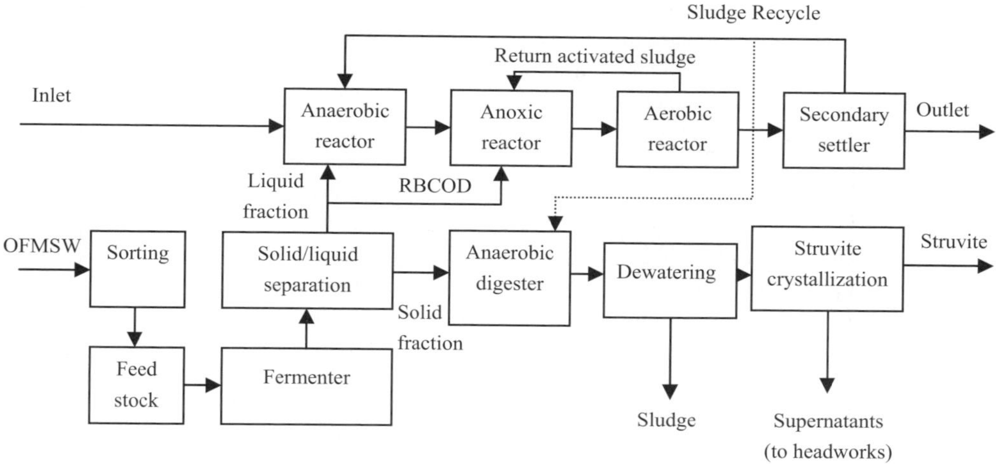

Figure 9.5. Scheme of the integrated wastewater treatment system (AF-BNR-SCP) (Cecchi et al., 1994).

The single processes characterizing the whole approach are:

(a) anaerobic acid fermentation of the organic fraction of municipal solid waste (OFMSW) in continuous stirred tank reactor (CSTR) in mesophilic conditions (T = 35–37 °C);

(b) anaerobic co-digestion of the solid residue from the OFMSW fermentation and of waste-activated sludge (WAS) and primary sludge (if a primary settler is present) by mesophilic process (T = 35–37 °C) or thermophilic process (T = 55 °C);

(c) wastewater treatment for biological removal of carbon, nitrogen and phosphorus;

(d) phosphorus and ammonia removal from supernatants by struvite and hydroxyapatite crystallization.

The separately collected source OFMSW is first pre-selected to eliminate packages and shredded. The mixture is then stocked before being sent to the fermenter unit where mesophilic acidogenic fermentation occurs. The liquid phase coming from the anaerobic fermentation step is fed into the biological nutrient removal plant along with the incoming stream of wastewater. The intention is to provide sufficient readily biodegradable carbon substrate to allow the combined phosphorus–nitrogen biological removal process (BNR) to enhance the removal of nutrients from the influent wastewater.

The solid phase leaving the solid–liquid separation after the anaerobic fermentation is blended with the WAS from the BNR treatment process and undergoes a methanogenic step, to produce energy (heat and electricity) which can contribute to the achievement of the power autonomy of the plant (Mata-Alvarez & Cecchi, 1989). In this unit, the biomass is broken down, thereby releasing also the phosphorus stored during the BNR process. This phosphorus, which otherwise should be sent to the plant headworks, leading to a significant feedback, can be removed and recovered by the struvite crystallization process (SCP) without any chemical addition.

In synthesis, the aims of the integrated process are:

- to reduce and stabilize the organic matter in the OFMSW together with the waste activated sludge from the wastewater treatment plant since it is mainly composed by water (more than 80%);
- to allow high performances in BNR plants by the use of acid fermentation effluent after OFMSW liquid fraction separation from the solid fraction;
- to perform the energy recovery thanks to biogas overproduction from anaerobic co-digestion of solids from fermentation processes and mixed or waste activated sludge;
- to perform the recovery of a non-renewable resource like phosphorus into a usable form.

9.4.2. Mass balance of the anaerobic fermentation of the organic fraction of municipal solid waste

The mass balance of the anaerobic fermentation of the OFMSW (Figure 9.6) accounts for the following contributions:

- the feed comprises the OFMSW shredded and homogenised in a quantity equal to 300 g/PEd. This specific production is typical for European countries (EEA, 2002) and can be here considered supposing a good efficiency for the source collection and for the sorting operation in the plant;
- after anaerobic fermentation the effluent is split into two fractions by screw-pressing:
  (1) the liquid fraction which is sent to the wastewater biological treatment section;
  (2) the solid fraction which is co-digested with the waste sludge from wastewater treatment.

The mass balance computations have been performed over average values which are included in the following ranges:

Parameter	Range
OFMSW entering the fermenter	
TS, g/kg	64.0–188.0
TVS, %TS	55.7–98.1
TCOD, g/kg	49.5–141.5

More details can be found also in Cecchi et al. (1994) and Pavan et al. (1994, 2000).

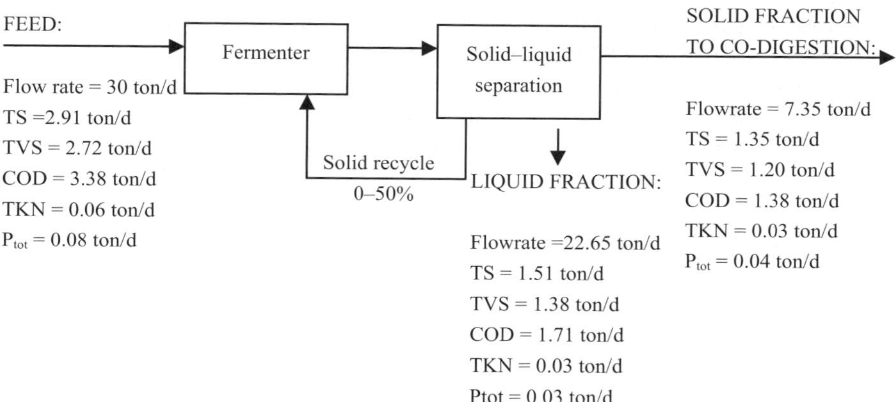

Figure 9.6. Schematic mass balance of the anaerobic fermentation of the OFMSW on a 100,000 PE basis (daily OFMSW production: 300 g/PE). (Mass balances are closed with an approximation of ±10%.)

9.4.3. Mass balance of the biological removal of carbon, nitrogen and phosphorus with the addition of the fermented OFMSW

The mass balance of the biological removal of carbon, nitrogen and phosphorus (Figure 9.7) accounts for the following contributions:

- the inlet is composed both by the wastewater generated by 100,000 PE on a 250 l/PEd basis (Metcalf & Eddy, 1991) and by the OFMSW fermentation liquid effluent flowrate (also on 100,000 PE basis) as described in paragraph 9.4.2.
- the effluent is finally discharged, after tertiary treatment, in natural waters or reused;
- the gaseous nitrogen leaves the system after denitrification;

- the waste sludge is sent to co-digestion with the solid residual of the OFMSW fermentation.

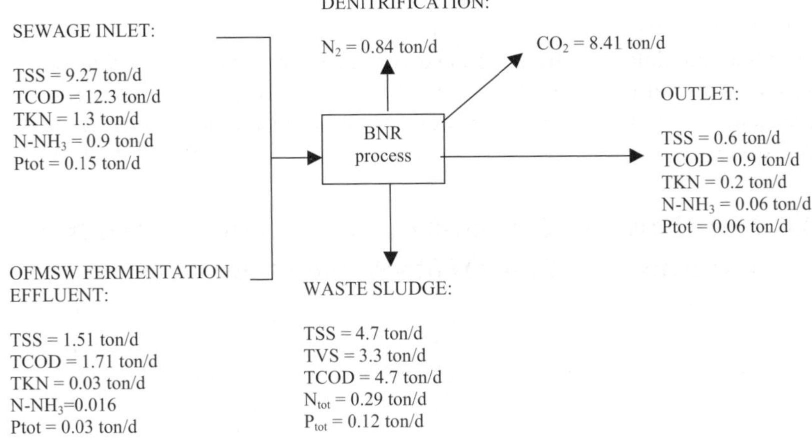

Figure 9.7. Schematic mass balance of the biological removal of carbon, nitrogen and phosphorus on a 100,000 PE basis (daily OFMSW production: 300 g/PE; daily wastewater production: 250 l/PE) (Pavan et al., 2000). (Mass balances are closed with an approximation of ±10%, CO_2 is expressed as ton O_2/d.)

The mass balance calculations are based on literature data (Battistoni et al., 1998) for a medium untreated domestic wastewater:

Contaminant	Concentration
Inlet wastewater	
TSS, mg/l	309
TCOD, mg/l	410
TKN, mgN/l	42.6
$N\text{-}NH_3$, mgN/l	30.2
P_{tot}, mgP/l	6

The nutrients content in the waste sludge was computed on the basis of literature data for BNR processes, where percentages of 7% (Beccari et al., 1993) and 3.5% (Pöpel & Jardin, 1993) of total N and total P respectively are reported with reference to TS.

The waste activated sludge production is based on simulation results with Activated Sludge Model No. 2 (Battistoni et al., 1998) applied to wastewater treatment plants without primary sedimentation and with the addition of liquid fermented OFMSW, considering the scenario where 250 l wastewater and 0.3 kgOFMSW are produced per capita per day. The reported values are the average of two simulations performed at two different temperatures (12 °C and 16 °C) and using SRT of 16 and 20 days respectively.

The values obtained for the gaseous products (nitrogen and carbon dioxide) have been calculated as the difference between the inlet contributions of total nitrogen and total carbonaceous matter (expressed as COD) and the loss from the system of the same compounds considering both the liquid effluent and the waste sludge.

9.4.4. Mass balance of the co-digestion of residual solid from fermentation of the OFMSW and of the WAS

The mass balance of the co-digestion of the residual solid from OFMSW fermentation and of WAS (Figure 9.8) accounts for the following contributions:

- the feed is composed by two different flux lines: the first is represented by the waste activated sludge produced in the wastewater treatment plant, the second by the residual solid deriving from OFMSW fermentation, after its separation by screw-pressing from the liquid phase;
- the effluent is composed by:
    1. the biogas which can be burnt to provide the system's heating and the generation of electrical power in co-generation units for the plant's requirement and for extra uses;
    2. a solid residual which must be stabilised and/or disposed (composting and land application, landfill, etc.).

The computations were performed according to the previous mass balance for the OFMSW fermentation and to the mass balance for a BNR wastewater treatment plant. An average value of 0.5 m^3/kgTVS is assumed as biogas production in these conditions (Cecchi et al., 1994). The waste activated sludge production has been calculated as described in the previous paragraph.

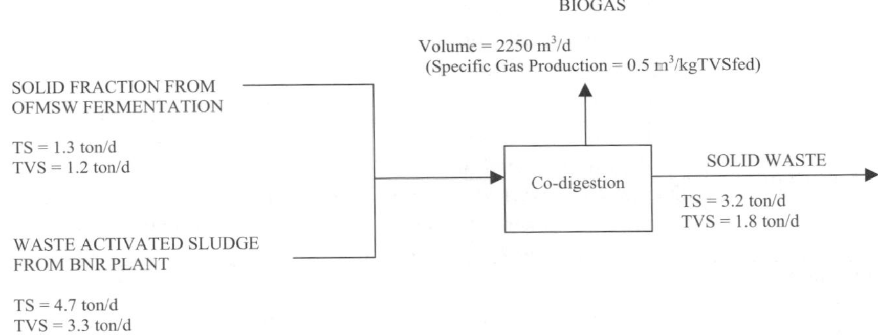

Figure 9.8. Schematic mass balance of the anaerobic co-digestion of residual solid from OFMSW fermentation on a 100,000 PE basis (daily OFMSW production: 300 g/PE; daily wastewater production: 250 l/PE). (Mass balances are closed with an approximation of ±10%.)

9.4.5. Economic balance for the integrated process

The aim of this paragraph is to make a comparison and to analyse the managing costs of two integrated processes: the first is a conventional plant for C–N removal with the addition of methanol as external carbon source (as representative status of the applied engineering at this moment); the second is a BNR plant working following the integrated approach AF-BNR-SCP where the external carbon source is the fermented OFMSW.

The calculations are made on a 100,000 PE basis and only the varying parameters are taken into account, leaving aside all those aspects that do not vary (e.g. electromechanical structures, process monitoring, analytical controls).

The parameters that have been considered are:

- OFMSW handling: it indicates the disposal of the total amount of the OFMSW coming from the considered basin (conventional approach) or the disposal of the sole residue from selection line (integrated approach, assuming it as 10% of the total collected);
- chemicals for process correction: mainly iron sulphate ($FeSO_4$ for phosphorus chemical precipitation in case the organic carbon implementation is not enough for the enhanced biological phosphorus removal process) and methanol for the conventional CN process;
- biogas as overproduction due to external carbon addition as a negative cost;
- oxygen demand;
- waste sludge disposal;
- personnel;
- amortization.

An economic analysis of the technical management of the conventional CN approach and the integrated BNR approach following the CEC Dir. 91/271 for 100,000 PE is given by the following scheme.

Operation	**Conventional approach CN process without primary sedimentation**			**Integrated approach BNR process**		
	Quantity	***Unit cost (Euro)***	***Total cost (Euro)***	***Quantity***	***Unit cost (Euro)***	***Total cost (Euro)***
OFMSW selection residue disposal	30 ton/d	77.5	+ 2325	3 ton/d	77.5	+ 232.5
Methanol	1140 kg/d	0.14	+ 160	—	—	—
$FeSO_4$	1303 kg/d	0.25	+ 326	81 kg/d	0.25	+ 20.3
Biogas production	990 m^3/d	0.18	− 178	1510 m^3/d	0.18	− 272
O_2 requirements	8640 kg/d	0.09	+ 778	9600 kg/d	0.09	+ 864
Sludge disposal	12.5 ton/d	77.5	+ 969	11.4 ton/d	77.5	+ 886
Number of people	10	106	+ 1060	11	106	+ 1167
Amortization C addition, selection and fermentation section			+ 19			+ 186
Total			**+ 5458**			**+ 3083**

9.4.6. Some guidelines for design

When designing a plant following the integrated approach, here are some important parameters that should be taken into account both for the OFMSW treatment plant and for the wastewater treatment plant:

- The PE number which are served by the wastewater treatment plant must be equal to civil and productive units connected by sewers existing or foreseen.
- The hydraulic loading must be forecast or calculated on the basis of the actual data regarding the sewers.
- The mass loading must be computed by the sum of sewers loading and acid fermented OFMSW loading.

- The OFMSW mass loading must be calculated on the basis of the plant potentiality and on the specific OFMSW production (150–300 g OFMSW/PE depending on local situation);
- The plant's capacity should be the same for both wastewater treatment and OFMSW collection;
- If the fermentate production is greater than the one needed in the plant it is possible:
    - to sell the acid mixture as external carbon source for other wastewater treatment plants. In that case particular attention must be paid to the prevention of any odour nuisance during stocking and transportation;
    - to use this material as additional biogas source in the co-digestion step, improving the energy balance of the whole plant.
- The line for secondary treatment of wastewater can be either a conventional one (carbon and nitrogen removal) or an advanced one (biological removal of carbon, nitrogen and phosphorus). Whatever the configuration, the design must account for the addition of the acid fermented OFMSW since this can lead to a small hydraulic overloading and a massive overloading of carbon. All these aspects are quantifiable over the fermentate main composition.
- The main effects of the fermentate and of the relative mass overloading are an increase in oxygen requirement in the biological treatment and a larger sludge production. All these effects are computable with consolidated design methods and parameters.
- The OFMSW can be supplied only by separate collection operations or by source separate collection in markets, canteens, etc. The collection can be performed in a loose way or in completely biodegradable bags. If polyethylene bags are used, an effective separation unit must be provided.
- The plant's location must be chosen to reduce transport pathways and to allow easy OFMSW discharging; furthermore, an adequate operating area must be designed both for trucks and for container lodging.

In particular, when designing the fermentation section it must be considered that:

- The aims of acid anaerobic fermentation are the hydrolysis and fermentation of the organic substrates with the consequent production of a liquid effluent to send to the wastewater treatment plant and a solid effluent to be sent to anaerobic co-digestion.
- The fermenter design can be performed on the HRT basis since no other method exists nowadays.
- The hydraulic retention time (HRT) to be applied must stand in the range 1–5 days with no chemical addition for pH control. If methanogenic conditions should set up for temporary and protracted feed deficiency, the flux must be deviated towards the anaerobic mesophilic digester.

- The fermenter must be provided with a footbridge and connected to a fast emptying device and to the anaerobic digester.
- The choice of the stirring mode could be a mechanical system inside the reactor. The power density to guarantee a complete mixing will be calculated according to biochemical engineering principles.
- The design must be comprehensive of the thermal balance and the fermenter temperature control devices, relative both to heat exchangers and to their control system. A maximum fluctuation of $\pm$2°C would be appropriate.
- The main operational parameters (pH, T, level, VFAs, RBCOD, degree of dissolution, etc.) should be monitored and recorded.
- Disposal of sludges must occur considering their characteristics. The pipes should always have a diameter larger than 4 inches (even for small operations) without any elbow or bottle-neck, pumping devices must be the volumetric kind (membrane, etc.) with a diameter equal to the pipes, security valves must be set on the main lines together with anti-clogging devices.

As regards the anaerobic digestion step it must be underlined that the aim of this operation is the energy recovery thanks to the biogas overproduction during the anaerobic co-digestion of the solid residual from the fermentation process together with the waste-activated sludge and the primary sludge (if primary settlers are used). Following the conventional methods for anaerobic digesters working with primary, mixed or waste-activated sludge, some aspects should be accounted for:

- mass loadings must account for the mass overloading of the solid residual of the OFMSW fermentation;
- the hydraulic retention time (HRT) can be assimilated to the one used for the anaerobic digestion of only wastewater treatment plant waste sludge;
- the maximum admissible loading in mesophilic digesters with medium loadings is equal to 3 kgTVS/m^{3}d;
- the biogas production can be calculated on a mass basis considering the overloading by the solid residual from the OFMSW fermentation;
- the solid residual from the co-digestion process can be calculated assuming a volatile solid removal percentage of 40–45%;
- the digested sludge and the residue from the OFMSW treatment follow the same disposal pathway as conventional digested waste sludge.

9.5. CASE STUDY: TREVISO CITY WASTEWATER TREATMENT PLANT

The new wastewater treatment plant of Treviso city (Veneto Region, Italy) use enhanced nutrient removal to protect Sile river protected area from eutrophication.

The logic followed in the plant's realization is to use the pre-existing conventional plant and to expand it with new sections to allow a potentiality of 70,000 PE. The pre-existing section is reserved to oxidation processes and serves 20,000 PE; the new section (50,000 PE) operates in a BNR mode with modular characteristics and the possibility of various process configurations avoiding primary sedimentation to preserve more COD for nitrogen and phosphorus removal steps. The process scheme is reported in Figure 9.9.

The domestic wastewater entering the plant contains low concentrations of readily biodegradable chemical oxygen demand (RBCOD, about 10% of the incoming COD) which is needed for extensive nutrient removal.

The blocks for RBCOD production and phosphorus removal from supernatants consist in two demonstrative areas serving the 50,000 PE (new line): the first operates the source separated OFMSW fermentation (OFMSW pilot area) the second the phosphorus removal/recovery as struvite from the supernatant coming from waste sludge anaerobic stabilization (SCP pilot area).

The demonstrative area for OFMSW fermentation is based, in addition to separate collection, on a refining step of OFMSW characteristics through a simplified line for substrate automated selection, in order to eliminate the rawest material. The selected OFMSW is then shredded and sent to an anaerobic fermenter (50 m^3 controlled CSTR) that works with an HRT within 3 and 6 days, depending on the substrate availability and the yields to be obtained. The fermenter effluent, after solid–liquid separation in a screw-press, is split into two fractions: the liquid flux, which is sent to wastewater treatment and the solid flux, which is sent to anaerobic mesophilic co-digestion section, together with the waste-activated sludge.

The second demonstrative area has the aim of removing the amount of phosphorus released in the supernatants during anaerobic digestion of waste activated sludge and solid fraction of OFMSW. The description is reported elsewhere (Battistoni et al., 2001).

The main volumetric and operative conditions of the new line in Treviso wastewater treatment plant are reported in Table 9.8.

Table 9.8. Sections volume and some operative conditions for the new line in Treviso wastewater treatment plant.

New section characteristics		
Wastewater treatment		
Influent average flow rate (Q_a), m^3/d		14,000
Maximum flow rate (Q_m), m^3/d		21,600
Peak flow rate (Q_p), m^3/d		21,000
Degritting, m^3		181
Pre-treatments, m^3		628
Pre-anoxic zone, m^3		400–1,200
Anaerobic zone	Volume, m^3	700–1,200
	Sludge recycle ratio, %	up to 150%
Anoxic zone	Volume, m^3	1,600–2,200
	Mixed liquor recycle ratio, %	up to 250%
Oxidation/nitrification zone, m^3		5,500
Chlorination, m^3		250
MLSS, kg/ m^3		3.47
Sludge load, kg BOD/ kg MLSS d		0.125
Total volume biological reactor, m^3		9,000
Secondary settler surface, m^2		1,300
Hydraulic loading rate, m/h		0.45
Peripherical height, m		2.15
Sludge treatment		
Thickener, m^3		210
Anaerobic digester, m^3		2 200
Digester organic loading rate, kgTS/ m^3d		1.75
HRT, d		23

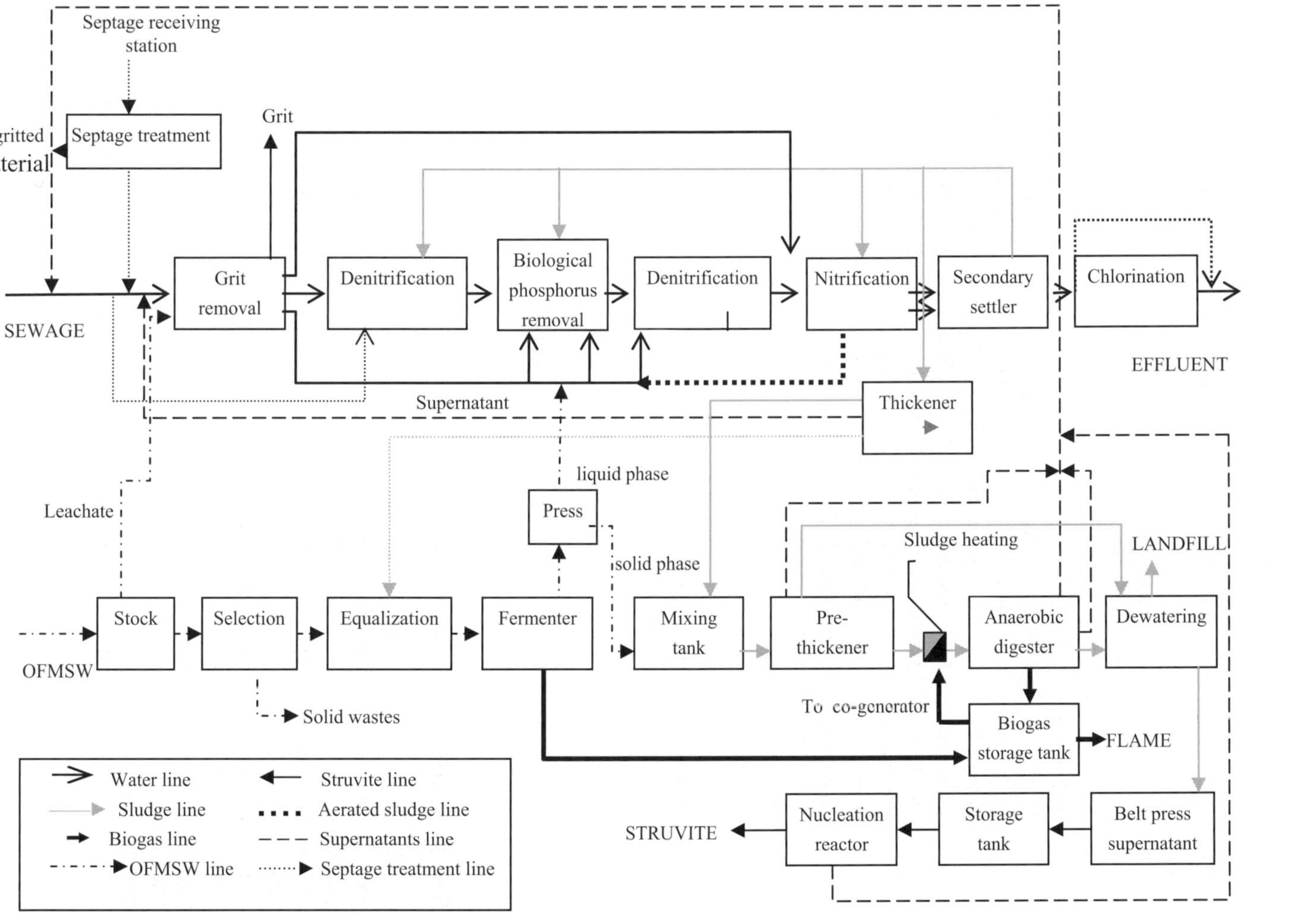

Figure 9.9. Logic scheme for the new wastewater treatment plant of Treviso city (Veneto Region, Italy).

The BNR plant and sludge treatment sections have the following characteristics.

1. BNR section: the 50,000 PE new line is able to biologically remove, besides carbon and nitrogen, also phosphorus following the configuration three-stage Phoredox (modified Johannesburg). The influent is firstly degritted and the sand is wasted while the organic fraction is recycled at the headworks. After, the influent is directly sent for biological treatment. This section has been designed with modular characteristics in order to be able to change the single tanks volume. This allows a high operating flexibility both respect to influent load and to its characteristics. The initial operative conditions consist in a pre-anoxic zone where only recycle activated sludge is sent for nitrates denitrification. The following sections are the anaerobic and anoxic ones where the fermented OFMSW is added. More precisely, the addition is planned in the anaerobic zone but it can be carried out also in the anoxic and pre-anoxic ones. Eventually, the oxidation/nitrification tank has been designed to operate at low loading conditions.
2. Sludge treatment section: in this section the anaerobic mesophilic single-stage digestion takes place. The sludge is sent for digestion after thickening and blending with the solid fraction coming out from the fermentative OFMSW area. The sludge coming out from anaerobic digestion is added with a polyelectrolyte and sent to mechanical dewatering. The supernatant from anaerobic digestion recycle is treated in the struvite section.

9.6. PERSPECTIVES AND CONCLUSIONS

In this section are summarised the advantages offered by the AF-BNR-SCP process in terms of its immediate application, which means a proper use of BNR processes for nutrient removal, OFMSW final disposal and phosphorus reclamation and, in perspective, as an indispensable tool for sewage sludge minimization.

Integration today between the organic fraction of municipal solid waste treatment and the wastewater treatment, that is the AF-BNR-SCP process application, allows the passage from a straight-line logic (solid/liquid waste → final disposal) to a circular logic (solid/liquid waste → treatment integration → resource recovery).

The fermentation of the OFMSW leads to an integrated process between solid and liquid waste treatments: by means of this process, in fact, a diversion of the OFMSW to be sent to landfills is obtained, since nearly 75% of it is removed as liquid phase to be added in the wastewater treatment plant and 25% is stabilised by co-digestion together with sewage and primary sludge. Using this liquid effluent from the fermenter as external carbon source for the BNR process it is possible to have two advantages: to double the contribution of RBCOD, in terms of VFA and lactic acid, in respect to the available RBCOD in the wastewater (this result should be compared with sludge fermentation where the highest RBCOD increase obtained is about 80%); and to increase/correct the influent C:N ratio (which is impossible only with the sludge fermentation). That is a powerful tool for phosphorus and nitrogen removal, without external chemical addition or with zero/negative cost for external chemical addition, and the ability to obtain a treated water as clean as economically possible and more suitable for reuse. Furthermore, the possibility of reclamation of the phosphorus present in the influent water must be underlined, since it is a non-renewable source, again without addition of chemicals.

Summarizing, the substantial benefits of the AF-BNR-SCP process can be outlined as:

- the recovery of renewable energy by methanization of the carbonaceous substrates (solid fraction from the OFMSW fermentation and waste-activated sludge with or without primary sludge);
- the very efficient removal of nitrogen to gaseous N_2 through assisted pre-denitrification step;
- the reclamation of phosphorus or the recovery of phosphorus and ammonia for agricultural purposes;
- the environmentally friendly disposal of a significant and otherwise difficult to dispose fraction of the MSW;
- the economical reliability of the integrated process mainly thanks to the biogas production and to the cost reduction of the OFMSW disposal;
- the possibility to perform a reduction of waste sludge production since the abatement of released nutrients related to this goal can be well performed with the AF-BNR-SCP process.

The development of this last concept must be focused as a further step for the whole AF-BNR-SCP process application and optimization. This in the frame of the solutions of the main critical problems to face in forthcoming years as much as concerns the wastewater treatments:

- The reduction or entire elimination of the amount of activated sludge produced in the wastewater treatment process so as to reduce the environmental and health impact of sludge and the cost of disposal which are still a problem nowadays.

- The achievement of higher standard quality treated water through enhanced biological nutrient removal and reclamation of phosphorus present in the influent water without any external chemical addition.

Many studies are at the moment ongoing about the reduction in waste-activated sludge which can be achieved by reducing the growth rate of the organisms present in the sludge and by chemical, physical or biological lysis of waste sludge produced (Cecchi et al., 2000; van Lier et al 2000). In this latter case, the non-converted products of lysis have to be fed back to the head-work to remove mainly nitrogen and phosphorus. To do that it needs a suitable process like the AF-BNR-SCP.

9.7. ACKNOWLEDGEMENTS

The authors thank Dr. Laura Innocenti for her important collaboration in the preparation of the manuscript, and Prof. Mario Beccari for his precious observations in revising the paper.

9.8. REFERENCES

Abu-Ghararah, Z. H. and Randall, C.W. (1991). The effect of organic compounds on biological phosphorus removal. *Wat. Sci. Tech.* **23**(4–6), 585.

Aesoy, A. and Ødegaard, H. (1994). Nitrogen removal efficiency and capacity in biofilms with biologically hydrolysed sludge as a carbon source. *Wat. Sci. Tech.* **30**(6), 63–71.

Battistoni, P., Pavan, P., Cecchi, F., Mata-Alvarez, J. and Majone, M. (1998). Integration of civil wastewater and municipal solid waste treatments. The effect on biological nutrient removal processes. In: *European Conference on New Advances in Biological Nitrogen and Phosphorus Removal for Municipal or Industrial Wastewaters*, 12–14 October, Narbonne, France, pp. 129–137.

Battistoni, P., Beccari, M., Cecchi, F., Majone, M., Musacco, A., Pavan, P. and Traverso, P. (1999). Una gestione integrata del ciclo dell'acqua e dei rifiuti. Fondamenti, stato dell'arte, ingegneria di processo. Ed. Franco Angeli.

Battistoni P., De Angelis A., Pavan P., Prisciandaro M. and Cecchi F. (2001). Phosphorus removal from a real anaerobic supernatant by struvite crystallization. *Water Research* **35**(9), 2167–2178.

Beacham, A., Seviour, R. and Lindrea, K.C. (1992). Polyphosphate accumulating abilities of Acinetobacter isolated from a BNR pilot plant. *Wat. Res.* **26**(1), 121–122.

Beccari, M., Passino, R., Ramadori, R. and Vismara, R. (1993). Rimozione di azoto e fosforo dai liquami. Hoepli.

Bhattacharya, S.K., Madura, R.I., Walling, D.A. and Farrell, J.B. (1996). Volatile solids reduction in two phase and conventional anaerobic sludge digestion. *Wat. Res.* **30**(5), 1041–1048.

Bolzonella, D., Innocenti, L., Pavan, P. and Cecchi, F. (2001). Denitrification potential enhancement by addition of the anaerobic fermented of the organic fraction of municipal solid waste. *Wat. Sci. Tech.* **44**(1), 187–194.

Burrell, P., Keller, J. and Blackall, L.L. (1999). Characterization of the bacterial consortium involved in nitrite oxidation in activated sludge. *Wat. Sci. Tech.* **39**(6), 45–52.

Cecchi, F., Battistoni, P., Pavan, P., Fava, G. and Mata-Alvarez, J. (1994). Anaerobic digestion of OFMSW and BNR processes: a possible integration. Preliminary results. *Wat. Sci. Tech.* **30**(8), 65–72.

Cecchi, F., Bolzonella, D., Innocenti, L., Pavan, P. and Battistoni, P. (2000). Excess sludge reduction and advanced biological nutrient removal wastewater treatment processes In: *Proc. 1st World Water Congress of the Int. Water Association – Poster presentation*, 3–7 July 2000, Paris, France.

Choi, E., Lee, H.S., Lee, J.W. and Oa, S.W. (1996). Another carbon source for BNR systems. *Wat. Sci. Tech.* **34**(1–2), 363–369.

Comeau, Y., Hall, K.J., Hancock R.E.W. and Oldham, W.K. (1986). Biochemical model for enhanced biological phosphorus removal. *Wat. Res.* **20**, 1511–1521.

Cuevas-Rodriguez, G., Gonzalez-Barcelò, O. and Gonzalez-Martinez, S. (1998). Wastewater fermentation and nutrient removal in sequencing batch reactors. *Wat. Sci. Tech.* **38**(1), 255–264.

EEA (2002). Review of selected waste streams. Technical report by the European Environment Agency, January 2002, pp. 48.

Ekama G.A. and Marais G.v.R. (1984). Biological nitrogen removal. In: *Theory, Design and Operation of Nutrient Removal Activated Sludge Processes*, Water Research Commission. Published by the Water Research Commission, Pretoria.

Ekama, G.A. and Wentzel, M.C. (1999). Difficulties and developments in biological nutrient removal technology and modeling. *Wat. Sci. Tech.* **39**(6), 1–11.

Ghosh, A., Conrad, J.R. and Klass, L. (1975). Anaerobic acidogenesis of wastewater sludge. *J. Wat. Poll. Control Fed.* **47**(1), 30–44.

Horan, N.J. (1990). *Biological Wastewater Treatment Systems. Theory and Operation.* John Wiley & Sons, Chichester, UK.

IAWQ Task Group on Mathematical Modeling for Design and Operation of Biological Wastewater Treatment processes (1995). Activated Sludge Model No. 2. *IAWQ Scientific and Technical Report* **3**, 32 pages.

Isaacs, S.H. and Henze, M. (1995). Controlled carbon source addition to an alternating nitrification–denitrification wastewater treatment process including biological P removal. *Wat. Res.* **29**(1), 77–89.

Knight, G., Seviour, R., Soddell, J., McDonnell, S. and Bayly, R. C. (1995). Metabolic variation among strains of Acinetobacter isolated from activated sludge. *Wat. Res.* **29**(9), 2071–2084.

Kristensen, G.H., Jorgensen, P.E., Strube, R. and Henze, M. (1992) Combined pre-precipitation, biological sludge hydrolysis and nitrogen reduction – a pilot demonstration of integrated nutrient removal. *Wat. Sci. Tech* **25**(5–6), 1057–1066.

Lotter, L. H. and Pitman, A.R. (1992). Improved biological phosphorus removal resulting from the enrichment of reactor feed with fermentation products. *Wat. Sci. Tech.* **25**(5–6), 943–953.

Mata-Alavrez, J. and Cecchi, F. (1989). Joint anaerobic digestion of sewage sludge and sorted organic fraction of municipal solid waste to attain the energetic autonomy in wastewater treatment plants. *Workshop of the FAO-CNRE: Biogas production technologies,* 10–13 April, Saragossa, Spain.

Metcalf & Eddy (1991). *Wastewater engineering: treatment, disposal, reuse*. McGraw-Hill.

Mino, T., Arun, V., Tsuzuki, F. and Matsuo T. (1987). Effects of phosphorus accumulation on acetate metabolism in the biological phosphorus removal process. In: *Advances in water pollution control. Biological phosphate removal from wastewaters*. (ed. R. Ramadori), pp. 27–38. Pergamon Press Oxford.

Moser-Engeler, R., Udert, K. M. and Siegrist, H. (1998). Products from primary sludge fermentation and their suitability for nutrient removal. *Wat. Sci. Tech.* **38**(1), 265–273.

Pavan, P., Battistoni, P., Musacco, A. and Cecchi, F. (1994). Mesophilic anaerobic fermentation of SC-OFMSW: a feasible way to produce RBCOD for BNR processes. In: *Proc. Int. Symp. On Pollution of the Mediterranean Sea*, 2–4 November, Nicosia, Cyprus. pp. 561–570.

Pavan, P., Battistoni, P., Amoroso, E., Cecchi, F. and Mata-Alvarez J. (1998). La triturazione sottolavello della frazione organica dei RSU. Una strategia attuale per potenziare la rimozione dei nutrienti in impianti per il trattamento di acque reflue civili. In: *Acque Reflue e Fanghi. Settimana Ambiente*, February, 1998.

Pavan, P., Battistoni, P., Bolzonella, D., Innocenti, L., Traverso, P. and Cecchi, F. (2000). Integration of wastewater and OFMSW treatment cycles: from pilot scale to industrial realization. The new full scale plant of Treviso (Italy). In: *Proc. 4th Int. Symp. On Environmental Biotechnology*, 10–12 April, Noordwijkerhout, The Netherlands.

Pitman, A.R., Ventre, S.L.V. and Nicholls, H.A. (1983). Practical experience with biological P removal plants in Johannesburg. *Wat. Sci. Tech.* **15**, 233–259.

Pitman, A.R., Lötter, L.H., Alexander, W.V. and Deacon, S.L. (1992). Fermentation of raw sludge and elutriation of resultant fatty acids to promote excess biological phosphorus removal. *Wat. Sci. Tech.* **25** (4–5), 185–194.

Pöpel, H.J. and Jardin, N. (1993). Influence of enhanced biological phosphorus removal on sludge treatment. *Wat. Sci. Tech.* **28**(1), 263–271.

Rabinowitz, B. and Oldham, W.K. (1985). Excess biological P removal in the activated sludge process using primary sludge fermentation. In: *Proc. Annual Conf. Of the Canadian Society for Civil Engineering*, Saskatoon, pp. 387–397.

Rabinowitz, B., Koch, F.A., Vasso, T.D. and Oldham, W.K. (1987). A novel operational mode for a primary sludge fermenter for use with the enhanced biological phosphorus removal process. In: *Proc. IAWPRC Spec. Conference on Biological Phosphate Removal form Wastewater*, pp. 349–356.

Randall, C.W., Barndard, J.L. and Stensel, H.D. (1992). *Design and retrofit of wastewater treatment plants for biological nutrient removal.* Water Quality Management Library, Vol. 5. Technomic Publishing Co. Inc. Lancaster, USA.

Rozzi, A., Bortone, G., Canziani, R., Andreottola, G., Ragazzi, M., Pugliese, M. and Tilche A. (1995). Mesophilic and psychrophilic fermentation of primary sludge for RBCOD production in a Phostrip® plant. In: *Mediterraneanchem – Int. Conf. On Chemistry and the Mediterranean Sea*, May 23–27, Taranto, Italy.

Sans-Mazon, C., Mata-Alvarez, J., Bassetti, A., Pavan, P. and Cecchi F. (1992). Pilot-scale demonstration for volatile fatty acid production by anaerobic digestion of municipal solid waste. In: *Waste Management International*, (ed. K. J. Thomé-Kozmiensky), Vol. 1, pp. 645–652.

Skalsky, I. and Daigger, G.T. (1995). Wastewater solids fermentation for volatile acid production and enhanced biological phosphorus removal. *Wat. Env. Res.* **67**(2), 230–237.

Traverso, P., Pavan, P., Innocenti, L., Mata-Alvarez, J. and Cecchi F. (2000). Anaerobic fermentation of source separated mixtures of vegetables and fruits wasted by supermarkets. In: *Proc. 4th Int. Symp. On Environmental Biotechnology*, 10–12 April, Noordwijkerhout, The Netherlands.

U.S. Environmental Protection Agency (1975). *Process design manual for nitrogen control*. Office of Technology Transfer, Washington DC.

Van Lier, J.B., Tilche, A., Ahring, B.K., Macarie, H., Moletta, R., Dohanyos, M., Pol, L.W.H., Lens, P. and Verstraete, W. (2000). New perspectives in anaerobic digestion. In: *Proc. 1st World Water Congress of the Int. Water Association*, 3–7 July 2000, Paris, France.

Virtutia, A.M., Mata-Alvarez, J., Sans, C., Costa, J. and Cecchi, F. (1992). Chemicals production from wastes. *Env. Tech.* **13**, 1033–1041.

10

Products, impacts and economy of anaerobic digestion of OFMSW

W. Edelmann

10.1. INTRODUCTION

The aim of digesting the OFMSW is to treat organic wastes at reasonable costs while closing ecological cycles. The anaerobic digestion (AD) of organic solid wastes produces two main products, i.e. biogas and compost. But additionally, there are some by-products, such as surplus water or gaseous emissions from compost and burnt biogas, which have also to be taken into consideration.

This chapter deals on the one hand with the products of AD including their environmental impacts. On the other hand, an economic comparison with competing technologies to treat organic wastes will show the ranking of AD as well as some details about investments and specific treatment costs.

Edited by J. Mata-Alvarez. ISBN: 1 900222 14 0

10.2. BIOGAS UTILIZATION

In general, digestion plants for the treatment of OFMSW show a minimal treating capacity of 10,000 Mg per year. Roughly, 100 kg of at-source separated biogenic waste may be expected yearly per population equivalent. Digestion plants for OFMSW will therefore treat the 'green' waste of at least 100,000 inhabitants producing per year around 1 million m^3 of biogas or more according to the plant treating capacity and depending on the properties of the incoming waste.

If burnt, 1 million m^3 of biogas will set free an enormous amount of heat. There will be a surplus of about 200 MW after covering the heat demand of the plant itself. This surplus could be used to supply big buildings with hot water in the plant surroundings. To take the heat for heating of buildings shows the disadvantage of generating waste heat in summer. And, in general, there are unfortunately no industries in the immediate neighbourhood of an AD plant with a correspondingly high heat demand also during summer season.

That is the reason why biogas is usually converted into electricity using the waste heat of the generator to cover the (relatively small) heat demands of the plant. The surplus electricity is fed into the national grid and sold to consumers. In Switzerland, electricity from cogeneration on OFMSW biogas plants has got the label 'naturemade star' due to an excellent performance in the environmental impact assessment of its generation (Schleiss and Edelmann, 2000). This label allows the electricity to be sold at a significantly higher market price than when selling fossil or nuclear current.

There are different types and sizes of cogenerator available. Today, they may convert up to more than 30% of the biogas energy into electric power at thermal efficiencies of about 60% (Piccini et al., 2000). An average composition of biogas from OFMSW is shown in Table 10.1. In addition, there may be traces of ammonia (NH_3). Water, carbon dioxide, hydrogen sulphide and ammonia may be eliminated by different scrubbing technologies depending on the biogas use.

Table 10.1. Typical composition of biogas from OFMSW.

Components		**Concentration (Vol, %)**
Methane	CH_4	55–60 (50–75)
Carbon dioxide	CO_2	35–40 (25–45)
Water	H_2O	2 (20 °C) – 7 (40 °C)
Hydrogen sulphide	H_2S	20–20,000 ppm (2%)
Nitrogen	N_2	< 2
Oxygen	O_2	< 2
Hydrogen	H_2	< 1

The critical temperature of methane is −82.5 °C; even with very high pressures it is not possible to liquefy methane at higher temperatures. This is probably the most important bottleneck in biogas utilization: biogas cannot be stored over long periods at reasonable costs, i.e. it has to be used immediately or within only a few hours. The conversion into electric current shows the advantage that it can be converted into expensive electricity during hours with high electricity demand. Thus this solution needs a gas storage capacity of about half a day to store the night production.

The CO_2 content of the biogas may depend on different factors, such as substrate composition, but also pH, reactor geometry/pressure, temperature, HRT, process design (plug flow/CSTR), etc. Carbon dioxide is an inert ingredient of biogas which may act slightly acidic, i.e. corrosive, within the condensing liquid. For thermal use CO_2 elimination is not necessary. On the contrary, for feeding the gas into a public gas pipeline and for its storage under high pressure (fuel for cars) CO_2 has to be cleaned off.

Water vapour creates condensation in the tubing. To prevent clogging and corrosion, it has to be condensed in water traps. This is especially important at high pressures, when higher amounts of corrosive components, such as CO_2 and H_2S, are dissolved, making the gas more aggressive.

Hydrogen sulphide is the component mainly responsible for corrosion. Gas scrubbing may be necessary, especially if the biogenic waste contains large portions of protein. More than about 100 ppm are not suited for use in burners or cogenerators. For humans, concentrations of more than 1000 ppm may be lethal when inhaled. Today, cheap and environmentally safe devices are available for biological sulphur elimination by adding small amounts of oxygen to the biogas in fixed-film gas washers (Hoejme Tecknik, 2000).

Free ammonia (NH_3) is in chemical equilibrium with the ionic ammonium (NH_4^+) depending on pH and temperature. An increase in the pH of the slurry by one unit will increase the ammonia concentration by a factor of ten. Ammonium is freed during the anaerobic breakdown of compounds, such as proteins and nucleic acids. The digestion of OFMSW normally takes place at rather elevated pH values of 7.5 or higher. Especially at thermophilic temperature ammonia may be set free, causing corrosion as well as the formation of NO_x while being combusted. Table 10.2 summarises where gas cleaning seems to be reasonable in addition to a simple condensation of the water vapour.

Table 10.2. Necessity for gas cleaning.

	H_2S	H_2O	CO_2
Combustion in burners	> 0.1%	No	no
Cogeneration	> 0.05%	No	no
Fuel for cars (compressed in bottles)	yes	Yes	yes
Feed into public gas pipeline	yes	Yes	yes

Different principles may be applied for gas cleaning, such as dry cleaning with different materials for absorption or adsorption (Shoemaker and Visser, 1999). Adsorption is a reversible process. In general, there are two columns: one for adsorption and one for desorption, i.e. regeneration by reversion of pressure, elevation of temperature and/or stripping with an inert gas. Other possibilities for cleaning the gas are cooling and compression for water and carbon dioxide removal or the use of membranes. Recently, technologies using molecular sieves have been developed to clean biogas to natural gas quality (for driving vehicles and feeding into public gas pipelines). Cleaning biogas to natural gas quality requires about 8% of the energy content of the biogas being treated (Egger, 1997). Actually, this amount has been reduced to less than 5%, however (Kompogas, 2001).

As already mentioned above, biogas from OFMSW is normally used for cogeneration of electricity and heat. Because large digestion plants produce a large amount of gas, it may be interesting to upgrade the biogas for driving gasoline/biogas hybrid cars. Depending on national laws, this may be the most financially interesting solution: If there is a reduction on fuel taxes for renewable energies, the (eventually) additional costs for converting a car into a dual fuel (or pure natural gas) vehicle are paid back very quickly. A button on the instrument panel enables the driver to switch from gasoline to biogas while driving. This ensures a high level of flexibility: The biogas increases the range of a conventionally powered vehicle.

A middle-class car requires the energy set free by 100 kg of organic waste to drive 100 km. Gas-powered cars are gaining ground continuously. In Switzerland, cars like BMW, Fiat, Honda, Mercedes, Opel, Renault, Volvo, VW, etc. as well as heavy trucks run on biogas (Kompogas, 2000). The gas is stored at 250 bar in bottles. Fuelling takes only about 2–3 minutes, allowing the car to be driven 200–250 km (Figure 10.1). Pure natural gas cars, such as the Fiat *Multipla*, drive more than 500 km without refuelling. Actually, it is intended to convert electro/gasoline hybrid cars (e.g. Nissan *Tino* or Toyota *Prius*) for additional biogas use or to use fuel cell technology to convert biogas directly into electricity. This would increase the range of action significantly – in perfect comfort – due to the lower fuel consumption of modern hybrid cars.

Figure 10.1. Fueling of biogas into a hybrid car (biogas/gasoline).

The use of biogas for driving vehicles shows an especial advantage, if there is no demand for the surplus heat of cogeneration. For the same reason, the processed biogas is fed into the public natural gas pipelines at some digestion plants (*ibidem*). This allows the refuelling of cars throughout the country. At the same time, biogas may be used for cooking or for industrial processes valorizing its total energy content.

10.3. PRODUCTION AND MARKETING OF COMPOST

The performance of AD technology depends on the quality of the substrate input (Saint-Joly et al., 1999). For the output, the first and most important rule is: the compost never shows better quality than its input, i.e. the OFMSW. While digesting, the inorganic compounds of the organic solid waste, such as heavy metals, are concentrated due to the volatilization of a considerable part of the dry matter converted into biogas. However, a good compost quality is the precondition for a successful compost marketing. For this reason Switzerland has already forbidden by law in 1986 to apply on agricultural fields those organic fertilizers that previously have been in contact with other garbage (Bundeskanzlei, 1986).

Properties and qualities of organic fertilizers are regulated by national laws (e.g. FAL, 1999). As shown in Table 10.3, there may be significant differences between countries or even between similar substrates. Further differences concerning biogas comparing legislation of different countries were published by Colleran (1999) and Nordberg (1999).

The compost quality consists of at least four aspects: (i.) plant nutritional value, i.e. inorganic nutrients as well as content of organic compounds for improving the structure and the humus content of the soil; (ii.) content of undesired goods, such as plastic, metal, glass etc.; (iii.) content of toxic compounds, such as heavy metals or AOX; and (iv.) good hygienic stabilization, i.e. absence of (phyto-)pathogens. On the one hand, pathogens can be destroyed by the process itself, i.e. by a high temperature over a period long enough to reduce them to a negligible number (therefore: preferably thermophilic MSW-digestion and plug flow design). On the other hand, the first three aspects are strongly correlated with the input quality. Therefore, it is strongly recommended to digest only source-separated OFMSW whenever possible. If the OSW comes into contact with the 'grey' waste fraction, toxic substances are mobilised by pH variations, by changing humidity contents and by organic acids set free by the microorganisms already present during collection. Those will convert metals into ionic forms, which afterwards will pollute the compost. A later technical separation of non-source-separated wastes will bring considerable additional contaminations (Bilitewski, 1989), such as pollution by heavy metals originating from destroyed batteries, etc.

In Switzerland, dumping of biologically non-stabilized OSW has been forbidden by law since 2000; Germany will follow in 2002. Therefore in some European countries such as Germany the so-called 'cold treatment of the grey fraction' ('kalte mechanisch-biologische Restmüllbehandlung') is being discussed and also realised in competition with thermal treatment (incineration, pyrolysis) (Wiemer and Kern, 1999). The organic fraction of the grey MSW is sorted out mechanically for a subsequent biotechnological treatment by AD and/or composting in order to reduce its biological activity as much as possible. However, after an extraction of the woody fraction for thermal use, the remaining compost has to be dumped because of its poor quality (Six, 1999).

Table 10.3. Legal limits for heavy metals in composts and sewage sludge in Germany (Bundesrepublik Deutschland, 1992; Bidlingmaier, 1994) and Switzerland (BUWAL, 1990; Bundeskanzlei, 1986).

Heavy metal	**Sewage sludge Germany** (mg/kg DM)	**Compost Germany** (mg/kg DM)	**Sewage sludge Switzerland** (mg/kg DM)	**Compost Switzerland** (mg/kg DM)
Lead (Pb)	900.00	150.00	500.00	120.00
Chromium (Cr)	900.00	100.00	500.00	100.00
Nickel (Ni)	200.00	50.00	80.00	30.00
Zinc (Zn)	2500.00	400.00	2000.00	400.00
Copper (Cu)	800.00	100.00	600.00	100.00
Mercury (Hg)	8.00	1.00	5.00	1.00
Cadmium (Cd)	10.00	1.50	5.00	1.00
Cobalt (Co)	—	—	60.00	—
Molybdenum (Mo)	—	—	20.00	—

The cold biotechnological treatment of the 'grey' fraction does not substitute for a separate collection of the OFMSW. As already mentioned, with biogenic waste separated at the source, great attention has to be paid before digestion to sort out the (small) percentage of undesired goods. As far as possible, metals are kept back by magnets. In Swiss digestion plants, undesired goods, such as glass, flower pots, plastic bags etc., are sorted out manually on a belt including a device for suction of the waste air (Figure 10.2). In Germany, for hygienic reasons (Strauch, 1993), automated solutions for sorting are generally preferred (supposed danger for sensitive workers of infection with *Salmonella* sp. or fungi such as *Aspergillus fumigatus*). Good results are obtained by a (second) classification at the very end, i.e. after digestion and composting, with a rotating vibration sieve (for large and heavy undesired goods), wind classification for small plastic pieces (not sorted out before digestion) as well as manual classification of remaining small goods like plugs of bottles or aluminium fragments (Müller, 1999).

Today, general criteria for definition as well as methods for determination of the different compost qualities are still missing on an international level (IEA, 1997a). However, the solid and liquid outputs of anaerobic digestion are considered to show different advantages, such as odour reduction or better fertilizing and soil improving qualities, in comparison to untreated manures and wastes (Klingler, 1999; Ortenblad, 1999). Digested slurries may be used directly in agriculture. However, an aerobic post-treatment is recommended improve further the quality of the digested matter.

In Switzerland, the digested material has to undergo a minimal aerobic post-treatment of two weeks fulfilling different criteria, if there is the intention to market it under the label 'Compost'. Without aerobic treatment it has to be called 'Gärgut', i.e. digested matter. Actually, different criteria have been formulated to define different (aerobic and anaerobic) compost qualities (Fuchs et al., 2001), such as:

- different criteria to guarantee good hygienic conditions;
- degree of degradation (for compost, the origin of the substrate must not be recognisable any more; exception: pieces of wood);
- ammonia content: maximum of 300 ppm NH_4^+–N per kg of FM (fresh matter);
- advanced humification (for higher qualities);
- plant growth tests to determine the suppressive activity on phytopathogens (for higher qualities); as well as
- ratio $NO_3^- / NH_4^+ > 2$ (> 20 for top qualities);
- salt content < 4 ms/cm (< 2.5 ms/cm respectively).

Figure 10.2. Manual sorting out of undesired goods of OFMSW before digestion. Waste air is sucked into the covering for the protection of the worker.

Looking at the results of the environmental impact assessment (section 10.5), for AD the methane emissions of the post-treatment play the dominant part in pollution. The digestat (digested matter) extracted by solid/liquid separation is saturated with anaerobic bacteria. If only stored on heaps, AD will continue to go on inside the warm material, causing a greenhouse effect by most effective methane emissions. Therefore it is most important to interrupt as quickly as possible the anaerobic breakdown after extraction of the digestat from the digester. Structural material, such as wood chips and branches, have to be added to the digestat after de-watering (if it is not already present within the digested matter). This will loosen the compact structure of the digestat for better and quicker access of the air. Technical composting is an artificial process that needs a lot of care (and energy input) for adequate aeration, moistening and turning over of the windrows. It seems to be reasonable to inoculate the digestat with an active aerobic compost for a quicker change in the composition of the microbiological populations. Sieving is recommended only at the very end of the process, because large particles enhance aerobic composting performance.

Standard quality compost is usually delivered at very low prices or free of charge to the farmers in the neighbourhood of the AD plant. The treatment costs are covered by the state or by direct taxes paid by the waste producers. Higher quality compost will be sold.

10.4. TREATMENT OF SURPLUS WATER

Even with so-called dry anaerobic digestion processes, some surplus water is generated by the solid/liquid separation after digestion.. The quantity of surplus water depends on the substrate properties (dry matter content) as well as on the digestion technology applied (wet or dry). Normally, the solid/liquid separation includes a first step, e.g. by a screw press, where the future compost is separated from a liquid still containing more than 10% of dry matter. This first 'liquid' may be separated in a second step by centrifugation into surplus water and a sludge respectively, which is added to the compost fraction.

With plug flow or batch processes, a part of the 'liquid' may be used for inoculation of the fresh substrate. With technologies for combination of composting and digestion, a large part of the liquid, which contains a high amount of dissolved inorganic nutrients, may be used for moistening and fertilization of the mainly woody compost windrows (Edelmann and Engeli, 1992). The most logical way of valorisation of still-existing surplus liquid is its agricultural use for fertilization and irrigation. However, in some situations this seems not to be possible because of long transport distances or even, as described by Loll (1999), because of too-high AOX-contents (adsorbable, organic bound halogens). (In this context, it seems to be wrong to forbid the recycling of surplus water from AD of OFMSW because of high AOX contents; it seems to be more reasonable to reduce the use of pesticides, because the same AOX are not only present in the liquid of AD, but also in our food on our table!)

Table 10.4 shows the composition of some liquids of dry digestion processes (Schleiss 1998). The liquid is rich in nutrients and DM. The heavy metal contents (g/ton DM = ppm) are somewhat higher than in the compost, but do not reach the low limits given by the law. The content of inorganic nutrients limits agricultural application since in several European countries farmers have to obey maximal limits of nutrient import per surface.

If there remains a surplus of liquid, it has to be further treated before disposal. This is mainly due to high loading of COD and ammonia nitrogen. Thus the aim of a wastewater treatment is to reduce COD as well as to achieve nitrification and denitrification. Different biological and physical/chemical systems have been studied, such as rotating disc reactor, fixed film reactor, ultrafiltration and reverse osmosis (Wilderer, 1999).

Table 10.4. Typical composition of liquids after solid/liquid separation of thermophilic-digested, source-separated OFMSW (Schleiss, 1998). FM: fresh matter, DM: dry matter.

General parameters 13 samples	**dry matter % FM**	**org. matter % DM**	**pH**	**C/N**	**NH_4–N kg/t DM**	**N–Min kg/t DM**
Average	14.2	44.8	8.2	10.2	11.24	11.25
Median value	13.9	44.9	8.2	10.1	10.41	10.40
Minimum	8.3	34.9	7.7	8.7	5.12	5.14
Maximum	20.4	53.3	8.6	11.9	22.88	22.88
Standard deviation	2.9	4.4	0.2	1.1	4.81	4.81

Nutrients 13 samples	**N-tot Kg/t DM**	**P_2O_5 (tot.) kg/t DM**	**K_20 (tot.) kg/t DM**	**Ca (tot.) kg/t DM**	**Mg (tot.) kg/t DM**
Average	21.0	12.8	31.6	36.4	9.7
Median value	21.4	13.2	31.4	36.3	9.8
Minimum	13.8	9.2	23.6	25.3	8.7
Maximum	26.1	14.6	42.8	52.8	11.8
Standard deviation	3.6	1.6	6.5	7.8	0.8

Heavy metals 13 samples	**Cd g/t DM**	**Cu g/t DM**	**Ni g/t DM**	**Pb g/t DM**	**Zn g/t DM**	**Cr g/t DM**	**Hg g/t DM**
Average	0.62	77.9	28.2	59.5	269.7	36.7	0.18
Median value	0.62	75.3	28.8	53.6	263.0	37.0	0.16
Minimum	0.49	70.5	19.9	36.8	229.6	26.2	0.12
Maximum	0.90	90.3	41.4	111.0	336.0	53.0	0.31
Standard deviation	0.10	6.3	6.7	21.9	29.0	8.2	0.06

Experiments carried out with biological systems showed that it is hardly possible to achieve COD reduction >60% in the liquid (Edelmann et al., 1999a). This is due to COD being partly recalcitrant. In contrast to COD, BOD_5 is fully degradable. Nitrogen removal was almost complete. No significant differences in performance could be observed among the different biological systems tested. This shows that the limits of microbiological breakdown are not strictly dependent on the type of biological system. But the remaining COD concentrations between 3.5 and 7 g/l are clearly too high to release the effluent into the sewer. To increase the biological breakdown of COD, wet oxidation with ozone or hydrogen peroxide has been tested. Ozone treatment was too expensive and UV-activated hydrogen peroxide oxidation was not effective enough. An adequate solution was found in the combination of a fixed-film reactor for nitrification/denitrification with a reverse osmosis unit (Figures 10.3 and 10.4).

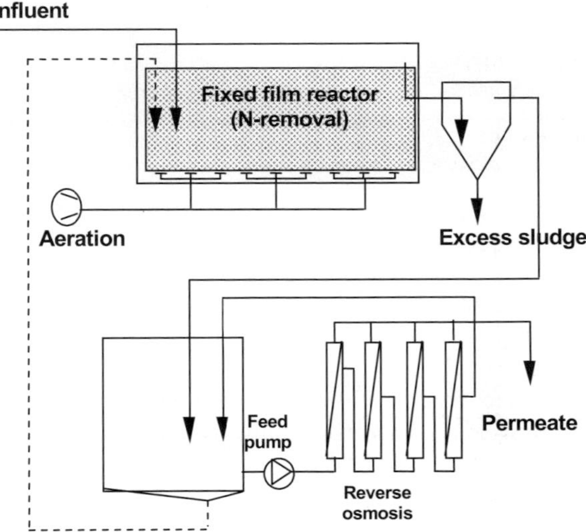

Figure 10.3. Flow sheet of the treatment of surplus water while digesting OFMSW (Edelmann et al., 1999a).

Figure 10.4. Reverse osmosis installation for the treatment of surplus water while digesting OFMSW (Edelmann et al., 1999a).

Results are given in Table 10.5. Two years of operation have shown the process to be reliable and effective. The experiences showed that a micro- or ultrafiltration put normally in front of the reverse osmosis is not compulsory. Instead, the fixed-bed reactor in front of the RO was of great support by lowering the ionic strength. Increased fouling and scaling was observed, if the wastewater was not pretreated biologically. Excess fouling and scaling were inhibited by an automated washing cycle.

Table 10.5. Composition of input and output of a reverse osmosis device in Otelfingen, Switzerland.

Parameter	Units	Influent	Permeate
Flow rate	m^3/d	40	32
COD	mg/l	12,000	30
BOD5	mg/l	3,000	15
Ammonia-Nitrogen	mg/l	1,200	40

10.5. ENVIRONMENTAL IMPACT ASSESSMENTS OF BIOWASTE TREATING

10.5.1. Compared objects and assumptions

For the biological breakdown of biogenic wastes, both aerobic and anaerobic technologies exist. So far, some comparisons of the technologies had been made (IEA, 1997b, Membrez and Glauser, 1997). However, most of them focused on single aspects such as economy or on environmental impacts of a few parameters. Edelmann and Schleiss (1999b) compared the different treating methods from ecological, energetic and economic points of view and in a more holistic way by comparing as many parameters as possible for standardised plants with treating capacities of 10,000 tons/a. Five different biotechnologies plus treatment in a modern incineration plant were examined. This work is summarised in this section and that following (economy of biowaste treatment). To improve understanding, information is also given on the methodology of environmental impact assessments or life cycle assessments, respectively.

Practical data were sampled on existing Swiss treatment plants. However, these installations differ in several ways: For example, the treating capacities of the plants observed vary from 5,000 to 18,000 t/year. To get comparable data, all data were standardised: data, such as construction materials, investment costs or salaries, were calculated for plant sizes of 10,000 t/year. It was assumed that all plants were constructed in the same suburban area. This allowed calculation with identical ground conditions as well as identical transporting distances for all biotechnological treatment methods while collecting the source-separated biogenic waste. It was further assumed that there was no possibility of using the waste heat of the cogeneration externally at this theoretical site. Similar to this example, assumptions were generally not made in favour of biogas, to be sure that the real biogas plants will perform in reality even better than those described in this study.

The compared plants mainly differ in (i.) process technology, (ii.) construction costs (money, energy and environmental factors) and (iii.) running costs including energy and emissions. The following process technologies have been compared:

•**EC**: fully **E**nclosed and automated **C**omposting plant with waste air treatment in a biofilter: The data were derived from a fully enclosed channel composting plant (IPS) with a compost biofilter;

•**OC**: **O**pen **C**omposting in boxes covered by a roof and in open windrows: COMPAQ-Boxes protected against rainfall followed by composting in open, low windrows reversed frequently and covered by gas permeable textile sheets,

•**DP**: fully enclosed, thermophilic, one step plug flow dry **D**igestion (horizontal Kompogas-digester) with aerobic **P**ost-treatment in an enclosed building equipped with compost biofilters,

•**DE**: combination of thermophilic dry **D**igestion combined with fully **E**nclosed, automated composting in boxes (BRV-technology), digestion of (only) 40% of the raw material before its addition to the compost line. The air is cleaned by bio-washers.

•**DO**: combination of multiple stage, thermophilic batch **D**igestion (romOpur-technology) combined with **O**pen windrow composting: digestion of 60% of the raw material before its addition to the compost line,

•**IS**: **I**ncineration in a modern incineration plant including enhanced **S**crubbing of the exhaust gas streams.

For the waste incineration, a plant with a treatment capacity of 100,000 t/a of mixed wastes was chosen. The effect of burning 10,000 tons of biogenic wastes in addition to mixed wastes was determined after Zimmermann (1996). The larger incineration plant causes longer transporting distances compared with biotechnological methods. On the other hand, there is not an additional green waste collection necessary. Identical impacts by waste transporting were assumed for all processes (for a detailed discussion see Edelmann and Schleiss, 1999b). For all plants, exactly the same waste composition was assumed (60% material relatively rich in kitchen wastes from public collection and 40% material rich in lignin derived from private suppliers). Table 10.6 shows the elementary composition of the waste. The waste composition corresponds to mean values of separately collected biogenic wastes (median values of large own database as well as data taken from literature).

Table 10.6. Assumption of the elementary composition of the biogenic waste.

Substance/element	Content (%)	element	Content (%)	Element	Content (%)
Water	60.00000	Arsenic	0.00020	Tin	0.00080
Oxygen (without O of water)	12.74000	Cadmium	0.00001	Vanadium	0.00030
Hydrogen (without H of water)	2.00000	Cobalt	0.00050	Zinc	0.00582
Carbon (organic plus inorganic)	16.24000	Chromium	0.00084	Silicon	3.99804
Sulphur	0.14993	Copper	0.00178	Iron	0.06000
Nitrogen	0.40000	Mercury	0.00001	Calcium	2.18000
Phosphorus	0.11300	Manganese	0.00043	Aluminium	0.99951
Boron	0.00102	Molybdenum	0.00004	Potassium	0.35000
Chorine	0.40000	Nickel	0.00054	Magnesium	0.28200
Fluorine	0.01999	Lead	0.00186	Sodium	0.15000

As shown in Figure 10.5, it was assumed that all the biotechnological treatment technologies are capable of causing a 50% loss of the organic matter (OS) by biological activities. The gaseous emissions, which were measured on site or taken from databases, were distributed according to the assumptions of Figure 10.5 as well as to the different percentages of material treated by aerobic and/or anaerobic methods, respectively (see above).

The functional unit, which was the base of all calculations, consisted in the treatment of 1 kg of waste such as defined in Table 10.6. However, all data presented further down refer to the yearly treating capacity of 10,000 tons of fresh substance of biogenic waste. It was assumed that 1% of the waste was undesired (e.g. glass, metal, plastic) and had to be sorted out and burnt in an incineration plant. (The emissions of this waste stream were taken into account.) The whole life cycle of the biogenic wastes was compared including all environmental impacts caused by the different treating methods as well as those while producing goods and equipment necessary for the specific treatment as well as its final disposal and recycling. The assessment starts at the moment when the waste leaves the household, and includes all steps to the final application on the field (compost) or to its storage in an adequate landfill (ashes of incineration).

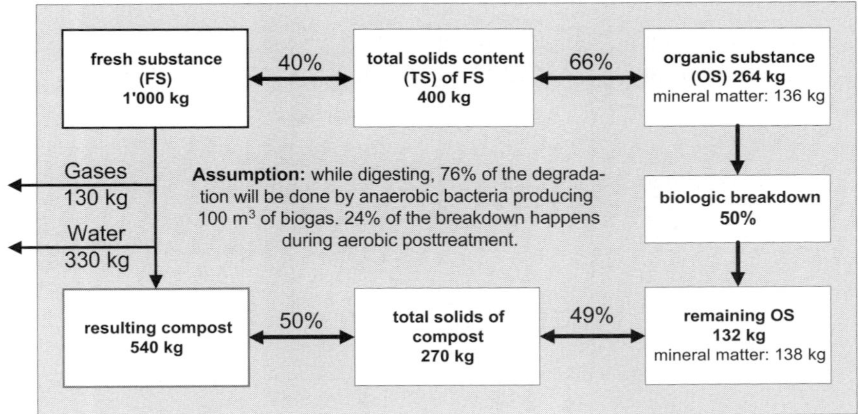

Figure 10.5. Assumptions for mass fluxes while degrading biogenic wastes with biotechnological methods.

Energy plays a very important role while treating biogenic wastes. As far as electricity had to be bought externally, it was calculated with the European electricity mixture (UCPTE) for the electricity needs of the processes (Frischknecht, 1996). It seemed to be reasonable to use the European electricity mixture, because in Europe all countries are connected with each other; additional electricity need in Switzerland will cause additional (mostly) thermal production in EC or will reduce the export of renewable water power, respectively—which shows reduction of thermal power production in EC.

Figure 10.6 shows the system borders: There are external additions to the system, such as material and energy necessary for the construction of the plant (infrastructure). The conversion of the renewable energy feed while degrading biogenic matter still is within the system borders (e.g. emissions of the motor burning the biogas). The emissions by compost application on the field and those caused by storing the ashes of incineration in a landfill are integrated into the consideration. Credits were taken into account for the benefits of the renewable energy and of some of the mineral fertilising substances within the compost.

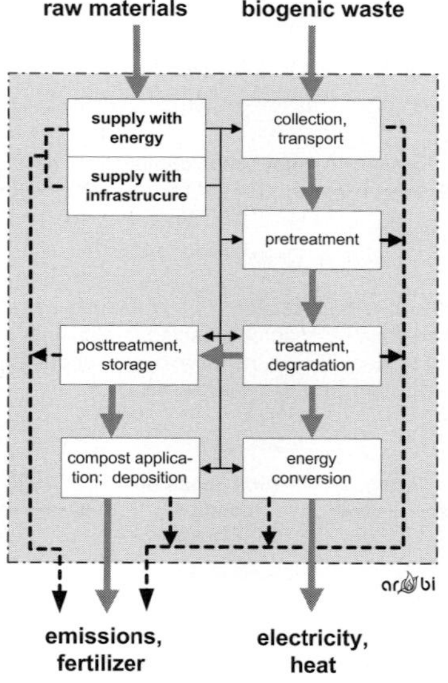

Figure 10.6. Definition of the borders of the observed systems: grey, thick arrows: mass and energy fluxes entering and leaving the system; dotted black arrows: emissions; black arrows: internal connections.

It was assumed that the fully enclosed biotechnological plants can be run without an excess of process water with this kind of waste (40% dry matter). The biologically generated compost heat is significantly higher than the heat needed to evaporate the water in Figure 10.5. In Switzerland, the anaerobic plants generally have no problems with surplus water (for a detailed discussion see Edelmann and Schleiss, 1999b). However, while composting in open windrows, which were exposed to the rain, the production of leachate had to be taken into account.

All data refer to 10,000 tons of biogenic waste. A description of each process included the evaluation of the infrastructural needs, such as buildings, asphalt surfaces, machines, infrastructure for pre- and post-treatment, etc. The materials needed to provide the treating infrastructure were divided by the span of their lifetime so as to obtain the yearly amounts of cement, metals, asphalt, etc. necessary to treat the 10,000 tons (assumptions: lifespan for mobile machines: 5 years; stationary engines: 10 years; buildings: 25 years). The ecological pre-investments to produce the building, the construction materials and the ecological costs for a future destruction of the plants were included by using data from ECOINVENT, a large database developed by the Swiss Federal Institute of Technology (Zimmermann, 1996).

The ecological running costs of the plant included energetic and material parameters such as energy fluxes, parts replaced because of attrition, commodities, etc. as well as the emissions into air and into water caused by the process.

For the methane emissions of composting sites there exist only few data (e.g. Marb et al., 1997; Ketelsen and Cuhls, 1999). To get appropriate values, the gaseous emissions were measured three times over the year by the closed chamber method (see Figure 10.7). Because the degree of organic matter degradation is defined (see Figure 10.5) and its carbon content is known (definition of elementary composition), the moles of emitted gas molecules containing carbon could be calculated. Because CO_2 and CH_4 both contain just one carbon atom and show a similar volume requirement, it is possible to calculate their total emissions, as soon as their relative ratio is known.

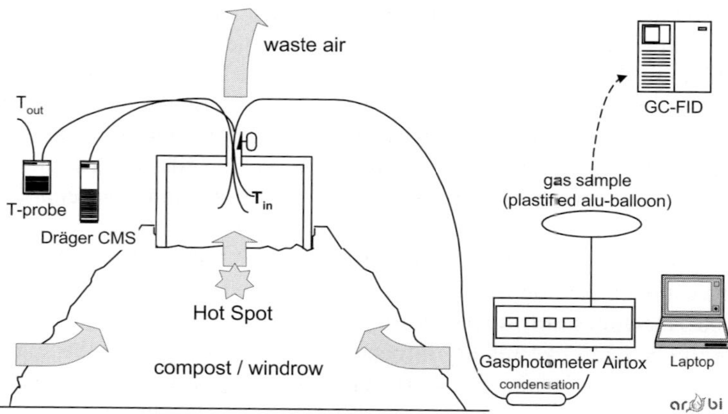

Figure 10.7. The closed chamber method to measure the gaseous emissions of compost: A box is placed over the hot spot of the compost windrow, where it will be filled by the (warm) ascending gases. The ratio of CO_2 to CH_4 to O_2 is monitored on line. For each process, measurements of one half up to two hours were made at different places on several composts of different ages or on the biofilters. The box was insulated and provided with heating facilities to prevent condensation.

10.5.2. Methodology of impact assessments, sensitivities

After the determination of the materials involved, the emissions of the mass fluxes were determined and weighted. All processes, such as raw material extraction, distribution and manufacturing were included up to the building and later destruction of the plants. Two tools were used for the weighing of the impacts: an improved version of EcoIndicator 95 (Goedkoop, 1995) and UBP, a tool developed by Buwal (1998). Calculations as well as the work on waste incineration were done by S. Hellweg at the Laboratorium für technische Chemie, ETH Zürich, with data from Zimmermann (1996). The effects of incinerating 10,000 t of defined biogenic waste together with 'grey' waste were also determined including all needs for infrastructure (ratio of 'green' to 'grey' corresponding to the composition of not-separately-collected Swiss household waste).

In EcoIndicator 95+, ten impact categories (such as greenhouse effect, ozone layer depletion, acidification, etc.) are defined. All the impacts caused by the different activities of the life cycle of a waste-treating process are first sorted and attributed to the relevant categories. For each damage category, a reference substance has been defined. The impacts are brought to a comparable size by multiplying with a factor corresponding to their relative damage potential (e.g. in the greenhouse effect methane is weighted – depending on the observation period – 21 times stronger than carbon dioxide, which is the reference substance). Like this, the effect scores can be normalised for each category. The damages caused by the reference substances are weighted for causing mortality, damage to health and ecosystem impairment. For the damage weighting factors, subjective weighing would be possible. In this study, the default values of the software have been applied as proposed by the authors.

The method UBP of BUWAL was used to compare the results of EcoIndicator to a tool, where the target set value is the Swiss national policy objectives, i.e. UBP-points depend on limits for toxic substances given by Swiss laws and the relative danger of releasing these substances into the environment. 'Danger' is dependent on the difference between the actual real situation in the field and the legal limits; in UBP, normalized Ecopoints are distributed not to the damages, but to individual emissions, energy consumption and environmental scarcity.

Sensitivity calculations show the influence of a single impact on the total result. Because EcoIndicator 95+ only takes account of heavy metals leached from soil to water, sensitivities were calculated for different degrees of heavy metal washout from the ground when compost was applied. Additional sensitivities were established for the different amounts of NH_3, N_2O and H_2S emissions (data from literature), for an optimal reduction of methane emission while digesting and for not giving benefits to the fertilizer values while applying compost.

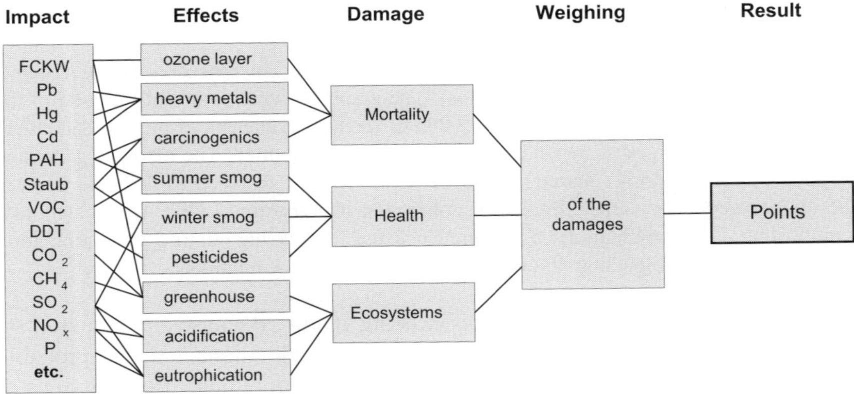

Figure 10.8. Methodical approach of Ecoindicator 95: Emissions of plant construction and running ('impacts') show effects in different impact categories. The three categories of damages are weighted and aggregated to a final ranking.

10.5.3. Results of the process comparison

Figure 10.9 shows the ratio of methane to carbon dioxide emission during biotechnological treatment. In digestion plants there is a considerable potential of methane emission during the 'aerobic' post-treatment, even if just a small percentage of the organic breakdown takes place outside the (enclosed) digester. On the other hand, there exist also significant methane emissions even in composts, which are reversed very often (OC; reversed daily during intensive composting period, windrow height only 1.2 m!). This confirms the suggestion that technical aerobic composting is a process that cannot be optimized because the optimizations of different parameters exclude one another (Edelmann, 2001).

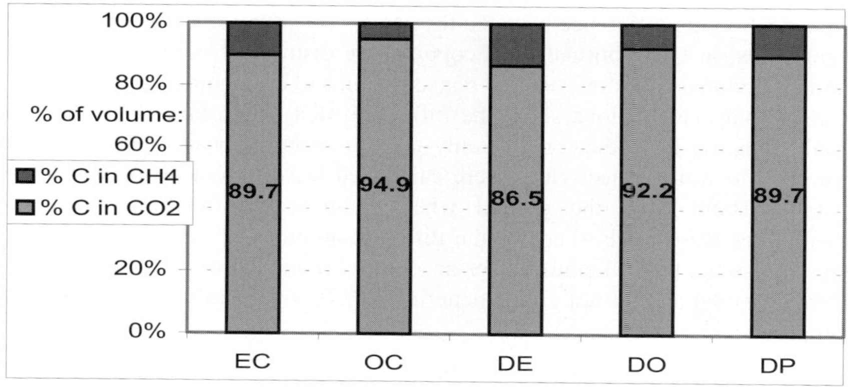

Figure 10.9. CO_2:CH_4 ratio of the different biotechnological processes (% of volume, weighted mean values of three campaigns). The graph shows the ratio of the total of the emissions, i.e. it took account of the fact that different percentages of the substrate were composted and/or digested depending on the technology applied. The methane generated by anaerobic digestion is counted as CO_2, because it will be oxidised while being burnt in the engine/generator. (EC: Enclosed automated Composting; OC: Open windrow Composting; DE: combination Digestion / Enclosed composting 60:40; DO: combination Digestion / Open composting 40:60; DP: Digestion with Post treatment).

The measurements were taken on existing plants. Because the situation on the real plants differed somewhat from the assumed situations, it seems probable that – taking mass streams such as defined in the assumptions – the emissions of DO will be rather higher and those of DP rather lower than shown in Figure 10.4. For detailed gas analyses and discussion of the emissions see Edelmann and Schleiss (1999b). The incineration plant emits no methane, but corresponding to the total oxidation of 100% of the carbon the double amount of gas in the form of CO_2.

The gaseous emissions of NH_3, N_2O and H_2S were very low compared with the emissions of CO_2 and CH_4 and difficult to measure accurately in the wet and warm gas emissions. Additional measurements have to be made to obtain data that are statistically significant. For this reason sensitivities were calculated with data taken from literature (Schattner- Schmidt et al., 1995; Gronauer et al., 1997; Hellmann et al., 1997; Smet et al., 1998).

Figures 10.10 and 10.11 show the sums of the Ecoindicator 95+ points for nine impact categories. For the nitrogen and phosphorus present in the compost, benefits corresponding to the savings due to less mineral fertilizer production were taken into account. In the incineration plant (IS) the nutrients are lost. In Figure 10.12, which shows the total Ecoindicator-points (EI-points), a sensitivity is calculated without fertilizer benefit.

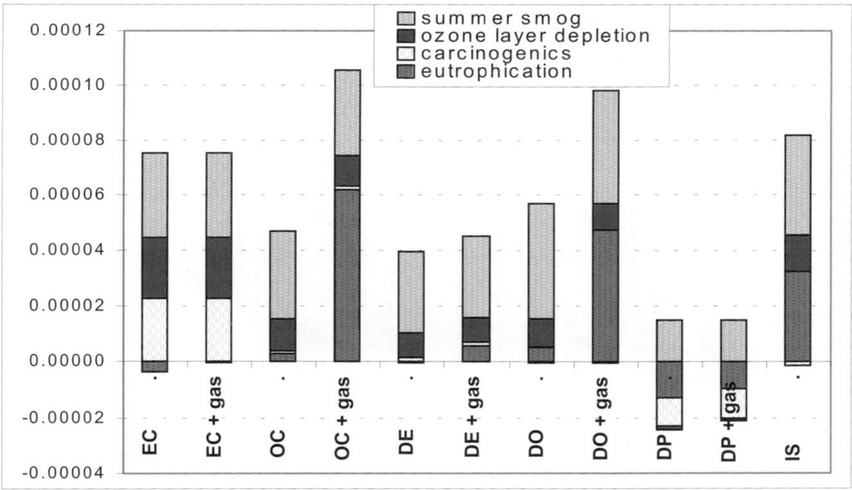

Figure 10.10. Ecoindicator 95+-points for the impact categories radioactivity, energetic resources, greenhouse effect, acidification and winter smog. The data '+gases' of the biotechnological processes were calculated including emission data (taken from literature) for NH_3, N_2O and H_2S into the air. NH_3 emissions are reduced to a large extent by the biofilters of the fully enclosed plants (EC, DP and DE). The radioactivity as well as much of the acidification is caused by the European electricity mixture (including nuclear plants as well as thermal plants running on fossil fuels). (EC: Enclosed automated Composting; OC: Open windrow Composting; DE: combination Digestion / Enclosed composting 60:40; DO: combination Digestion /Open composting 40:60; DP: Digestion with Post treatment; IS: Incineration including Scrubbing).

Sensitivities calculated with data from literature for gases, such as NH_3, N_2O and H_2S, show that the environmental impacts of these emissions play a very important role. Mainly because of acidification and greenhouse effect (Figure 10.10) as well as eutrophication (Figure 10.11), outdoor steps of the waste treatment increase the total impacts considerably (see Figure 10.12; +/- gas). The only considerable negative impact of DP is observed in the category of greenhouse effect (note: the scaling of Figure 10.10 differs from that of Figure 10.11!). This is mainly caused by the relatively large methane emissions of the material, which is (post-)composted after digestion. Here, there is a considerable improvement potential by creating as quickly as possible aerobic conditions immediately after digestion. The measurements cited above suggest that biofilters reduce only a relative small part of the methane emissions by methane oxidising biocenoses. (For detailed discussions see Edelmann and Schleiss, 1999b.)

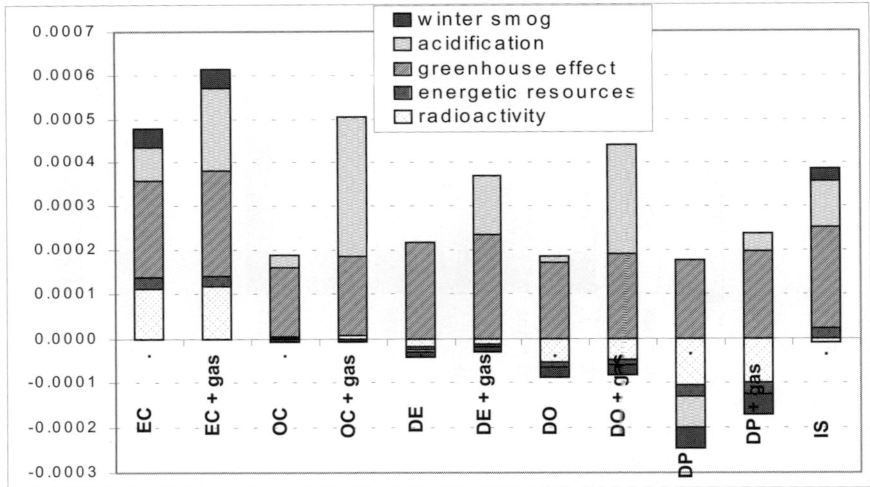

Figure 10.11. Ecoindicator 95+-points for the impact categories eutrophication, carcinogens, ozone layer depletion and summer smog. The data '+gas' of the biotechnological processes were calculated including emission data for NH_3, N_2O and H_2S into the air. For the sensitivity with gas emission, the eutrophication becomes especially large in OC and DO, where ammonia is able to escape into the open air. Similar to Figure 10.10, DP shows some negative values which are caused by the high net surplus production of renewable electricity that substitutes for all UCPTE-electricity. (EC: Enclosed automated Composting; OC: Open windrow Composting; DE: combination Digestion / Enclosed composting 60:40; DO: combination Digestion / Open composting 40:60; DP: Digestion with Post treatment; IS: Incineration including Scrubbing).

Figure 10.12 shows the overall performances of the processes. High scores correspond to high environmental impacts. Energy plays a predominant role: taking all the necessary activities into account, the anaerobic digestion (DP) causes more or less no environmental impacts, i.e. for the impacts is compensated by the generation of a surplus of renewable energy which substitutes for fossil and nuclear energy. The benefit of re-using the fertilising value of the green waste is shown by the sensitivity ‘no nutrients’ in Figure 10.12: for example, if it is not necessary to produce ammonia by the Haber-Bosch synthesis, the use of fossil energy including a lot of negative environmental impacts may be prevented.

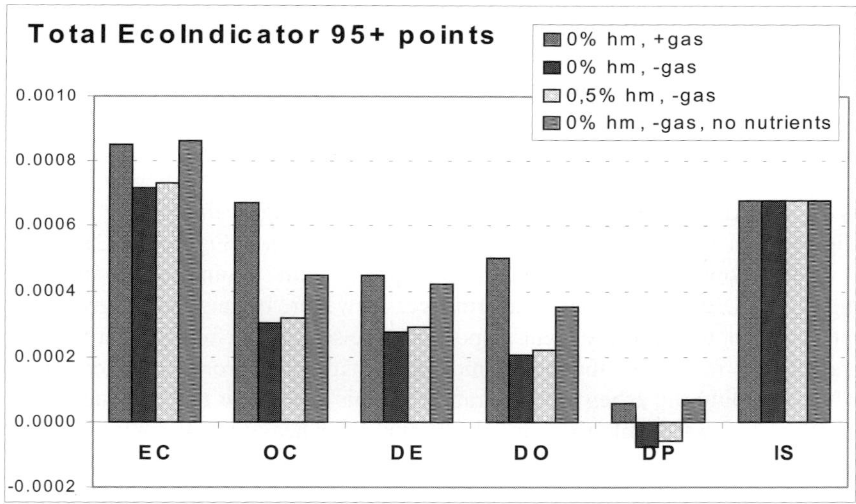

Figure 10.12. Total sums of Ecoindicator 95+-points for four sensitivities: +/- gas: with/without emission of NH_3, N_2O and H_2S into the air; no nutrients: no benefit for fertilizer substitution; 0/0,5% hm: amount of heavy metal export from soil by water. (EC: Enclosed automated Composting; OC: Open windrow Composting; DE: combination Digestion / Enclosed composting 60:40; DO: combination Digestion / Open composting 40:60; DP: Digestion with Post treatment; IS: Incineration including Scrubbing).

For incineration, which was calculated with identical system borders and identical tools, no sensitivities are shown in Figure 10.12, because for incineration different sensitivities are relevant compared with biotechnological processes. For more details on incineration, see Hellweg (1999).

In addition to the energetic and environmental construction costs, the energy running costs of EC are more than 100 kWh electricity per ton of waste (and correspondingly more primary energy), which causes considerable negative impacts. Composting in a highly automated plant seems to be even more polluting than incineration in a modern incineration plant with advanced energy recovery. A comparison of EC with DP on the level of primary energy (energy necessary for the production of electricity, substitution of non-renewable energy by biogas, including all needs for plant construction and running etc.) showed a difference as large as 700 kWh per ton of waste treated (Edelmann and Schleiss, 1999b). The substitution of the plant nutrients present in compost with mineral fertilisers causes a primary energy need of nearly 90 kWh/t of compost, in addition to considerable environmental impacts in different impact categories (*ibidem*).

One heavy metal sensitivity is integrated in Figure 10.12. Ecoindicator 95+ only takes heavy metals into account, if they are exported from the soil into the water. Different sensitivities were calculated for the heavy metal leaching. Here data are shown for no leaching and for 0.5% heavy metal export into water. With 5% export into water – which is an extremely high value – only DP has a slightly lower sum of total EI-points than IS (not shown in Figure 10.12).

UBP showed a similar ranking compared with Ecoindicator: anaerobic digestion (DP) showed best performance. However, because UBP gives very much weight to the heavy metal import into the soil, IS performs better than OC and EC due to the fact that heavy metals are withdrawn from ecological cycles while dumping the ashes of incineration. In this context, it may be questioned, whether it is reasonable to blame the compost for heavy metal contents, which originate only for a negligible part from the treating process itself; most of the heavy metals are deposited by mineral fertilisers, wind and rain on the biomass. Considering the fact that the heavy metal load of the compost usually is far below the legal limits (Schleiss, 1999), it seems not to be a reasonable solution to incinerate precious organic matter to withdraw heavy metals from ecological cycles. The solution has to be a reduction of heavy metal imports into the environment.

Comparing the biotechnological solutions with incineration, they would even perform significantly better, if it were possible to give benefits for the compost quality, i.e. not only for substituting for plant nutrients, such as N or P, but also for improving the structure of the ground by adding organic material, for suppression of phyto-pathogens, for improving the water retaining capacity of the soil, etc. Because of missing data, these aspects were not taken into account.

The data of this study were also taken for calculations with the newest version of Ecoindicator, i.e. Ecoindicator 99. With this improved tool, the ranking of the pure anaerobic solution DP is even more significantly better than the remaining technologies (unpublished). It may be concluded that, from an ecological point of view, future solutions to treat biogenic wastes have to include percentages of anaerobic digestion as high as possible.

10.6. ECONOMY OF BIOGENIC WASTE TREATMENT

The different plants for the yearly treatment of 10,000 tons of biogenic waste, which were compared in the previous section from an ecological point of view, were also compared on an economic level. For the abbreviations of the different technologies see 10.5.1. The investment and running costs of the different processes were inquired at existing plants and afterwards standardized for plant sizes with treating capacities of 10,000 t/a. For the cost comparison, all plants were calculated with identical methods (Prochinig, 1997) and with the following assumptions:

- costs for real estate: CHF 30.-/m^2 (industrial area) (about 20 Euro/m^2);
- costs per worker 100%: CHF. 100,000 /year (incl. social costs) (about 66,000 Euro/year);
- capital costs: 5%/a on 60% of initial investment (Ammann, 1997);
- linear deduction: mobile machines: 20%/a, stationary machines: 10%/a, constructions: 4%/a;
- servicing and maintenance: 1%/year of construction costs; 2%/year of investments for machines;
- insurances: 0.4%/year of total invest (fire, etc.).

Figure 10.13 visualises the investment costs. The investments of the fully automated, enclosed tunnel composting are very high; it may be suggested, however, that it could be possible to construct a new plant of similar design with some financial savings.

With all technologies except open windrow composting, the most important investment is caused by stationary machines. These costs also contribute most to the fixed plant running costs (exception with OC: 'mobile machines'). The incineration data are only divided into 'stationary machines' and 'other costs'. They refer to the treatment of 10,000 tons of biogenic wastes (in adequate proportion together with 'grey' waste) in an incineration plant with a capacity of 100,000 t/a (no separate collection the organic fraction). The costs for the incineration plant are derived from a project, which is under construction in Switzerland (plant with advanced gas scrubbing and selective catalytic reduction of NO_x). Detailed information on the cost comparison is given in Edelmann and Schleiss (1999b).

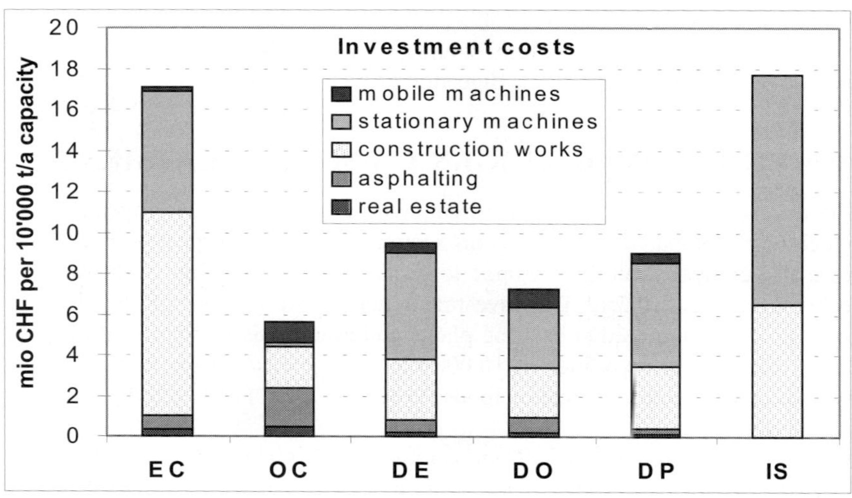

Figure 10.13. Investment costs of different processes for biowaste treatment. (EC: Enclosed automated Composting; OC: Open windrow Composting; DE: combination Digestion / Enclosed composting 60:40; DO: combination Digestion / Open composting 40:60; DP: Digestion with Post treatment; IS: Incineration including Scrubbing).

Figure 10.14 visualises the specific costs treating one ton of biogenic waste (for biotechnological treating methods: separated at source). In Figure 10.14, the costs for waste collection are not included. For the biotechnological treatments they can be assumed to be identical. With incineration it is not necessary to collect twice ('grey' and 'green'), but the transporting distances are significantly longer due to the larger radius of the collection area, caused by the higher treating capacity (250,000 population equivalents compared with 100,000 for the biotechnological treatment).

Open windrow composting (OC) shows the lowest treatment costs due to the relatively low investment. On the other hand, the running costs of OC are higher than those of the other biotechnological treatments because of lower automation. It has to be stressed, however, that open windrow composting is not suited for waste containing much of easily degradable (kitchen) waste for environmental reasons (see section 10.5). Digestion combined with open composting (DO) is slightly cheaper than the other digestion technologies because of lower investment (open composting without biofilter etc.), but it is not recommended for all applications due to gaseous emissions. Fully enclosed, automated composting (EC) shows the lowest variable costs, but very high investment costs causing high fixed costs.

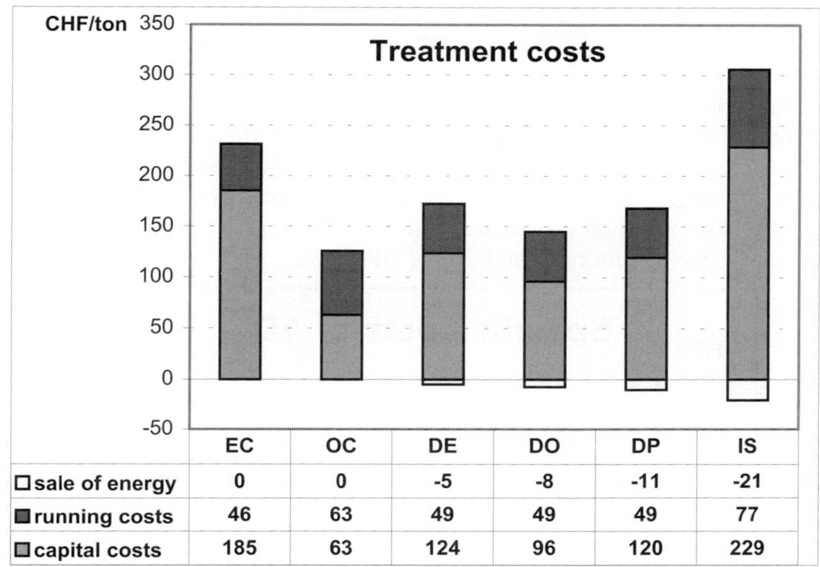

	EC	OC	DE	DO	DP	IS
sale of energy	0	0	-5	-8	-11	-21
running costs	46	63	49	49	49	77
capital costs	185	63	124	96	120	229

Figure 10.14. Specific treatment costs of the different processes. In contrast to anaerobic digestion, it was assumed that with IS waste heat could be sold for house and water heating purposes (therefore considerably higher sale of energy than AD). Effective costs: sale of energy has to be deduced from the total. (EC: Enclosed automated Composting; OC: Open windrow Composting; DE: combination Digestion / Enclosed composting 60:40; DO: combination Digestion / Open composting 40:60; DP: Digestion with Post treatment; IS: Incineration including Scrubbing.)

In Figure 10.14, it was assumed that – in contrary to digestion technologies – the waste heat of the incineration plant could be sold. This assumption is in favour of incineration and difficult to achieve in reality. Sensitivities were calculated for different costs for real estate (no significant effect except for OC) and for prolonging the plant pay-back time from 25 to 35 years, which favours to some extent the processes with high investment costs, but has no effect on the ranking.

Other publications confirm the data of Figure 10.14 (Auksutat, 1999; Widmann, 1999). AD is significantly cheaper than incineration with advanced cleaning technology for exhaust gas, and competitive with the other biotechnological treating methods.

In the economic comparison of Figure 10.14 it was calculated with a biotechnological treating capacity of 10,000 tons/a, because this seems to be a reasonable size for Swiss (semi-rural) conditions from an ecological as well as economic point of view: larger capacities seem to be too big, because the environmental impacts caused by the longer transporting distances grow too much (except in big cities). On the other hand, smaller solutions will be too expensive. Figure 10.15 shows the decrease of economic costs by increasing the AD plant size. From an economic point of view, larger treating capacities are important especially for two-stage AD of OFMSW.

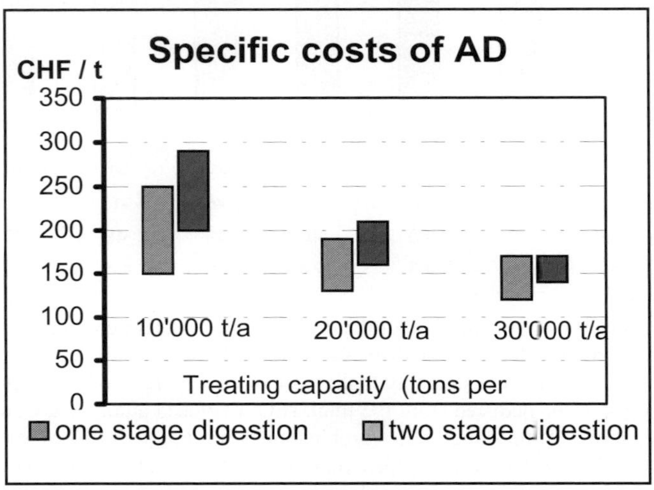

Figure 10.15. Ranges of the specific treatment costs per ton of biogenic waste for one- and two-stage digestion (Oetjen-Dehne and Ries, 1995, modified).

A significant cost reduction is possible by realising co-digestion plants: Denmark has much experience with large, centralized, agricultural biogas plants for co-digestion of agricultural and OFMSW (Holm-Nielsen and Al Seadi 2000). Economic break-even is reached when obtaining at the gate DKK 55/m^3 of waste (Hjort-Gregersen, 1999). In Switzerland, the feasibility of digesting organic solid wastes within the digester of a sewage treatment plant has been investigated (Edelmann, et al., 1999c). If it is possible to take advantage of a digestion infrastructure already present on the waste water treatment site, municipal co-digestion is a cheap solution for rural and tourist areas with small amounts of biogenic wastes; the treatment will cost only around Sfr 60–80 per ton at treating capacities of 500 to 1000 tons/year. Prerequisite for co-digestion on STPs is a good quality of the sewage sludge and its recycling in agriculture.

Recently, DP has become significantly cheaper: Kompogas (i.e. the process considered in the LCA and in the economic considerations above) has developed a modular plant which is optimised and reduced to the really necessary parts (e.g. the fermenter is not built inside a building, but placed outside, etc.) (Kompogas, 2001). The investment costs are reduced by more than 50% and the specific treatment costs (Figure 10.14) have gone down by significantly more than one third; they are actually less than 100 CHF/t. Under these new conditions, AD of OFMSW is most competitive now compared with composting technologies.

10.7. PERSPECTIVES AND CONCLUSIONS

Ecological and economic considerations show that for biogenic waste treatment biotechnology is generally favourable with respect to incineration. The pure composting technologies appear to be less ecological than digestion. There is no doubt any more that anaerobic digestion is the best solution to treat biogenic wastes: the higher the percentage of digestion, the better the score.

The three categories of greenhouse effect, acidification and heavy metals play an important role in the environmental impact assessment. The greenhouse effect is caused mainly by CO_2 and CH_4. Carbon dioxide emission cannot be prevented, if biogenic matter is degraded. Methane, on the other hand, is freed in nature as soon as biomass is piled up into heaps. Because the aerobic degradation of OFMSW needs at the same time water, solid particles and air, i.e. liquid, solid and gaseous phases, it is not possible to prevent anaerobic zones, i.e. methane emissions in technical composting. For an aerobic post treatment after AD, there is the disadvantage that the organic matter is well inoculated with anaerobic bacteria. Even if just a very small share of the organic matter is degraded during composting after AD, methane emissions may be larger than those caused by pure composting. Here, there is a potential for further improvement of AD (see section 10.5.3).

When comparing the different technologies, energy plays a predominant role. Digestion plants are better from an ecological point of view, mainly because they do not need external fossil and nuclear energy. If only one quarter of the biogenic waste is digested, combined digestion and composting plants can be self sufficient in energy (Edelmann et al., 1998). The production of renewable energy has positive consequences on nearly all impact categories, because of savings in or compensation for non-renewable energy. This reduces the impacts of parameters such as radioactivity, dust, SO_2, CO, NO_x, greenhouse gases, ozone depletion, acidification or carcinogenic substances. Digestion plants could show an even better performance, if they were constructed near an industry which can use the waste heat of electricity production all year round.

Solar energy is fixed inside the chemical bonds of biomass compounds. This energy is freed not only when producing biogas, but also when composting. For technical reasons, however, it is nearly impossible to take advantage of waste heat while composting (Edelmann et al., 1993). Looking at the results of the impact assessment and the economic comparison, it is very difficult to understand that today composting plants are still constructed: in these plants high value fossil and nuclear energy is invested to *destroy* the precious renewable solar energy which has been fixed by photosynthesis in the chemical compounds of biomass and thus in the biogenic waste. Within a few decades, the 'post-fossil' inhabitant of the so called 'first' world will not be able any more to have meals with fruit juice from Central America, shrimps from Indonesia, lamb from New Zealand, dried beans from China and wine from South Africa, because the energy will be too precious for such polluting behaviour. But also for sure, he will not be able any more to afford the non-renewable energy necessary to destroy the solar energy of biomass by composting and by aerobic waste water treatments.

On 'space shuttle Earth', nature has not gone broke over 4 billion years because of closed ecological cycles (Vester, 1978). Nature is a fine network with countless links depending on each other. Unfortunately, modern man has not learned at school to recognize interdependencies, to think in a multi-disciplinary way. In our energy-spoiling civilisation we become aware of a problem, and we look for a solution without realising that we create several new problems somewhere else. For example, if we become aware that there is too much ammonium in the liquid of solid/liquid separation, we develop a new process to destroy (with non-renewable energy) the fertilising nitrogen salt, converting it to elementary nitrogen by nitrification/denitrification. Some miles away, we build up (with non-renewable energy) an industry converting (with natural gas and a lot of non-renewable energy) elementary nitrogen into ammonia. Both processes exhaust non-renewable resources and create greenhouse effect as well as a lot of other damage to the environment. This cannot be the solution for coming generations.

Anaerobic digestion, such as discussed in this book, is a first and big step in a better direction, because it is a solution that closes ecological cycles and is friendly to the environment. But solutions valid for a future mankind have to be even more integrated and based on holistic, long-lasting concepts. The Otelfingen Kompogas plant is an example of additional steps which in future will certainly become more important.

The Otelfingen plant treats the biogenic wastes of around 100,000 inhabitants of Zürich-North. Lignified matter is converted into wood chips for firing and for use in gardens (mulch). The remaining waste is digested in a thermophilic plug flow reactor with a treating capacity of 10,000 t/a ('dry' digestion). The compost is – after aerobic post treatment including air treatment by biofilters – distributed to the farmers in the neighbourhood (standard quality) or sold for application in private gardens. The surplus water of the solid/liquid separation after digesting is partly spread out on agricultural fields. Another part is used in a large-scale pilot project for aquaculture (Figures 10.16a and 10.16b).

The liquid surplus contains precious plant nutrients and organic compounds. But it cannot be delivered to agriculture all year round; it is forbidden to apply fertilizer when the soil is frozen. Therefore the owner of the Otelfingen biogas plant looked for additional possibilities for upgrading the waste water: The nutrients are recycled successively by different trophic levels in an aquaculture plant (Staudenmann and Junge, 2002). The plant includes 36 ponds of 360 m^2 on a surface area of about 450 m^2. About 300 m^2 of the water surface is located within greenhouses to increase the temperature by solar energy for better growth. The partly cleaned input water shows the following characteristics: TOC: 670 mg/l, NO_3–N: 150 mg/l, NH_4–N: 95 mg/l, P_{tot}: 50 mg/l (Staudenmann and Junge, 2000). The dissolved inorganic nutrients are converted into precious biomass: in a cascade of trophic levels there is first a production of macrophytic water plants, such as *Eichhornia*, *Pistia* and duckweed as well as plants for decoration, which are sold. Afterwards modules follow for algae, *Daphnia*, aquatic snails and further fodder for crabs (*Astacus astacus*) and fishes (carp, tench, rudd, *Tilapia*, etc.), which are kept partly outside the greenhouses. Rainwater is sampled on the roofs of the buildings to compensate for the evaporation within the greenhouses.

The water still containing nutrients (excrement of fishes, etc.) is used for growing vegetables in hors-sol-production inside greenhouses, which may be heated additionally by the warm exhaust gases of cogeneration, enriched with carbon dioxide to enhance photosynthesis. Finally, the water undergoes a final cleaning in a reed pond and is diverted into the nearby brook. The technical installation for water cleaning by membrane technology is only in operation when there is a surplus of waste water.

The biogas is converted partly into electricity and heat by cogeneration. The heat covers the need of the plant and the electricity surplus is sold to private consumers ('Ecolabel'). The biogas surplus is cleaned and sold at the integrated gas station for driving biogas cars or fed into the natural gas pipeline for use at further stations along the pipeline network, respectively. Figure 10.17 shows the products of the Otelfingen plant.

Figure 10.16. Views of the aquaculture in Otelfingen (inside greenhouses).

In the future, similar integrated concepts have to be developed. Then, anaerobic digestion of biogenic wastes will probably play a very important role in education, because it is an excellent example for interdisciplinary thinking while closing ecological cycles, reducing pollution and grading household wastes up to precious new products.

Figure 10.17. Products of the Otelfingen Kompogas plant for the treatment of OFMSW: 'green' electricity, heat, wood chips and mulch (not shown), compost, fuel for cars and trucks, vegetables, flowering plants, fishes, crabs and clean water as well as cleaned gas fed into the public pipeline for cooking purposes.

10.8. REFERENCES

Ammann, H. (1997). *Maschinenkosten 1998, Kostenelemente und Entschädigungsansätze für die Benützung von Landmaschinen*, FAT_Bericht Nr. 507, Eidg. Forschungsanstalt für Agrarwirtschaft und Landtechnik (FAT), Tänikon.

Auksutat, M. (1999). Kostenstrukturen der biologischen, mechanisch-biologischen und thermischen Abfallbehandlung, in: Wiemer K., Kern M. (Ed.) *Abfall-Wirtschaft, Proceedings des Kasseler Abfallforums: Bio- und Restabfallbehandlung III,* Kassel, Baeza Verlag, Witzenhausen, pp. 103.

Bidlingmaier, W. (1994). Bundesgütegemeinschaft Kompost e.V. und die RAL-Richtlinien, Gütekriterien für Kompost, in: *Müllhandbuch 6515*, Erweiterung 1/94, Erich Schmidt Verlag, Berlin.

Bilitewski, B. (1989). Beeinflussung der Schwermetallfrachten durch mechanische Sortierung und getrennte Sammlung, in: Thomé-Kozmiensky K. [ed]: *Biogas - Anaerobtechnik in der Abfallwirtschaft*, EF-Verlag Berlin, pp. 261–286.

Bundeskanzlei (1986). *Verordnung über umweltgefährdende Stoffe, StoV*, revisison: 1992, EDMZ, CH-3003 Bern.

Bundesrepublik Deutschland (1992). Klärschlammverordnung (AbfKlärV), 15.4.92, in: *Müllhandbuch 0526*, Erich Schmidt Verlag, Berlin.

BUWAL (1990). *Technische Verordnung über Abfall*, TVA, 10.12.1990, revision 1.1.93, EDMZ, CH-3003 Bern

BUWAL (1998). *Bewertung in Ökobilanzen mit der Methode der ökologischen Knappheit, Ökofaktoren 1997*, Schriftenreihe Umwelt #297, Bundesamt für Wald, Abfall und Lanschaft, Buwal, CH-3003 Bern.

Colleran, E. (1999). Hygienic and sanitation requirements in biogas plants treating animal manures or mixtures of manures and other organic wastes, in: AD-Nett (ed): *AD: Making energy and solving modern waste problems,* Ortenblad H., Herning municipal utilities, Denmark, pp. 77–86.

Edelmann, W. and Engeli, H. (1992). Combined digestion and composting of organic industrial and municipal wastes in Switzerland, in: F.Cecchi (ed.) *Proceedings of International Symposium on anaerobic Digestion of solid Waste*, pp. .225–241, Venezia, 14.-17.4.92 (published also in *Wat. Sci. Tech.*).

Edelmann, W., Engeli, H., Gradenecker, M., Kull, T. and Ulrich, P. (1993). *Möglichkeiten der Wärmerückgewinnung bei der Kompostierung,* Schriftenreihe Forschungsprogramm Biomasse BFE, arbi CH-6340 Baar.

Edelmann, W., Brotschi, H. and Joss, A. (1998). *Kompostier- und Gäranlage "Allmig" - Betriebsergebnisse und Energiebilanz*, Schlussbericht, BFE, CH-3003 Bern.

Edelmann, W., Engeli, H. and Joss, A. (1999a). Behandlung von Abwässern aus der Abfallvergärung mit der Membrantechnik, in: Wilderer P. (ed.): *Prozessabwasser aus der Bioabfallvergärung*, Berichte aus Wassergüte- und Abfallwirtschaft, TU München, pp. 123–132.

Edelmann, W. and Schleiss, K. (1999b). *Ökologischer, energetischer und ökonomischer Vergleich von Vergärung, Kompostierung und Verbrennung fester biogener Abfallstoffe*, Final report for BFE/BUWAL, BFE, CH-3003 Bern. (118 pages.)

Edelmann, W., Engeli, H. and Gradenecker M. (1999c) Co-digestion of organic solid waste and sludge from sewage treatment. *Wat. Sci. Tech.* **41**(3), 213–221.

Edelmann, W. (2001). Grundlagen der biochemischen Umwandlung, in: Kaltschmitt M., Hartmann H.: [ed.] Energie aus Biomasse, Springer Verlag Berlin/Heidelberg/New York, pp. 585, ISBN 3-540-64853-4.

Egger, K. (1997). *Kompo_Mobil: Biogasnutzung in Fahrzeugen*, Schlussbericht, P+D_Projekte, Bundesamt für Energie BFE, CH-3003 Bern, September 1997

FAL (1999). *Wegleitung zur Bewertung und Zulassung von Düngern und diesen geichgestellten Erzeugnissen, EDMZ* #730-960d (english: 730-960e), CH-3003 Bern.

Frischknecht, R. (1996). *Ökoinventare von Energiesystemen*, ISBN 3-9520661-1-7, ENET, Postfach, Bern dritte, überarbeitete Auflage). (Data on CD-ROM.)

Fuchs, J., Galli, U., Schleiss, K. and Wellinger, A. (2001). *VKS-Richtlinien 2001: Qualitätseigenschaften von Komposten und Gärgut aus der Grüngutbewirtschaftung,* Richtlinien des VKS, Verband Kompostwerke Schweiz, CH-3322 Schönbühl. (Download at: www.vks-asic.ch).

Gronauer, A., Helm, M. and Schön, H. (1997). *Verfahren und Konzepte der Bioabfallkompostierung - Vergleich - Bewertung - Empfehlungen*, Bayerische Landesanstalt für Landtechnik der TU München-Weihenstephan, BayLfu 139. (130 pages.)

Goedkoop, M. (1995). *The eco-indicator 95 - Weighting method for environmental effects that damage*, National Reuse of Waste Research Program, Pré-consultants, NL 3811 Amersfoort. ISBN 90-72130-80-4.

Hellmann, B., Zelles, L., Palojärvi, A. and Bai, Q. (1997). *Emission of climate relevant trace gases and succession of microbial communities during open windrow composting. App. and Env. Mikrobiol.* March 1997, pp. 1011.

Hellweg, S. (1999). Life cycle assessment of thermal waste processes, in: *Proceedings of R99*, Feb. 1999, Geneva

Hjort-Gregersen, K. (1999). *Centralized biogas plants - integrated energy production, waste treatment and nutrient redistribution facilities*, Bioenergy dept., University of southern Denmark, Esbjerg

Hoejme Tecknik (2000). Newsletter 1-11-2000, 5250 Odense SV.

Holm-Nielsen, J. and Al Seadi, T. (2000). *Danish centralized biogas plants - plant descriptions*, Bioenergy dept., University of southern Denmark, Esbjerg.

IEA (1997a). *Development of a protocol for assessing and comparing the quality of aerobic composts and anaerobic digestats*. Bioenergy AD activity, International Energy Agency.

IEA (1997b). *Life cycle assessment of AD - a literature review*. Resource Development Associates, Final Report for International Energy Agency.

Ketelsen, K. and Cuhls, C. (1999). Emissionen bei der mechanisch-biologischen Behandlung von Restabfällen und deren Minimierung bei gekapselten MBA-Systemen, In: Wiemer K., Kern M.. (Ed.) *Abfall-Wirtschaft, Proceedings des Kasseler Abfallforums: Bio- und Restabfallbehandlung III*, Kassel, Baeza Verlag, Witzenhausen, pp. 461.

Kompogas (2000). Information on the process and on vehicles, published on Internet: www.kompogas.com.

Kompogas (2001). D. Würgler; pers. comm.

Klingler, B. (1999). Environmental aspects of biogas technology, in: AD-Nett (ed): *AD: Making energy and solving modern waste problems,* Ortenblad H., Herning municipal utilities, Denmark, pp. 22–33.

Loll, U. (1999). Übersicht über Mengen und Beschaffenheit von Prozessabwässern aus der Vergärung biogener Abfälle, in: Wilderer P. (ed): *Prozessabwasser aus der Bioabfallvergärung*, Berichte aus Wassergüte- und Abfallwirtschaft, TU München, pp. 1–22.

Marb, C., Dietrich, G., Köberni, M., Neuchl, C. (1997). Vergleichende Untersuchungen zur Kompostierung von Bioabfällen in Reaktoren und auf Mieten: Emissionen, Qualität und Schadstoffe, Müll und Abfall, 10/97.

Membrez, Y. and Glauser, M. (1997). *Evaluation environementale du Degré de Centralisation d'installations de Méthanisation des Dechèts organiques au Moyen d'Ecobilans*, BFE, CH-3003 Bern.

Müller, A.G. (1999). Allmig-Kompost: Frei von Fremdstoffen, "Forum" #42, Alfred Müller AG, CH-6340 Baar.

Nordberg, A (1999). Legislation in different European Countries regarding implementation of anaerobic digestion, in: AD-Nett (ed). *AD: Making energy and solving modern waste problems,* Ortenblad H., Herning municipal utilities, Denmark, pp. 110–124.

Ortenblad, H. (1999). The use of digested slurry within agriculture, In: AD-Nett (ed): *AD: Making energy and solving modern waste problems,* Ortenblad H., Herning municipal utilities, Denmark, pp. 53–65.

Oetjen-Dehne, R. and Ries, G. (1995). Was kostet die biolgische Abfallbehandlung, in: Thomé-Kozmiensky K. [ed]: *Biologische Abfallbehandlung*, EF-Verlag Berlin, pp. 168–174.

Prochinig U. (1997). *Mittelflussrechnung*, 3. Auflage 1997, Verlag SKV, Zürich.

Piccini, S., Fabbri, C. and Riva, G. (2000). Sizing and installation of cogenerators with Italy as an example, In: AD-Nett (ed): *AD: Making energy and solving modern waste problems,* Ortenblad H., Herning municipal utilities, Denmark, pp. 142–159.

Saint-Joly, C., Desbois, S. and Lotti, J-P. (1999). Determinant impact of waste collection and composition an anaerobic digestion performance: industrial results. *Water Sci. Tech.* **41(**3), 291–297.

Schattner-Schmidt, S., Helm, M., Gronauer, A., Hellmann, F. and Hellmann, B. (1995). Kompostierung biogener Abfälle, Landtechnik, 6/95 pp. 364.

Schleiss, K. (1998). *Kompostier- und Vergärungsanlagen im Kanton Zürich - Jahresbericht 1997,* AWEL, Amt für Abfall, Wasser und Energie, 1998/4, Zürich.

Schleiss, K. (1999). *Grüngutbewirtschaftung im Kanton Zürich aus betriebswirtschaftlicher und ökologischer Sicht,* Dissertation ETH-Z 13'476, Library ETH, Zürich (on PDF-file) 177 pages.

Schleiss, K, and Edelmann, W. (2000). *Oekobilanz der Stromproduktion aus der Feststoffvergärung*, Biogas Forum, NovaEnergie, CH_8356 Ettenhausen.

Shoemaker A. and Visser A. (1999). Treatment of biogas, in: AD-Nett (ed): *AD: Making energy and solving modern waste problems,* Ortenblad H., Herning municipal utilities, Denmark, pp. 125-141.

Smet E., Van Langenhove H. and de Bo I. (1998). The emission of volatile compounds during the aerobic and the combined anaerobic/aerobic composting of biowaste, Preprint of a paper submitted to Atmospheric Environment.

Staudenmann J. and Junge-Berberovic, R. (2000). Treating biogas plant effluent through aquaculture: First results and experiences from the pilot plant Otelfingen (Switzerland) In: *Waste recycling and resource management in the developing world* (ed. Jana, B.B., Banerjee, R.D., Guterstam. B., and Heeb J.). pp. 51–57. University of Kalyani, India and Int. Ecol. Eng. Soc., Switzerland.

Staudenmann, J., Junge-Berberovic, R. (2002). Recycling nutrients from industrial wastewater by aquaculture systems in temperate climates (Switzerland). *Journal of Applied Aquaculture*. (In press.)

Strauch, D. (1993). Hygienefragen beim Umgang mit Bioabfällen, in: *ANS-Tagung 17.11.93 in Brugg,* Unversität Hohenheim, Stuttgart.

Six, W. (1999). Auftrennung und Vergärung von Restabfall mit dem Dranco-Verfahren, In: Wiemer K., Kern M. [ed.], (1999). *Bio- und Restabfallbehandlung III, biologisch - mechanisch - thermisch*, Baeza-Verlag, D-37213 Witzenhausen, pp. 909–916.

Vester, F. (1978). *Unsere Welt - ein vernetztes System*, Klett-Cotta Verlag, Stuttgart.

Widmann, R. (1999). Anaerobic procedures (fermentation) in: *Proceedings of Orbit 99*, Int. Conf. on biological treatment of the waste and environment, Bauhaus, Weimar, pp. 169 ff.

Wiemer, K. and Kern, M. [ed.], (1999). *Bio- und Restabfallbehandlung III, biologisch - mechanisch - thermisch*, Baeza-Verlag, D-37213 Witzenhausen (1005 pages).

Wilderer, P. [ed] (1999). *Prozessabwasser aus der Bioabfallvergärung*, Berichte aus Wassergüte- und Abfallwirtschaft, TU München (146 pages).

Zimmermann, P. (1996). *Ökoinventare von Entsorgungsprozessen: Grundlagen zur Integration der Entsorgung in Ökobilanzen*, ISBN 3-9520661-0-9, ENET, Postfach, Bern.

11

Anaerobic digestion of organic solid waste in bioreactor landfills

F.G. Pohland, A.B. Al-Yousfi and D.R. Reinhart

11.1. INTRODUCTION

Effective municipal solid waste management includes an integrated approach involving waste minimization, reuse/recycle, treatment and ultimate disposal. Of these options, ultimate disposal in landfills is often selected because of comparative economic advantage and a recognition of an inherent capacity for *in situ* attenuation of organic as well as inorganic waste constituents. The latter attenuation capacity is possible due to microbially mediated conversion of the organic waste fractions, and a concomitant physical–chemical conditioning of the reaction environment to permit both biotic and abiotic processes to be operative. Therefore, landfills are currently being designed and constructed as controlled systems, which with integrated leachate recirculation and gas management throughout the operational and postclosure periods, has led to the emergence of new generation bioreactor landfill technologies.

11.2. LANDFILL PROCESS ENHANCEMENT

The principal operational variable associated with process enhancement in bioreactor landfills is regulated leachate generation, containment, collection, recirculation, removal, and ultimate disposal (Pohland, 1996). Implicit in this tactic is moisture accumulation and management to optimize and accelerate the microbially mediated waste transformation reactions throughout the landfill complex, as each landfill cell or compartment progresses through the sequential phases of stabilization. These phases are depicted by changes in leachate and gas indicator parameters in Figure 11.1. Accordingly, organic waste constituents are converted into intermediate and end products that appear in the leachate and gas over a shortened and more predictable time increment than in landfills operated without leachate recirculation. With the exception of an initial relatively brief aerobic phase, most of these transformation reactions occur under anaerobic or anoxic conditions, with pronounced acidogenic and methanogenic phases common to anaerobic digestion processes. However, because of the comparatively massive waste loadings and large batch-type reaction zones developed as disposal operations proceed, the temporal and spatial distribution of these phases is more evident, magnified by the relatively long contact times and by leachate recirculation.

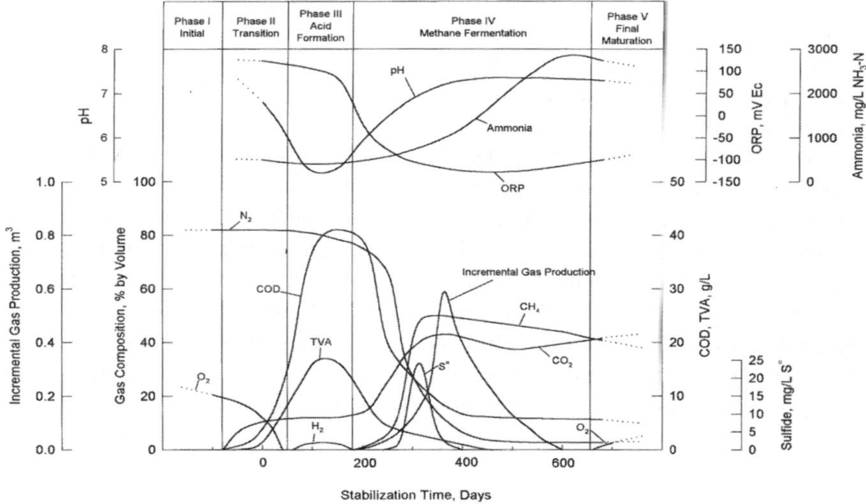

Figure 11.1. Stabilization characteristics within a bioreactor landfill unit.

The nature and function of the ecological milieu established during the sequential phases of landfill stabilization are reflected by the merger of leachate and gas production and quality from the multiple landfill zones, as affected by moisture availability and augmented by the frequency and loading intensity of the recirculated leachate. The associated conversion processes can be described in terms of recognized reaction mechanisms and a natural succession of microbial species, as the overall system progresses toward and establishes an equilibrium or steady-state condition. Therefore, a landfill cell or compartment exposed to leachate recirculation is essentially analogous to an attached growth anaerobic treatment process, with internally generated and usually diminishing substrate loadings, but which may be utilized and sustained by transfer and application of leachate from adjoining landfill areas. The loadings and retention times in this sequencing batch-type system are regulated by selected leachate recirculation and gas recovery and utilization schedules.

Drawing from an understanding of microbially mediated conversion of complex organic matter under anaerobic conditions, a series of interdependent oxidation-reduction sequences can be identified as indicated in Figure 11.2 and Table 11.1 (Pohland and Kim, 2000). Hence, organic waste polymers are converted by hydrolytic bacteria into monomers of sugars, fatty acids and amino acids. These conversion products are then further converted by fermentative bacteria into alcohols and organic acids characteristic of the acidogenesis phase of landfill stabilization as illustrated in Figure 11.1 (Phase III). Thereafter, these intermediate substrates are converted either by syntrophic bacteria to acetate, hydrogen (H_2) and carbon dioxide (CO_2), or directly by methanogenesis to methane (CH_4) as also illustrated in Figure 11.1 (Phase IV). Whether these sequences are working in harmony is a function of prevailing environmental conditions, reaction opportunities, and the relative conversion capacities and growth rates of the microbial consortia mediating the constituent reactions, as well as associated abiotic influences. For example, if a normal symbiotic association between the hydrogen oxidizing methanogens and aceticlastic methane fermenters is interrupted because of some imbalance or inhibition, hydrogen and organic acids could accumulate, interrupt the desired pattern of organic waste conversion, and lead to a condition analogous to a failed anaerobic digestion process. Moreover, the competition between sulphate-reducing and methane-producing bacteria for available electron donors could also intermit the normally desired anaerobic conversion sequence.

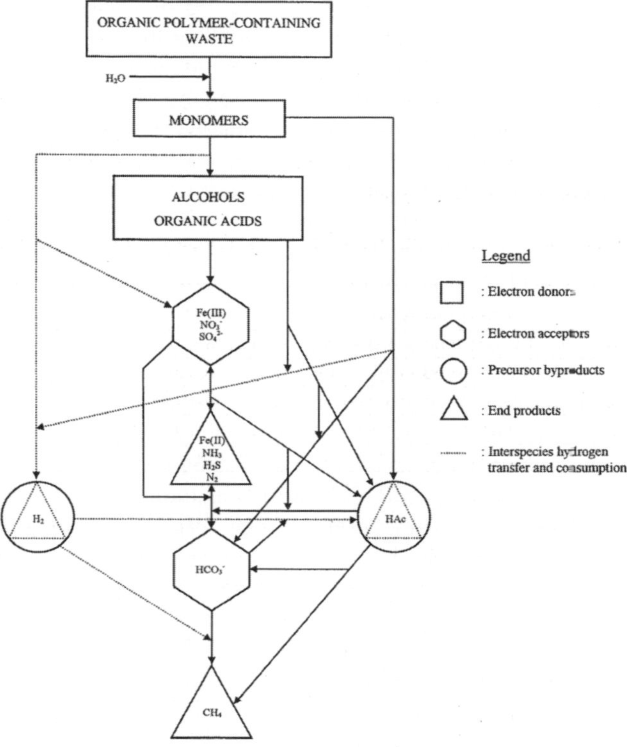

Figure 11.2. Waste conversion sequences in a bioreactor landfill.

As indicated by the dashed lines in Figure 11.2 and the redox half-reactions of Table 11.1, H_2 plays an important role in overall waste conversion. Although the indicated thermodynamic favourability can suggest reaction preference, accumulation of H_2 can shift and interrupt the normal conversion pattern. Thus, removal of H_2 through interspecies hydrogen transfer or consumption by other electron acceptors (e.g., sulphate, nitrate) is necessary in order to avoid accumulation of intermediate reaction products, including the lower fatty acid homologues and their condensation products. The availability of other electron acceptors within the landfill matrix can both alter and help preserve conditions favourable to waste stabilization. Accordingly, reduction of sulphates to sulphides is particularly important, both to retrieve excess H_2 and to attenuate heavy metals by *in situ* precipitation of sparingly soluble sulphides under reducing (low ORP) conditions. Retention by filtration and matrix capture is facilitated by leachate recirculation. Likewise, reduction of nitrates to ammonia provides a H_2 sink, but because the total nitrogen available usually exceeds microbial nutrient requirements, leachate ammonia may accumulate. Such an accumulation may raise the leachate pH, and possibly necessitate treatment before ultimate discharge unless it is subject to oxidation and evolved as N_2 through denitrification (Onay and Pohland, 1998). In addition, participation of H_2 in reductive dehalogenation reactions, together with abiotic transformations of heavy metals, are adjunct processes within landfills that serve to help detoxify otherwise detrimental waste constituents.

Table 11.1. Redox half-reactions during anaerobic stabilization in bioreactor landfills.

	Oxidation (electron donating) reactions*	ΔG^o (KJ)
Caproate → Propionate	$CH_3(CH_2)_4COO^- + 2H_2O \rightarrow 2CH_3CH_2COO^- + H^+ + 2.5\ H_2$	+48.3
Caproate → Acetate	$CH_3(CH_2)_4COO^- + 4H_2O \rightarrow 3CH_3COO^- + H^+ + 4H_2 + 2H$	+96.7
Caproate → Butyrate + Acetate	$CH_3(CH_2)_4COO^- + 2H_2O \rightarrow CH_3(CH_2)_2COO^- + CH_3COO- + H^+ + 2.5\ H_2$	+48.4
Propionate → Acetate	$CH_3CH_2COO^- + 3H_2O \rightarrow CH_3COO^- + HCO_3^- + H^+ + 3H_2$	+76.1
Butyrate → Acetate	$CH_3CH_2CH_2COO^- + 2H_2O \rightarrow 2CH_3COO^- + H^+ + 2H_2$	+48.1
Ethanol → Acetate	$CH_3CH_2OH + H_2O \rightarrow CH_3COO^- + H^+ + 2H_2$	+9.6
Lactate → Acetate	$CH_3CHOHCOO^- + 2H_2O \rightarrow CH_3COO^- + HCO_3^- + H^+ + 2H_2$	-4.2
Acetate → Methane	$CH_3COO^- + H_2O \rightarrow HCO_3^- + CH_4$	-31.0
	Reductions (electron accepting) reactions*	
HCO_3^- → Acetate	$2HCO_3^- + 4H_2 + H^+ \rightarrow CH_3COO^- + 4H_2O$	-104.6
HCO_3^- → Methane	$HCO_3^- + 4H_2 + H^+ \rightarrow CH_4 + 3H_2O$	-135.6
Sulfate → Sulfide	$SO_4^= + 4H_2 + H^+ \rightarrow HS^- + 4H_2O$	-151.9
	$CH_3COO^- + SO_4^= + H^+ \rightarrow 2HCO_3^- + H_2S$	-59.9
Nitrate → Ammonia	$NO_3^- + 4H_2 + 2H^+ \rightarrow NH_4^+ + 3H_2O$	-599.6
	$CH_3COO^- + NO_3^- + H^+ + H_2O \rightarrow 2HCO_3^- + NH_4^+$	-511.4
Nitrate → Nitrogen gas	$2NO_3^- + 5H_2 + 2H^+ \rightarrow N_2 + 6H_2O$	-1120.5

*pH 7, 1 atm, 1 kg/mol activity, 25°C

Based on results of laboratory, pilot-scale and full-scale landfill investigations, both readily degradable organic substrates as well as more recalcitrant inorganic and organic waste constituents are amenable to *in situ* attenuation. As indicated in Table 11.2, coupled biochemical and physico-chemical processes are often required, again influenced by prevailing environmental conditions and reaction opportunity.

Table 11.2. Bioreactor landfill attenuation of representative recalcitrant organic and inorganic constituents*.

Constituent	Dominant Attenuation Mechanisms during Landfill Stabilization Phases
*Organic Compounds***	
Halogenated Aliphatics (PCE, TCE, DBM)	Volatilization and mobilization in gas and leachate due to high vapor pressure and solubility with low K_{ow} prior to abiotic and biotic reductive dehalogenation under methanogenic, methanotrophic, sulfate reducing and denitrifying conditions.
Chlorinated Benzenes (HCB, TCB, DCB)	Volatilization and sorptive matrix capture due to relative volatility with low solubility and high K_{ow} prior to partial reductive dechlorination.
Phenolics and Nitroaromatics (DCP, NP, NB)	Mobilization in leachate but with low volatility due to low K_{ow} and vapor pressure prior to dechlorination or nitro-group reduction, biodegradation and complexation.
Phthalate Esters (BEHP), Polynuclear Aromatics (NAP) and Pesticides (LIN, DIEL)	Low volatility and mobility in gas and leachate due to low vapor pressure and solubility with high K_{ow} prior to complete or partial biodegradation.
Heavy Metals: (Cd, Cu, Cr, Fe, Hg, Ni, Pb, Zn)	Conversion to a reduced oxidation state with complex formation (Fe, Cr, Hg). Mobilization in leachate by organic (e.g., aromatic hydroxide carboxcylic acid, aromatic amine humic and fulvic acid) and inorganic (e.g., chloride, sulfate, carbonate) ligands. Formation of sparingly soluble hydroxides (Cr) and sulfides (Cd, Cu, Fe, Hg, Ni, Pb, Zn) Physical sorption and ion exchange within the waste matrix. Filtration and retention within stagnant pools of interstitial water. Precipitation and encapsulation in localized alkaline environments.

*Pohland and Kim (2000); Pohland, et al. (1993); Reinhart and Pohland (1991)

**Tetrachloroethene (PCE), Trichloroethene (TCE), Dibromomethane (DBM), Hexachlorobenzene (HCB), Trichlorobenzene (TCB), Dichlorobenzene (DCB), Dichlorophenol (DCP), Nitrophenol (NP), Nitrobenzene (NB), Bis(2-ethylhexyl)phthalate (BEHP), Naphthalene (NAP), Lindane (LIN), Dieldrin (DIEL); includes daughter products.

11.3. EMERGING DEVELOPMENTS AND IMPLEMENTING STRATEGIES

The controlled bioreactor landfill has received global attention and acceptance as an appropriate, cost-effective, and superior technological innovation. Selected examples of pilot- and full-scale bioreactor landfill installations are included in Table 11.3. A range of sizes and a variety of operational techniques have been used to provide enhanced stabilization, including leachate recirculation and other optimization approaches. However, to fully benefit from the associated potential for enhanced waste attenuation and biogas recovery and utilization, traditional landfill designs and operational protocols, with relatively small and isolated discrete cells and lifts, need to be modified to accommodate planned development and operation of leachate and gas management systems, synchronized with landfill construction and waste loading schedules. Waste placement, without excessive compaction or isolation, should be practiced to facilitate controlled leachate (and gas) movement throughout the landfill complex.

A prime example of the evolution of bioreactor landfill technology is the continuing development at the landfills managed by the Delaware Solid Waste Authority since the original start up of an 11 ha landfill in 1980. As detailed in Table 11.4, these facilities are located at two sites and included two small test cells, with and without leachate recirculation, to field demonstrate the prior laboratory investigations on accelerated stabilization (Pohland et al., 1993). The large-scale landfills were variously designed and operated to provide intermittent leachate recirculation through vertical wells, sequentially constructed of perforated manhole sections filled with stone and perforated PVC pipe, augmented by plastic surface 'infiltrators' located under the final cover; a combination of vertical gas/recharge wells and fields with each well connected horizontally at selected lift intervals using stone trenches; horizontal gas/leachate trenches operated under pressure; and parallel horizontal trenches containing perforated HDPE pipe surrounded by stone or tire chips at the bottom, and with a second pipe at the end of the trench near the landfill side slopes for gas extraction. This latter innovation was incorporated into the operation of the newest landfill extensions at each site, with initiation of leachate recirculation in summer of 2000.

Such bioreactor landfill innovations, conceptually depicted in Figure 11.3, are intended to promote unimpeded leachate recirculation throughout the waste mass, coupled with efficient gas extraction, using inter-connected wells and trenches or transport grids, possibly augmented by alternative or temporary daily and intermediate cover systems (Pohland and Graven, 1993). Increased leachate flow rates throughout the active phases of waste stabilization need to be accommodated in the leachate management system with appropriately designed and operated leachate containment, collection and recirculation facilities (Reinhart and Townsend, 1998). Landfill designs should also include plans to recover the increased landfilling space generated as enhanced waste stabilization proceeds. Because increased moisture content leading to waste saturation can create positive internal pressure and a reduced angle of friction, and changes in shear strength due to waste conversion, construction techniques need to address changes in waste properties that affect landfill structural stability.

Table 11.3. Description of selected pilot-scale and full-scale bioreactor landfills

Location (Source)	**Size**	**Start up date**	**Leachate recirculation technique**	**Bioreactor cost***	**Comments**
Kootenai Co., ID (Miller and Emge, 1996)	2.83 ha	1993 (landfill operation) 1995 (leachate re-circulation)	Surface spray (summer only); trenches 24.4-m spacing; wells	$1,035,000 amortized + operating costs = $449,600/yr	First lined landfill in Idaho.
Bluestem SWA, Linn Co. IA (Hall, 1998)	0.20 ha, 7700 tons waste, divided into 2 subcells	1998	Trenches 4.6-m spacing; 10,670 l/d	$959,000 (cell construction)	Experimenting with bag opening; biosolids addition.
Milwaukee, WI (Viste, 1997)	61 m x 12.2 m	1999	Trenches	NA	No compaction; shredded; biosolids added.
Keele Valley LF, Toronto, Canada (Mosher et al., 1997)	Pilot	1990	Vertical wells - 1.2 wells/ha; ~ 190 - 400 l/m	NA	Well water added to adjust moisture content (not leachate).
Eau Claire, WI, 7 Mile Creek SL (Magnuson, 1998)	720 tpd landfill, (Phase I at 180 tpd)	1998	Trenches 7.6-m spacing; 73 lpd/m^2	NA	Tire chips acceptable in trenches; gas production increased by 25 % in wells near recirculation.
Yolo County, CA (Yolo Co., 1998)	Two 930 m^2 cells, 4080 kg MSW each, 12 m deep	1995	14 infiltration trenches at surface	$563,000 (cell construction)	Enhanced gas production, settlement. Shredded tires successful in gas collection.
Lower Spen Valley LF, West Yorkshire, UK (Blakey et al., 1997)	Two cells, ~ 860 tons waste, ~ 890 m^2 each, ~ 5.5 m deep	1991	Trenches	NA	Biosolids and wastewater addition. Low temperature prevented maximum gas production.
Crow Wing MSW LF, MN (Doran, 1999)	5.18 ha	1997	11 trenches, 15-m spacing, filled with shredded tires; 310 l/d/m	$290,000; $72,500 savings/yr (1997-8)	No off-site hauling of leachate in 1998; recirculation operated 3 mo/yr.
Worcester Co. LF, MD (Kilmer and Tustin, 1999)	6.9 ha, 24 m deep	1990	Vertical wells surrounded by 7.6-m of gravel blanket	$50,000; Net benefit $3.2 million per 6.9-ha cell (after mining)	Average 65% of leachate recirculated. Upper layers did not degrade extensively.

Table 11.3. (*continued*).

Location (Source)	Size	Start up date	Leachate recirculation technique	Bioreactor cost*	Comments
Lyndhurst LF, Melbourne, Australia (Yuen et at., 1995)	1.3 ha	1995	Recharge wells and trenches	NA	Complete instrumentation for monitoring leachate, temperature, gas, climate, moisture distribution, head on liner.
VAM Waste Treatment, Wijster, The Netherlands (Oonk and Woelders, 1998)	7062 m^2	1997	Trenches 10-m horizontal, 3-m vertical spacing (plus surface infiltration at 5-m spacing)	NA	Gas collection in wood chips at the top liner. Filled with mechanically separated organic fractions < 45-mm diameter.
Baker Rd LF, Columbia County, GA (Hudgins and Marks, 1998)	3.24 ha, 3 m	1996	20 vertical wells	$25 - 30,000 capital; operating and maintenance costs not reported	Air injected into leachate collection system; settlement increased by 4.5%, bio-degradation rate increased by > 50%.
Live Oak LF, Atlanta, GA (Johnson and Baker, 1999)	1.01 ha, 9 m deep	1997	27 vertical wells, 1.5 - 4.6 m deep, 18 air injection wells	NA	Air and liquid injection into same well improved fluid distribution.
Shin-Kamata LF, Fukuoka City, Japan (Fukuoka City Environmental Bureau, 1999)	NA	1975	Horizontal pipes	NA	Semi-aerobic process using large leachate collection pipes that draw in air.
Trail Road LF, Ontario, Canada (Warith et al, 1999)	270 m x 500 m	1992	Infiltration lagoons	NA	Lagoons were moved around;~ 50% of field capacity achieved.

Table 11.4. Bioreactor landfill developments at the Delaware Solid Waste Authority.

Location	Size	Start up date	Re-circulation start date	Leachate recirculation Technique	Leachate recirculation cost	Comments
CSWMC Test Cells	0.4 ha each	1989	1990	Leachate fields constructed under the cap. The fields were made from stone and plastic "infiltrators".	Approximately $1,000,000 design/construction cost for 2 double lined cells	CSWMC Test Cells
CSWMC Area A/B	10.9 ha	1980	1983	Vertical wells with stone; manhole sections and fields on the top and within the waste mass.	No additional construction cost, since existing gas wells were utilized. Some fields constructed with tire chips.	First landfill using recycled leachate by the Delaware Solid Waste Authority. Also experimented with ponding the leachate and spray irrigation on the surface.
CSWMC Area C	7.7 ha	1988	1990	Vertical wells constructed of concrete manhold sections with stone and perforated pipe.	No additional construction cost since existing gas wells were utilized.	Increased the size of stone in the wells to reduce clogging.
CSWMC Area D	8.9 ha	1993	1994	Combination of vertical wells connected at various lifts, and fields installed below the cap.	$70,000 cost to construct fields and control system.	Connected the wells together using stone trenches. Fields were added after the cell was completed.
CSWMC Area E	13 ha	1999	2002	Horizontal injection trenches installed at every other lift of waste.	Approximately 60 horizontal trenches @ $5,000 ea.	Changed from vertical wells and fields (gravity systems) to horiozntal pressure system.
SSWMC Cells 1 & 2	17 ha	1986	1987	Vertical well injection.	Approximately 30 vertical meters of well constructed ($40,000 cell 1)	Odor problems associated with recirculation events let to ban of recirculation in February 1993.
SSWMC Cell 3	9.7 ha	1996	2002	Horizontal injection trenches.	Approximately ~ $26/linear m ($70,400 1st lift)	

*NA: Not available.
CSWMC: Central Solid Waste Management Center.
SSWMC: Southern Solid Waste Management Center.

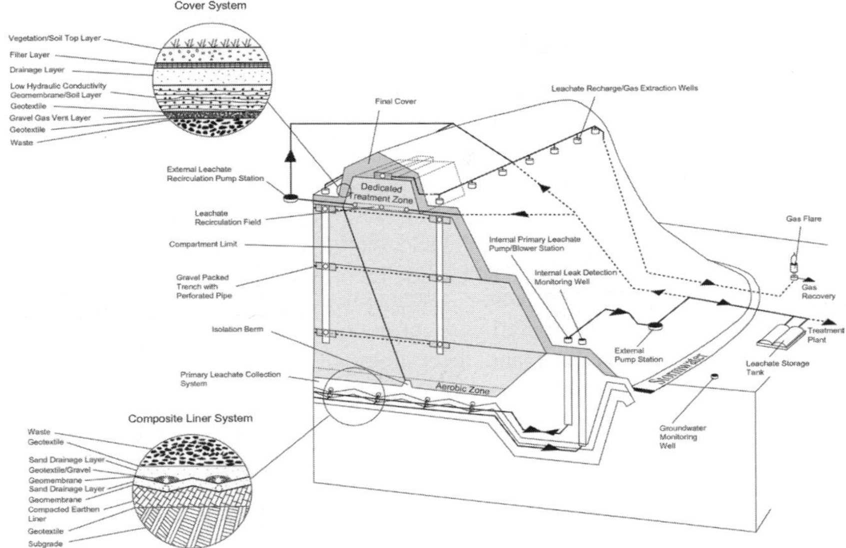

Figure 11.3. Construction and operational features of bioreactor landfill system.

Management of a bioreactor landfill complex, including the operation of the leachate recirculation system, must rely on routine inspection and monitoring so that controlled stabilization can be optimized and maintained, e.g., monitoring for leachate and gas characteristics (Figure 11.1) can direct leachate loading schedules to help promote and sustain the desired treatment outcome, whether principally focused on optimizing anaerobic methanogenic activity, or providing opportunities for aerobic/semi-aerobic oxidation, nitrification, co-metabolism, or anoxic denitrification. Flexibility to allow interchange of leachate loadings between dedicated treatment zones should be provided, possibly utilizing older, more stabilized areas to provide treatment capacity for excess leachate from newer areas. The total quantity of leachate allowed to accumulate should be restricted to that required to operate the system at the desired loading levels, to comply with regulatory restrictions, and to minimize the eventual residual quantity requiring removal and ultimate disposal, with or without post-treatment, and after active methanogenesis (Phase IV) is essentially complete. Post-treatment requirements for residual leachate from a bioreactor landfill, whether by on-site or off-site management, should be simpler and of lesser complexity because the amount and character of the stabilized leachate would be established sooner and more reliably. Moreover, because removal of residual leachate from a stabilized landfill basically deprives it of much of the moisture and nutrients necessary for continuation of active microbially mediated conversion of remaining and less readily degradable waste constituents, the overall system should become relatively dormant, thereby establishing more realistic requirements, reducing the extent and frequency of monitoring and maintenance, and providing potential cost savings (Pohland and Al-Yousfi, 1994).

11.4. CONCLUSIONS

A new generation of controlled bioreactor landfills has evolved largely due to recognition of the advantages associated with their operation as an essential element of integrated solid waste management. These advantages include formalized design and operational protocols based on a fundamental understanding of the science and technology of microbially mediated waste conversion processes and the capacity of engineered systems for optimization and control. Implicit in these advantages is the opportunity to exploit the attenuation capacity of landfills for enhanced *in situ* waste conversion and leachate treatment as well as regulated gas production and energy recovery potential. Collectively, these attributes include more efficient and cost-effective landfill management, recovery of landfill space, reduced leachate treatment and disposal costs and postclosure care requirements, avoidance of adverse environmental impacts, and increased societal and regulatory acceptance.

Further development needs to include more standardization and better integration of design and operational protocols with training and certification, innovations in construction techniques for cell interconnectivity and leachate and gas containment and management, classification by type with consensus standards of practice, and enabling policies and regulations. Resolution of these needs will help establish the technical and societal framework from which the next generation of bioreactor landfills will evolve.

11.5. REFERENCES

Blakey, N.C., Bradshaw, K, Reynolds, P. and Knox, K. (1997). Bio-reactor landfill-a field trial of accelerated waste stabilization. *Proc. Sardenia* 97, 6th *Intl. Landfill Symp.* **I**, 375–386.

Doran, F. (1999). 'Lay leachate lay', *Waste Age*, April 4–79.

Fukuoka City Environmental Bureau (1999). The Fukuoka Method – What is the semi-aerobic landfill? Fukuoka City.

Hall, T. J. (1998). Implementing low-cost bioreactor cell technology in Iowa. *Proc. SWANA's 3rd Annual Landfill Symp.* Palm Beach Gardens, FL, 1–6.

Hudgins, M. and Marks, J. (1998). In-situ municipal solid waste composting using an aerobic landfill system. American Technologies, Inc., Aiken, SC, USA.

Johnson, W.H. and Baker, J. (1999). Operational characteristics and enhanced bioreduction of municipal waste landfill mass by a controlled aerobic process. *Proc. SWANA's 4th Annual Landfill Symp.* Denver, CO., 127–141.

Kilmer, K.S. and Tustin, J. (1999). Rapid landfill stabilization and improvements in leachate quality by leachate recirculation. *Proc. from SWANA's 4th Annual Landfill Symp.* Denver, CO., 71–99.

Magnuson, A. (1998). Leachate recirculation. *MSW Mngmt.* March/April, 24–31.

Miller, D.E. and Emge, S.M. (1996). Unique design and operations approaches enhance leachate recirculation system performance at the Kootenai County (Idaho) landfill. *Proc. SWANA's 1st Annual Landfill Symp.* 157–175.

Mosher, F.A., McBean, E. A., Crutcher, A. J. and MacDonald N. (1997). Leachate recirculation to achieve rapid stabilization of landfills theory and practice. *Proc. SWANA's 2nd Annual Landfill Symp.* 121–134.

Onay, T.T. and Pohland, F.G. (1998). *In situ* nitrogen management in controlled bioreactor landfills. *Wat. Sci. & Tech.* **32**(5), 1383–1392.

Oonk, H. and Woelders, H. (1998). Full-scale demonstration of treatment of mechanically separated organic residue in a bioreactor at VAM in Wijster, *Proc. 3rd Swedish Landfill Res. Symp.* Lulea, Sweden, October 6–8.

Pohland, F.G. (1996). Landfill bioreactors: fundamentals and practice. *Wat. Qual. Intl.* Sept./Oct., 18–22.

Pohland, F.G. and Kim, J.C. (2000). Microbially mediated attenuation potential of landfill bioreactor systems. *Wat. Sci. & Tech.* **41**(3), 247–254.

Pohland, F.G., Cross, W.H., Gould, J.P. and Reinhart, D.R. (1993). Behavior and assimilation of organic and inorganic priority pollutants codisposed with municipal refuse. *Natl. Tech. Info Serv.* Springfield, VA , PB 93-227198A5, Vol. **I**, 197 pp. and Vol. **II**, 329 pp.

Pohland, F.G. and Graven, J.T. (1993). The use of alternative materials for daily cover at maniple solid waste landfills. *Natl. Tech. Info. Serv.* Springfield, VA PB93-227197, 210 pp.

Pohland, F.G. and Al-Yousfi, A.B. (1994). .Design and operation of landfills for optimum stabilization and biogas production. *Wat. Sci. & Tech.* **30**(12), 117–124.

Reinhart, D.R. and Pohland, F.G. (1991). The assimilation of organic hazardous wastes by municipal solid waste landfills. *Jour. Ind. Microbiol.* **8**, 193–200.

Reinhart , D.R. and Townsend, T.G. (1998). Landfill bioreactor design and operation. Lewis Publishers, CRC Press LLC, Baca Raton, FL, 189 pp.

Viste, D.R. (1997). Waste processing and biosolids incorporation to enhance landfill gas. *Proc. Sardinia 97, 6th Intl. Landfill Symp.* **I**, 369–374.

Warith, J.A., Smoklin, P.A. and Caldwell, J.G. (1999). Effect of leachate recirculation on the enhancement of biological degradation of solid waste: case study. *Proc. SWANA's 4th Annual Landfill Symp.* Denver, CO. 65–70.

Yolo County Division of Integrated Waste Management (1998). Trash to cash – controlled landfill bioreactor project. Public Technology, Inc. Energy Program, Order No. 98/97-317, 28 pp.

Yuen, S.T.S., Styles, J.R. and McMahon, T.A. (1995). Active landfill management by leachate recirculation: a review and an outline of a full-scale project. *Proc. Sardinia 95, 5th Intl. Landfill Symp.* Cagliari, Italy, 403–418.

Index